For a complete listing of titles in the
Artech House Microwave Library,
turn to the back of this book.

The VNA Applications Handbook

Gregory Bonaguide

Neil Jarvis

BOSTON | LONDON
artechhouse.com

Library of Congress Cataloging-in-Publication Data
A catalog record for this book is available from the U.S. Library of Congress.

British Library Cataloguing in Publication Data
A catalogue record for this book is available from the British Library.

Cover design by John Gomes

ISBN 13: 978-1-63081-600-1

685 Canton Street
Norwood, MA 02062

10 9 8 7 6 5 4 3 2 1

To my mentors:
Bob Kenney and Stu Atkinson

"If I have seen a little further it is by standing on the shoulders of Giants."
—Sir Isaac Newton, 1675

—GB

To my father, rest in peace
Thank you for all of the motivation, education, inspiration, and patience.
I was certainly a handful.

—NJ

Contents

Preface

The modern vector network analyzer (VNA) is one of the most loved and least understood pieces of radio frequency (RF) test equipment in use today. Conceived in the pre-dawn light of the microprocessor revolution, the automated VNA provided an important tool for dispersing the long shadows of the arcane black art of RF design. The VNA burst into the modern communications age with dazzling measurement speed and accuracy that has set the standard for RF and microwave measurements ever since. Today, we are beneficiaries of the astounding technological advancements made possible by these amazing instruments. However, there are many new engineers and technicians encountering the VNA for the first time: those who have yet to turn their fear of the unknown beast into awe and respect, and those seeking to broaden their RF knowledge, improve their measurements, or simply make fewer mistakes.

New arrivals to the VNA world tend to pose the same question in different ways: How do I get there from here? While there are many excellent books and conference papers treating S-parameter calibration and measurement theory and plenty of marketing material extolling VNA virtues and capabilities, there is still a rather broad gap in between that has mostly been filled by personal experience or tutelage under the watchful eye of a seasoned VNA guru. The former path is expensive (think trials, tribulations, and an occasional puff of very expensive smoke), while the latter may be unavailable (or unavoidable). So this book provides an important bridge in one's personal journey from clumsy newbie to competent contributor. Our objective in writing this book is to help engineers and technicians get the most out of their VNAs by providing a short learning curve gleaned from more than 30 years of combined applications experience to where we have helped people avoid costly mistakes and get the best measurements possible in the shortest times possible.

If reading this book assists you in achieving any of these goals, we will consider the preparation time and effort well spent. Best of luck on your journey!

1

Architecture of the Modern Vector Network Analyzer

1.1 What Is a Vector Network Analyzer?

The best tool to accurately determine the amplitude and phase response of a radio frequency (RF) or microwave circuit is a vector network analyzer (VNA). The VNA was designed originally with the goal of measuring two-port S-parameters. While virtually all modern VNAs still have this basic capability, they now have to ability to measure and display many other things and are often used in high-speed digital design. Examples measurements include:

- Time domain reflectometry (TDR);
- Multiport measurements with a large number of ports;
- Differential amplitude and phase response;
- Skew;
- Nonlinear measurement;
- Gain compression;
- Amplitude to phase conversion (AM to PM);
- Noise figure;
- Frequency conversion;
- Load pull;

- Material defect analysis;
- Radar cross-section analysis.

Each of these measurements has different challenges. To really understand how to make these measurements accurately and repeatedly, a fundamental understanding of the architecture of a modern VNA is necessary. This chapter will begin by explaining the high-level blocks in a VNA. Next, the concept of a reference plane will be introduced and how it is used in conjunction with errors and calibration.

1.2 Wave Quantities and S-Parameters

High-frequency devices will see reflections when transmission lines or circuits see impedance mismatches or physical discontinuities. To characterize these high-frequency devices, rather than voltages and currents, wave quantities are most often used. This is because power and energy are much easier to measure at these frequencies. Wave quantities are used to differentiate between incident waves and reflected waves while avoiding the challenges of measuring voltages and currents. Figure 1.1 shows the signal flow and wave quantities for a one-port device.

1.2.1 One-Port Measurements

The input power delivered to the device is $|a|^2$, while the power reflected from the device is $|b|^2$. A very common characteristics of high-frequency devices is the reflection coefficient Γ, the ratio of the reflected wave to the incident wave. Γ is typically a complex quantity that can be calculated as a ratio of the device impedance Z to the reference impedance Z_0 or $\frac{Z}{Z_0}$. This normalized impedance ratio can be used to calculate Γ.

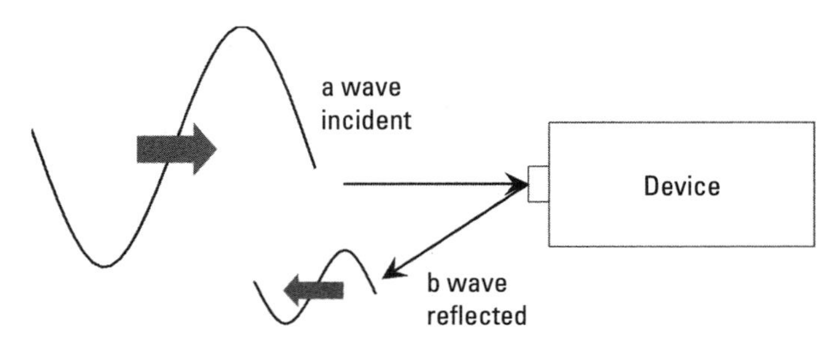

Figure 1.1 Wave quantities of a one-port device.

$$\Gamma = \text{Type equation here.} \frac{Z-1}{Z+1}$$

Measurements of Γ versus frequency are commonly plotted in the complex Γ plane, referred to as a Smith chart, developed by Phillip H. Smith. In the Smith chart, the circles represent the real component of the complex impedance and the curves represent the complex component of the complex impedance.

Figure 1.2 shows a Smith chart with key reference impedance points. The VNA has the ability to display a Smith chart for matching circuits. The Smith chart is an important tool for an RF engineer, particularly when trying to match impedances between non-50Ω devices.

1.2.2 Two-Port Measurements

More than one port can be measured similarly, with two-port devices being the most common. When measuring two-port devices, each port will include a transmission measurement in addition to the reflected measurement, but will be in both directions. Figure 1.3 shows the respective S-parameters and how the measurements are determined.

When measuring two-port devices, four S-parameters exist: S_{11}, S_{12}, S_{21}, and S_{22}. Each of these respective S-parameters is calculated by taking the complex ratio of the wave quantities, as shown in Figure 1.3. In each of these measurements, the unused *a* wave is equal to 0.

The signal flow in Figure 1.3 of a two-port device under test (DUT) with an ideal couple is as follows:

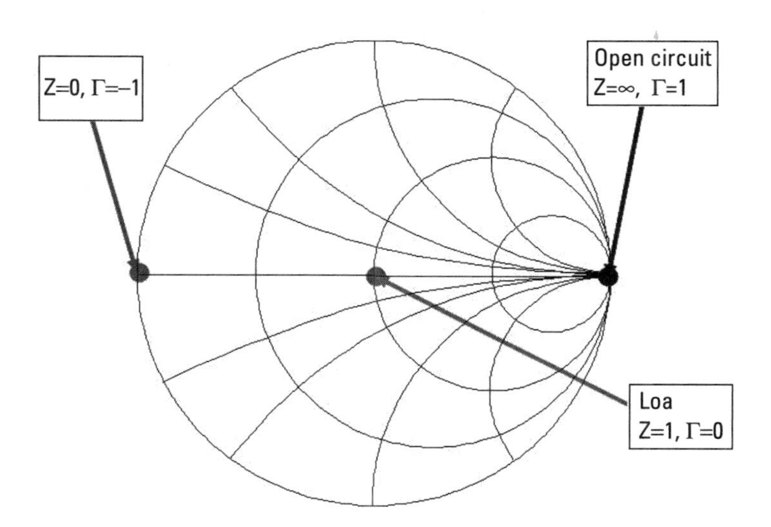

Figure 1.2 Smith chart showing reference points for key impedances.

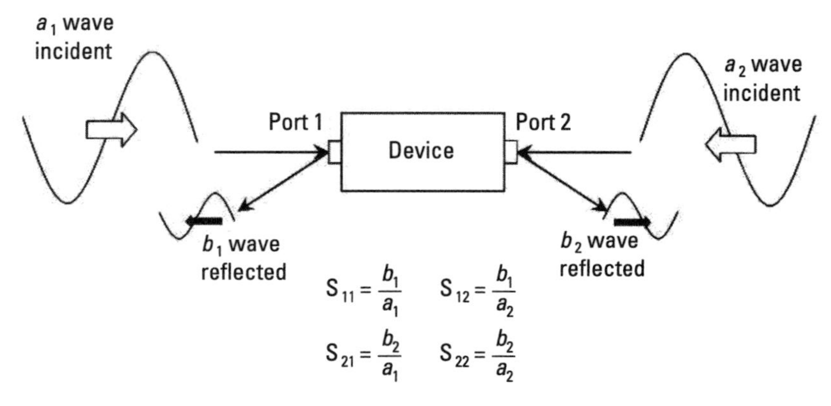

Figure 1.3 Wave quantities for a two-port device measurement.

- All reflection measurements will consist of stimulating a single port with an *a* wave and measuring a single *b* wave at the same port:
  - Wave a_1 is first used as a stimulus incident at DUT port 1 moving left to right.
  - Any mismatch between port 1 and the DUT input will cause a wave to be reflected from the DUT right to left, wave b_1.
  - The resulting ratio gives S_{11} and is representative of the input impedance of the DUT.
  - S_{22} is similarly measured by stimulating the DUT with wave a_2 moving right to left.
  - Any mismatch between port 2 and the DUT output will cause a wave to be reflected from the DUT left to right, wave b_2.
- The gain or loss (forward and reverse) will consist of stimulating a single port with an *a* wave and measuring a single *b* wave at the alternate port (reverse gain is also referred to as isolation):
  - Wave a_1 is first used as a stimulus incident at DUT port 1 moving left to right.
  - The device's transfer function will alter the wave and give a resulting output at port 2.
  - The resulting output wave b_2 is measured at port 2 moving left to right.
  - The resulting ratio gives S_{21}.
  - The final S-parameter to be measured is representative of the reverse isolation or reverse gain of the DUT.
  - Wave a_2 is used as a stimulus incident at DUT port 2 moving right to left.

  - Any wave traveling through the device from right to left will be measured as b_1.
  - The final resulting ratio gives S_{12}.
- This gives a complete representation of the DUT with the resulting S-parameters.

$$S_{11} = \frac{b_1}{a_1} \quad S_{12} \frac{b_1}{a_2}$$

$$S_{21} = \frac{b_2}{a_1} \quad S_{22} = \frac{b_2}{a_2}$$

As the number of ports increase, the number of waves increases as a square of the number of ports. For example, a three-port device will require nine waves.

1.3 Architecture of an N-Port Network Analyzer

Next we delve into the architecture of a modern VNA and how we measure these scattering or S-parameters. Figure 1.4 shows a block diagram of a VNA.

In some network analyzers, crosstalk can be minimized by using the alternate sweep mode instead of the chop mode (the chop mode makes measurements on both the reflection (A) and transmission (B) receivers at each frequency point, whereas the alternate mode turns off the reflection receiver during the transmission measurement).

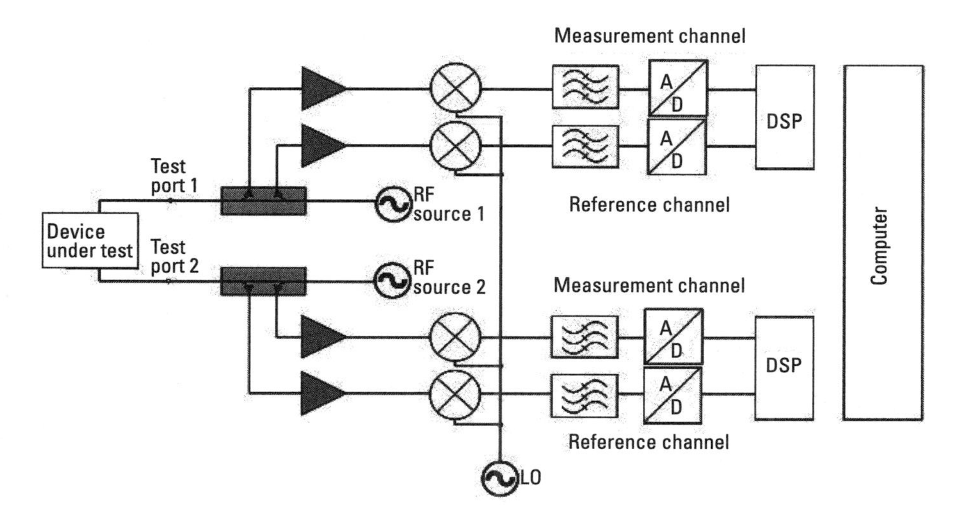

Figure 1.4 Detailed architecture of a modern VNA.

1.3.1 Main Blocks

The primary functional blocks in a VNA include the following:

- Sweeping source or generator;
- Test set or directional coupler;
- Reference receiver;
- Measurement receiver;
- Computer or processor.

The key blocks that will be focused on are shown in Figure 1.5. Each will be discussed in the following sections.

1.3.1.1 Source or Signal Generator

Figure 1.5 shows two sources or RF generators (the circles on the left and the right). The function of the source is to generate a clean RF signal that can be swept in frequency or amplitude and applied to the DUT as the stimulus. Sometimes the source is shared among ports with a switch.

1.3.1.2 Test Set or Directional Coupler

The test set establishes directionality of signals. It separates the incoming source waves from the waves reflected from the DUT at the test port. This separation is necessary to measure the reflection coefficient or Γ of the DUT. The test set is commonly either a dual directional coupler or a directional bridge. In comparing the two, the coupler has better high-frequency performance while the bridge has better low-frequency performance. The directional coupler often is more broadband.

The incident wave is fed into the reference receiver and the response signal is simultaneously fed into the corresponding measurement receiver for amplitude and phase measurement.

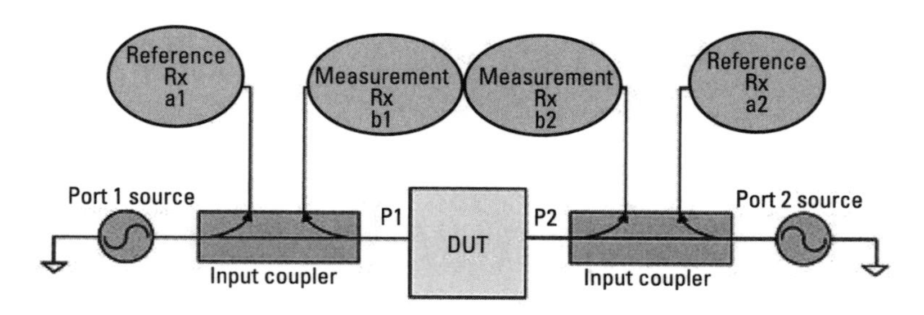

Figure 1.5 Simplified VNA block diagram.

1.3.1.3 Reference Receiver

As the signal is passed through the coupler, it is sampled by the reference receiver. This establishes the phase and amplitude reference of the input signal to the device. There is a reference receiver at each port for forward and reverse measurements. This reference measurement is used for calculating ratios with the measurement results from the measurement receiver's ports. Therefore, rather than calculate the ratios using the calculated phase and amplitude of the source, they are actually measured very accurately by the reference receiver. Understand that this is a reference signal from the analyzer and is not influenced by the DUT. It provides a very precise measurement of the outgoing signal in both phase and amplitude. As an example, Figure 1.6 shows the signal leaving the source on the left and entering the directional coupler on the right. The coupler directs the wave traveling from left to right into reference receiver *a*, while rejecting signals entering in the opposite direction. Each port will have a similar architecture to measure the outgoing signals with no influence from the DUT.

The extra connections in the upper right corner of Figure 1.6 are for measurements requiring direct receiver access. This will be discussed in a later chapter.

1.3.1.4 Measurement Receiver

The input signal is applied to the input of the DUT, which results in a reflected wave returning from the DUT input moving right to left into the coupler. As the reflected signal enters the coupler from right to left, it is separated by the coupler and sampled by the measurement receiver *b*. The measurement receiver *b* then measures the amplitude and phase of the signal relative to the reference signal.

Each port will have a similar architecture to measure the outgoing signal *a* with no influence from the DUT. Similarly, it measures the incoming signal *b* with the influence of the DUT.

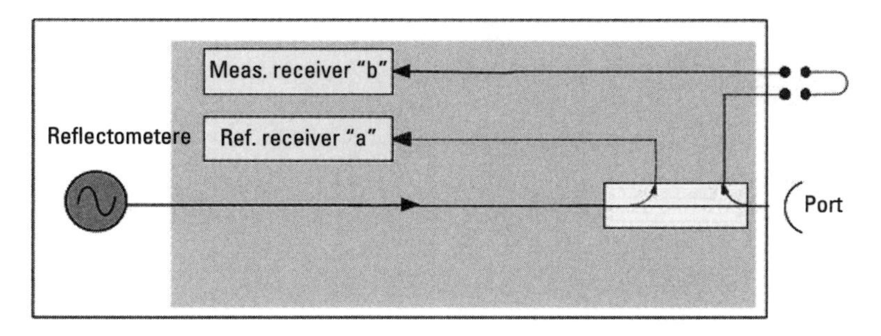

Figure 1.6 Forward measurement path.

In addition to the *b* receiver measuring the wave reflected from the DUT at the input, port 1, an additional receiver at the output of the DUT, will measure the transmitted wave at the output of the DUT port 2, as shown in Figure 1.8.

1.3.1.5 Computer

A computer is used to do the system error correction and to display the measurement data. It also provides the user interface and the remote control interfaces. The preinstalled software is known as the firmware.

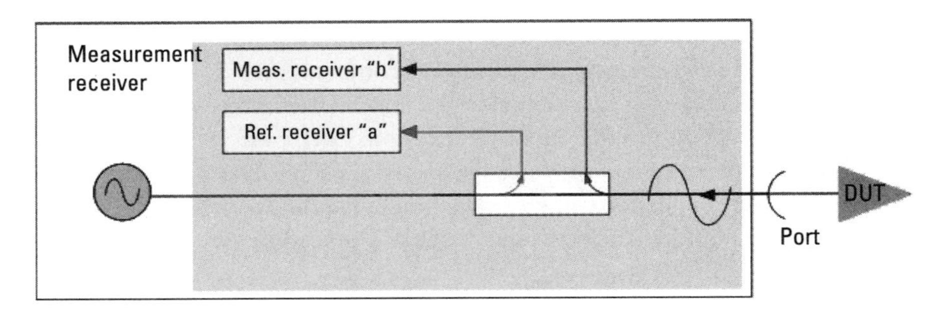

Figure 1.7 Reverse measurement path.

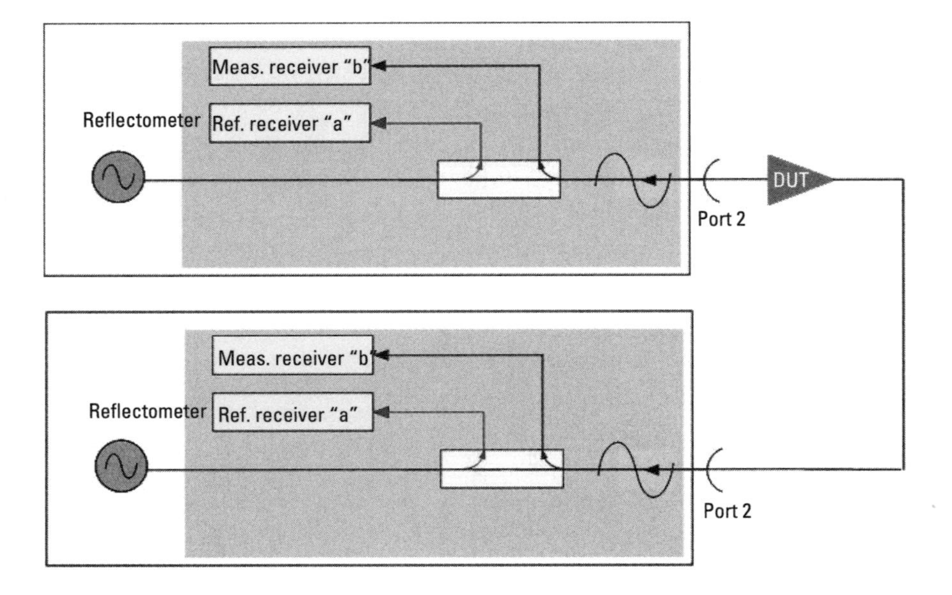

Figure 1.8 Transmitted measurement path.

1.3.2 Errors

When making measurements, there are typically three types of measurement errors: random errors, drift errors, and systematic errors. Random errors change with time in an unpredictable fashion. These errors might include thermal or environmental noise, connector repeatability, or cable flexion. Drift errors typically are changes in the measurement systems after calibration. The effect of temperature is the primary drift error. The final type of error, the most interesting, is the systematic error. Because systematic errors are time-invariant and fairly easy to measure, we can improve the measurement performance of our systems significantly with calibration. Calibration is very effective at correcting these errors and will be addressed in detail in Chapter 2.

One of the most common misconceptions about vector network analyzers lies in calibration. The concept is that any undesired response in the measurement setup can be removed. This includes cables adapters, connectors, launches, and perhaps even antennas. The first problem with this is that for errors to be calibrated out, they must be systematic. Engineers inexperienced with the use of a VNA commonly try to use low-cost, low-quality cables with the expectation that the VNA will remove any performance issues related to these interconnects. The fallacy with this lies in that poor-quality cables and connections rarely show systematic performance. A good experiment is to go measure the response of a cable and store the response in trace memory. Next, put some flex or bend on the cable and put some tension on the connections. You will see that with high-quality cables the response will return very closely to the prior measured trace stored in memory while poor cables will not.

1.3.2.1 Frequency Response Reflection Tracking

When making reflection measurements, reference signal a is compared to the measured signal b at the same port, either port 1 or port 2. For an ideal, complete reflection, the reflected wave will be equal in amplitude to the reference wave $\Gamma = 1$. In an actual measurement system, they are always different and this difference must be measured. This error results from the frequency dependence of everything in the measurement path. This will include phase and amplitude variations versus frequency of couplers or bridges internal to the VNA and any cables or connections internal and external to the measurement and reference paths. These errors are typically removed by the reflection standards in the calibration. This error will scale and rotate the Smith chart as shown in Figure 1.9. Calibration will correct this error and return the Smith chart to its proper scale and origin.

Figure 1.10 shows the signal flow for reflection tracking error with an open installed at the port. As shown, the wave from the source flows directly

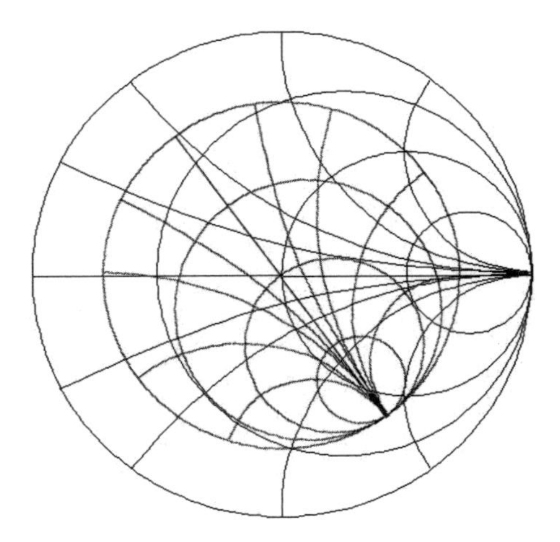

Figure 1.9 Scale and rotation effect on a Smith chart from reflection tracking.

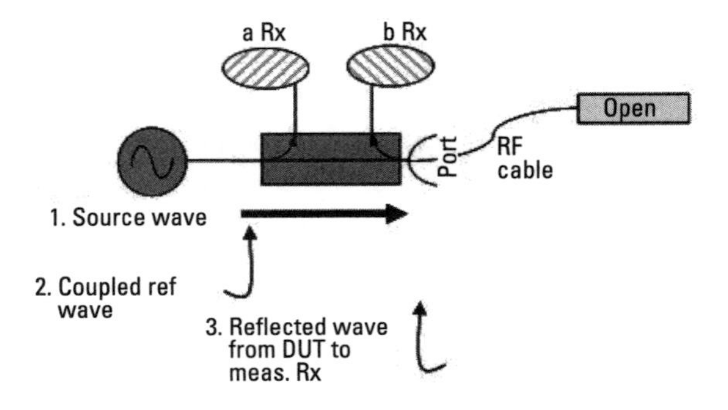

Figure 1.10 Reflection tracking error.

into the coupler. The wave is first sampled by the *a* receiver and the wave reflected from the open flows back through the coupler in the reverse direction to the *b* receiver. The difference in amplitude and phase of these two waves up to the reference plane is reflection tracking error. The calibration process will measure these differences and apply them to the measured data. It will rotate the Smith chart and scale it.

1.3.2.2 Frequency Response Transmission Tracking

When making transmission measurements, reference wave a_1 is compared to the measured wave b_2. For an ideal transmission, the a_1 wave will be equal in amplitude and phase to the b_2 wave.

1.3.3 Test Set Challenges

1.3.3.1 Directivity

One of the more challenging components to design in a VNA is the signal separation device, typically either a bridge or a coupler. Figure 1.10 shows a dual directional coupler used as a signal separation element. Directive elements are either a bridge or a directional coupler. Table 1.1 summarizes the advantages and disadvantages of the different directional elements.

In a perfect coupler, the forward path would only include signals traveling from coupler port 1 to port 2. The forward *a* wave would never leak into the *b* receiver. In a real directional element, it does but it is a systematic error and will be reduced by calibration.

When making reflection measurements, reference wave *a* is compared to the measured signal *b* with a load connected at the port. With a load applied, no signal should be reflected from the DUT back into the port. Most of the signal seen by the *b* receiver will be due to leakage of the forward traveling wave into the *b* receiver in the wrong direction. When the forward traveling wave is seen at the *b* receiver, this is a measure of directivity at a single port.

The same process will be carried out to measure the directivity at each port.

1.3.3.2 Port Match

In an ideal measurement system, when measuring reflections from a DUT, all of the reflected signal will be measured by the measurement receiver *b*. In reality, the VNA never has a perfect match. Due to this, reflections and rereflection will occur between the DUT and the analyzer. These signals will be measured at the *b* receiver, but will not be seen by the reference or *a* receiver. Figure 1.13 shows the signal flow for port match error at the input of a DUT. When performing a calibration, reflection standards will be connected to the port

Table 1.1
Comparison of Bridge Versus Directional Coupler

	Bridge	**Directional Coupler**
Pro	Low frequency perf. Port match Size Thermal Stability ESD Insensitive	Power handling Wide bandwidth Low source attenuation AC coupled to receivers
Con	Power handling Lower source power DC path (port to receivers) Thermal Drift	Low frequency performance Additional ESD protection Size

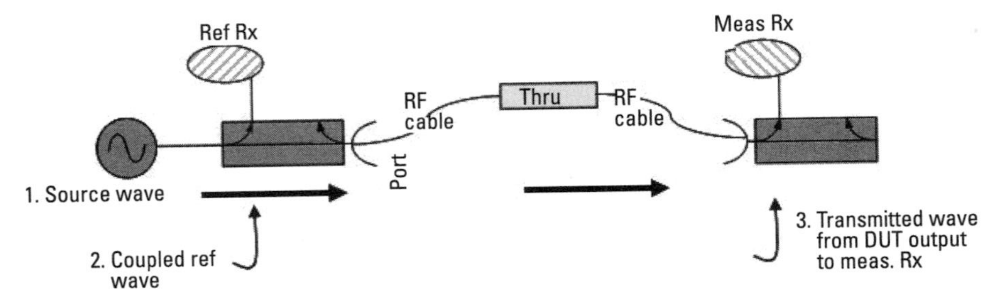

Figure 1.11 Transmission tracking error.

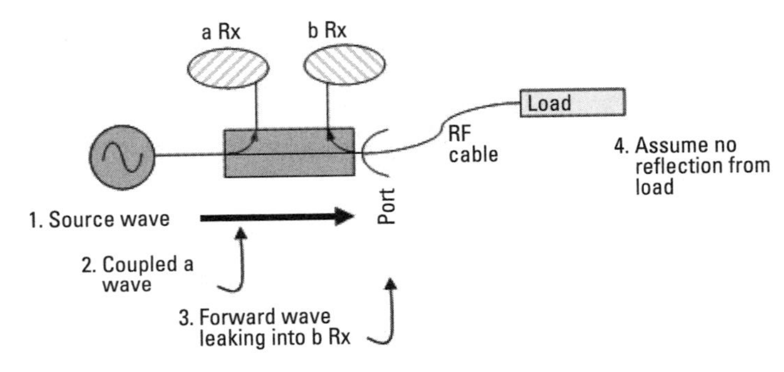

Figure 1.12 Single-port directivity error.

with known reflection coefficients. These standards will be measured by the *b* receiver and used to calculate the port match error. The VNA will subsequently use the error coefficients to remove match errors from each port.

1.4 Swept Versus Stepped Mode

When sweeping the frequency of a VNA, the source will start at the start frequency, step at the step size, and continue this algorithm until it reaches the stop frequency and then repeats. While this may seem like a simple and intuitive process, it becomes more complex when trying to measure devices with long delays in them. This is because when the output is delayed significantly, the *b* receiver may be measuring the response to the wave at the previous frequency if the delays are long enough. Stepped mode is designed to remedy this problem by enabling the user to define an associated settling time for the measurement receiver to wait before measuring. The default mode for sweeping the frequency is stepped mode. In the stepped mode, as the frequency is changed, the measurements are performed at each frequency point. The benefit of this is that the source is settled for a predetermined time to settle the phase, frequency, and amplitude. This enhances measurement precision with the drawback of

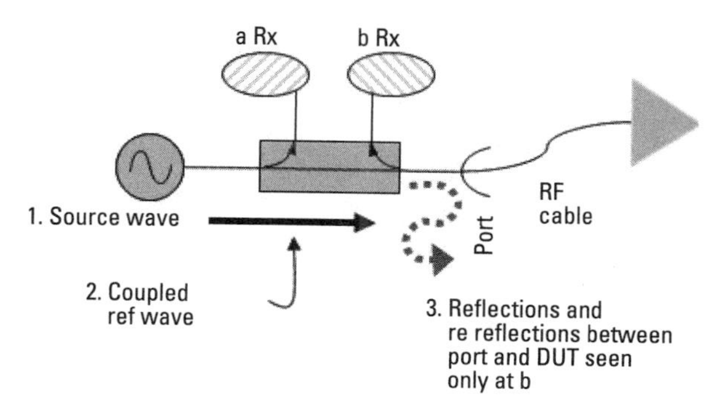

Figure 1.13 Source match error at input.

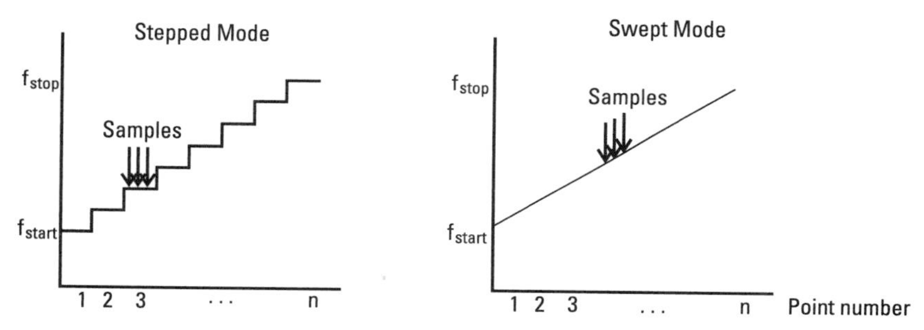

Figure 1.14 Comparison of the stepped and swept modes.

reduced speed. The swept mode is intended to reduce sweep time by continuing to sweep regardless of the receiver status. Therefore, there may be reduced precision while operating in the swept mode.

1.4.1 Chopped Versus Alternate Mode

Alternate and chopped modes are additional mechanisms for reducing measurement times and improving measurement performance. In the chop mode, the source will stimulate one port of a device and measure the resulting response waves and then switch measurement direction and measure in the opposite direction before moving to the next frequency point. It is the default mode of all of the VNAs on the market. In the alternate mode, a complete frequency sweep is performed in the forward direction, then another complete frequency sweep is performed in the reverse direction. Depending on the application and DUT characteristics, alternate or chopped mode will deliver a faster sweep or better measurement performance.

The alternate mode is also useful for longer devices with significant delay and where the delay can manually be set by the user between measurement points.

2

Calibration

The modern VNA has found its way into practically every RF laboratory. It is not hard to understand why: no other instrument provides the same level of measurement accuracy as the VNA for evaluating the important building blocks of modern radio technology, from the smallest chip components to complex multifunction subsystems. At the heart of this modern measurement marvel lies the remarkable S-parameter. Relatively easy to derive at high frequencies, it bears a direct relationship to many important parameters of interest for RF and microwave designers.

S-parameters and VNAs have become so entwined in technical circles that verification of the former invariably requires the latter (and vice versa). Because of this important connection, it is vital to gain a good understanding of S-parameters before embarking on VNA calibration and measurement topics.

2.1 VNA Measurements in an Ideal World

Our discussion on S-parameters begins with the humble wave quantity, which forms the basis for every S-parameter measurement. The VNA generator launches a wave towards a device under test (DUT) (Figure 2.1(a)). This is the incident wave, denoted by the letter a. Once this wave reaches the DUT, a portion is reflected back towards the generator. This is the reflected wave, denoted by the letter b. The true power of the incident wave is $|a|^2$, and likewise the true power of the reflected wave is $|b|^2$ [1]. The ratio of the reflected to incident wave is called the reflection coefficient and is given by:

$$\textit{reflection coefficient}, \Gamma = \frac{b}{a} \tag{2.1}$$

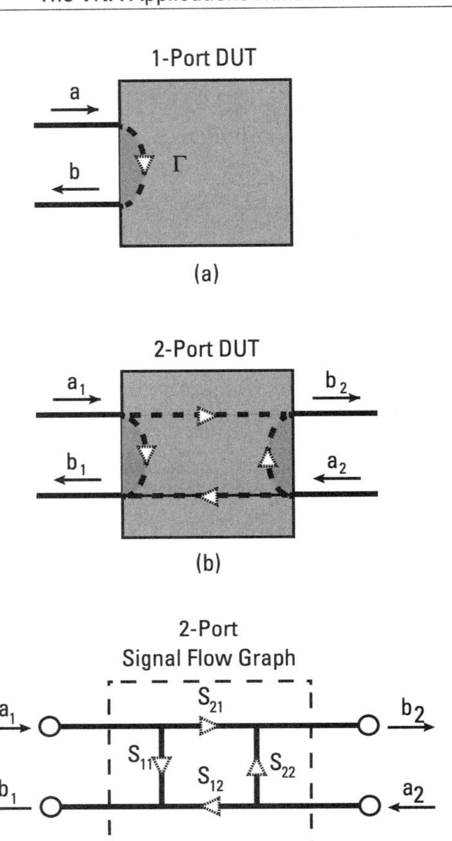

Figure 2.1 (a) Incident and reflected waves associated with a one-port device. (b) Incident and reflected waves associated with a two-port device. (c) Signal flow graph for a two-port device.

This is rather intuitive: the reflection coefficient simply measures the amount of energy reflected back relative to the amount sent forward.

The DUT generally has a complex impedance that varies over frequency, while the measurement instrument ideally has a constant impedance, usually 50Ω. The critical takeaway is that the reflection coefficient, Γ, is intimately related to the device impedance [2]. The basic relationship is given by:

$$\Gamma = \frac{(Z_l - Z_o)}{(Z_l + Z_o)} \tag{2.2}$$

If we divide both numerator and denominator by Z_o (the characteristic impedance of the system) and define a normalized impedance as:

$$z = \frac{Z_l}{Z_o} \tag{2.3}$$

then we get:

$$\Gamma = \frac{(z-1)}{(z+1)} \tag{2.4}$$

Solving for z, we obtain:

$$z = \frac{(1+\Gamma)}{(1-\Gamma)} \tag{2.5}$$

We went through this exercise to demonstrate that if you can measure Γ and know your instrument's system impedance, you can determine the unknown impedance of your device. VNAs make this type of calculation a cinch, because they are optimized for measuring incident and reflected waves at every port.

Next, let's take this wave quantity concept to the next level by considering a two-port device. In Figure 2.1(b), there are now incident and reflected waves at both ports. We use subscripts to keep these straight. Here, we have the a_1 incident at port 1. Is the reflected value b_1 caused by reflections from port 1 or signal feedthrough from port 2, or some combination of these? Enter the S-parameter concept, which clearly delineates reflections from different sources (incident waves). A signal flow graph for a two-port network is shown in Figure 2.1(c). The formula for the complete set of S-parameters for a two-port device is given by:

$$b_1 = s_{11}a_1 + s_{12}a_2 \tag{2.6a}$$

$$b_2 = s_{21}a_1 + s_{22}a_2 \tag{2.6b}$$

This set of equations presents the reflected waves measured at each port as a function of the incident waves applied at all other ports. This relationship is linear, with complex coefficients (the four S-parameters) multiplying each incident wave quantity (a_1 and a_2). So the composite b_1 reflection can be obtained by measuring the impact of each incident wave separately and simply adding them together (vectorially). For example, if we set $a_2 = 0$ in (2.6a) (implying there is no incident wave applied at port 2), we can isolate the s_{11} term:

$$s_{11} = \frac{b_1}{a_1}\Big]_{a_2=0} \tag{2.7a}$$

If you compare (2.7a) with (2.1), you will see that another name for s_{11} is the reflection coefficient.

We can use this same process to isolate the three other terms:

$$s_{12} = \frac{b_1}{a_2}\Big]_{a_1=0} \tag{2.7b}$$

$$s_{21} = \frac{b_2}{a_1}\Big]_{a_2=0} \tag{2.7c}$$

$$s_{22} = \frac{b_2}{a_2}\Big]_{a_1=0} \tag{2.7d}$$

Incidentally, this is the same divide-and-conquer strategy used by the VNA. It applied a stimulus (incident wave) at only one port at a time and measures the reflected waves from every other port. For our idealized two-port VNA, the process looks like this:

- *Forward sweep:* Apply stimulus (a_1) at port 1 and terminate port 2 in 50Ω (to force $a_2 = 0$), and then measure b_1 and calculate $b_1/a_1 = s_{11}$ and measure b_2 and calculate $b_2/a_1 = s_{21}$.
- *Reverse sweep:* Apply stimulus (a_2) at port 2 and terminate port 1 in 50Ω (to force $a_1 = 0$), and then measure b_1 and calculate $b_1/a_2 = s_{12}$ and measure b_2 and calculate $b_2/a_2 = s_{22}$.

Note that both a forward and a reverse sweep are required to obtain all four S-parameters.

2.2 Measurement Errors in the Real World

In the real world, things are a bit more complicated. As opposed to our ideal VNA (Figure 2.2(a)), measurements from a real VNA are degraded by both ran-

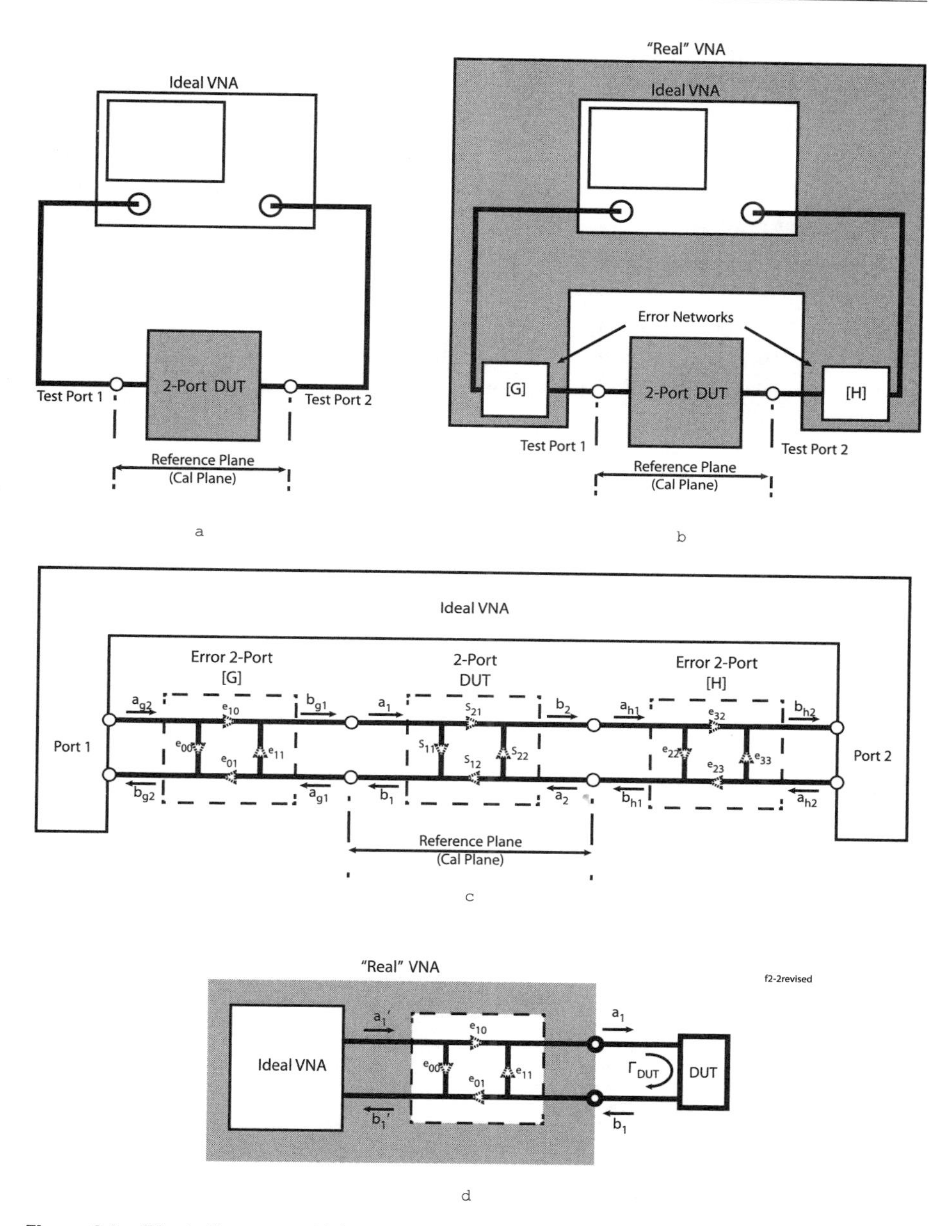

Figure 2.2 Block diagrams of (a) ideal and (b) real-world VNAs, showing how the error network models of a real-world VNA surround an ideal VNA. (c) Individual error paths are identified within each error network. (d) Error network for a one-port measurement.

dom and systematic errors [3]. Random errors are errors that can only be quantified statistically (at best), and include things like thermal drift, repeatability, and noise. Systematic errors are repeatable and can be completely eliminated (theoretically) through vector error correction. This process requires that both the magnitude and the phase of an error contributor be determined so that the correction algorithm can effectively remove its influence from the measured S-parameters. In practice, we consider a real VNA to consist of an ideal VNA with systematic errors concentrated into linear error network models at each port, as shown in Figure 2.2(b). Each error network contains its own two-port S-parameter signal flow graph, as shown in Figure 2.2(c). The error model includes all possible contributions of systematic inaccuracies, including test port coupler directivity, the match presented by each reflectometer, the frequency response of the reflectometers, transmission between ports, and crosstalk between ports. Chapter 3 discusses error contributions associated with directivity in Section 3.4.1 and source match in Section 3.4.2.

Notice in Figure 2.2(c) the additional introduction of the reference plane (sometimes called the calibration plane). All error correction is performed relative to this plane. So it is important (particularly at microwave frequencies) to understand exactly where this plane is defined with respect to coaxial connectors. Figure 2.3 shows this reference plane for both a male and female *N* connector: It is at the mating surface (or plane) of the outer conductors. Note that for many RF connectors, the outer conductor for the RF signal is different from the conductor that provides the nut and screw threads for making the mechanical connection.

2.2.1 Random Errors

Since random errors are statistical in nature, they cannot be eliminated by the mathematical techniques described above but must rely upon good VNA design and high-quality auxiliary components (cables and adapters) coupled with

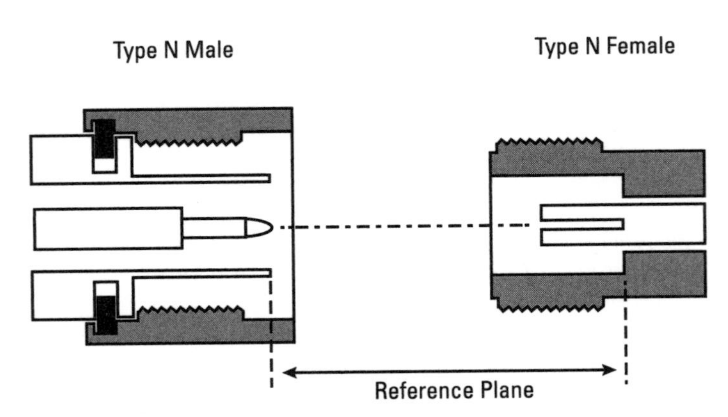

Figure 2.3 Reference plane for male and female N connectors.

good operator practice to reduce their influence as much as possible. Several sources of random errors are discussed below, along with mitigation strategies.

2.2.1.1 Thermal Drift

Temperature changes affect both trace stability and accuracy of VNA measurements. Search any VNA datasheet for "temperature" and you will see how this environmental factor plays a role. In general, sensitivity to temperature should be small in a well-designed instrument (e.g., "0.01 dB/K, 0.1°/K (typical)" for the ZNA). VNA performance is often specified for a fairly narrow operating range (+18°C to +28°C), with the expectation that the environment will remain within 1C of the calibration temperature for subsequent measurements.

Temperature changes in the lab not only affect the VNA, but also affect the conductors and dielectric material of the test cables. Thermal mitigation strategies include ensuring good airflow around the VNA and avoiding colocating a VNA with other intermittently used instrumentation/heat sources (e.g., power supplies that are only occasionally turned on to power DUTs). Also, avoid placing the VNA and its components directly under heating or air conditioning vents.[1]

2.2.1.2 Repeatability

Repeatability refers to the correlation between successive measurements over a short period of time under the same conditions, for example, measuring the same S-parameter on the same DUT by the same instrument with all the same VNA settings, using the same measurement procedure. High repeatability requires the use of high-quality test cables and connectors, and optimized measurement technique that minimizes cable movement between the calibration and measurement phase, and also between measurements. Repeatability is also highly dependent on the quality/cleanliness of the connection plane and the connection process. Mitigation strategies include keeping the connector interfaces clean and free of contaminants. Never use water, acids or abrasives. Cotton swabs moistened with isopropyl alcohol do a reasonably good job of cleaning the connectors. I eschew common ear swabs, which are much too bulky for the fine features of RF connectors, and prefer pointed swabs.[2] Common ear swabs often utilize a glue to keep the cotton in place. This glue may dissolve in alcohol and contaminate RF connectors during cleaning. Avoid completely saturating the swab in alcohol, and verify that no cotton threads remain in the connector after cleaning. For screw threads, I press an alcohol-moistened paper towel into the body end of the screw thread with my fingernail and rotate the body of

1. It can be eye-opening to correlate VNA measurements with the temperature profile of the VNA environment over a 24-hour period.
2. A good example is the CCT2425 MicroPoint swabs from Chemtronics.

the connector counterclockwise until I reach the end of the device. Use liquid compressed air or nitrogen to augment the cleaning process.

Use a pin depth gauge to regularly check the position of the center pin relative to the defined reference plane for that particular connector type. You do not want to be the person who damages the precision test port connector on your laboratory's VNA because of a cheap (or damaged) cable, adapter, or component.

Another very simple (but very important) mitigation strategy involves the simple act of making a coaxial connection. Align the male connector with the female, and turn only the nut on the male connector, never the body of the female connector. When you turn the nut, the male pin remains stationary with respect to the female receptacle throughout the connection process. If you make the mistake of turning the body, the female receptacle rotates with respect to the male pin. Once contact is initially made, this rotating motion will scrape the (gold) plating off of the pin or the end of the receptacle (or both), leading to excessive wear, intermittent or poorly repeatable electrical performance, and drastically shortened service life. (This is by far the most common "newbie" error.) Also, get into the habit of using the proper torque wrench for your connector type every time you make a connection. This habit will serve you well towards the goal of achieving better measurement repeatability.

2.2.1.3 Thermal Noise

Thermal noise is another source of random measurement uncertainty that can have a large or insignificant impact on results, depending on the measurement. Thermal noise may be calculated from the following equation [4]:

$$N_T = kTB \tag{2.8}$$

where N_T = thermal noise (in watts), k = Boltzmann's constant = 1.38×10^{-23} J/K (Joules/kelvin) or W/K-Hz, T = temperature in Kelvin, and B = bandwidth in hertz.

Typically, thermal noise is calculated in a 1-Hz bandwidth at room temperature (290K), which is 4.0×10^{-21} W. Converting to dBm yields $10\text{Log}_{10}(4e_{21}/0.001) = -174$ dBm in a 1-Hz bandwidth. The noise affecting a VNA measurement is influenced by four factors:

1. Bandwidth of the IF filter used for the measurement;
2. Filter shape (relative to an ideal rectangular filter);
3. VNA receiver noise figure;
4. Receiver temperature.

Of these four factors, the receiver temperature is often ignored because the equipment is maintained close to ambient room temperature. Equation (2.9) accounts for the remaining three factors in determining the noise level (or noise floor) of a VNA:

$$L_N = -174\,dBm + NF_{RX} + 10log10\left(S_f\right) + 10log10(B) \tag{2.9}$$

where L_N = VNA displayed noise level (in dBm), −174 dBm is the normalized value of thermal noise (1-Hz bandwidth, 290K), NF_{RX} = VNA receiver noise figure (in decibels), S_f = VNA receiver's IF filter shape factor (relative to an ideal rectangular filter), and B = VNA receiver's IF filter bandwidth.

This equation suggests that if either the bandwidth or the noise figure increases, so does the noise power (and thus the noise floor). High dynamic range (DR) is important in a VNA and is specified as the difference between the maximum output power and the noise floor measured in a 1-Hz bandwidth (in decibels). Similarly, high signal-to-noise ratio (SNR) is desirable in a VNA measurement. Anything that increases noise power decreases SNR. Let's quantify this with a measurement example. The objective is to attain a SNR of at least 45 dB at the VNA receiver (Figure 2.4) to maintain measurement uncertainty (due to noise) no greater than +/−0.05 dB.[3] Let's use two tables to keep track of signal and noise power at a VNA measurement receiver (b_2). To begin, we select a cable as the DUT, and assume this cable has an insertion loss of 0.5 dB. Further, we will use default VNA settings for power (−10 dBm) and bandwidth (10 kHz). In the signal level table, we start with −10 dBm, add the 0.5-dB insertion loss, and end up with −10.5 dBm at port 2 of the VNA. Inside the VNA, the port-2 directional coupler imposes an additional coupling loss of 13 dB, reducing the signal level at the b_2 receiver to −23.5 dBm.

In the noise level table, the thermal noise floor (−174 dBm) is impacted by the 10-kHz bandwidth of the IF filter (= 40 dB) and filter shape factor (2.6 dB). To this we add the noise figure of the receiver, which is 35 dB at 1 GHz [5].

The resulting SNR, 72.9, exceeds our objective (45 dB). This result agrees with real-world experience: it is easy to get a very stable (noise-free) measurement of a short cable with default VNA settings. You can see that lossy DUTs whittle away our SNR margin. For example, measuring a 30-dB attenuator reduces our SNR to only 43.4 dB, which means we will start seeing some trace noise. With increasing attenuation, the SNR becomes negative, implying that the desired signal is now buried below the thermal noise. For this situation we have two choices: either decrease the IF bandwidth, or increase the signal generator output power level (as long as the higher power does not overdrive the VNA receivers).

3. The quantitative influence of error signals is discussed in Section 3.4.1.

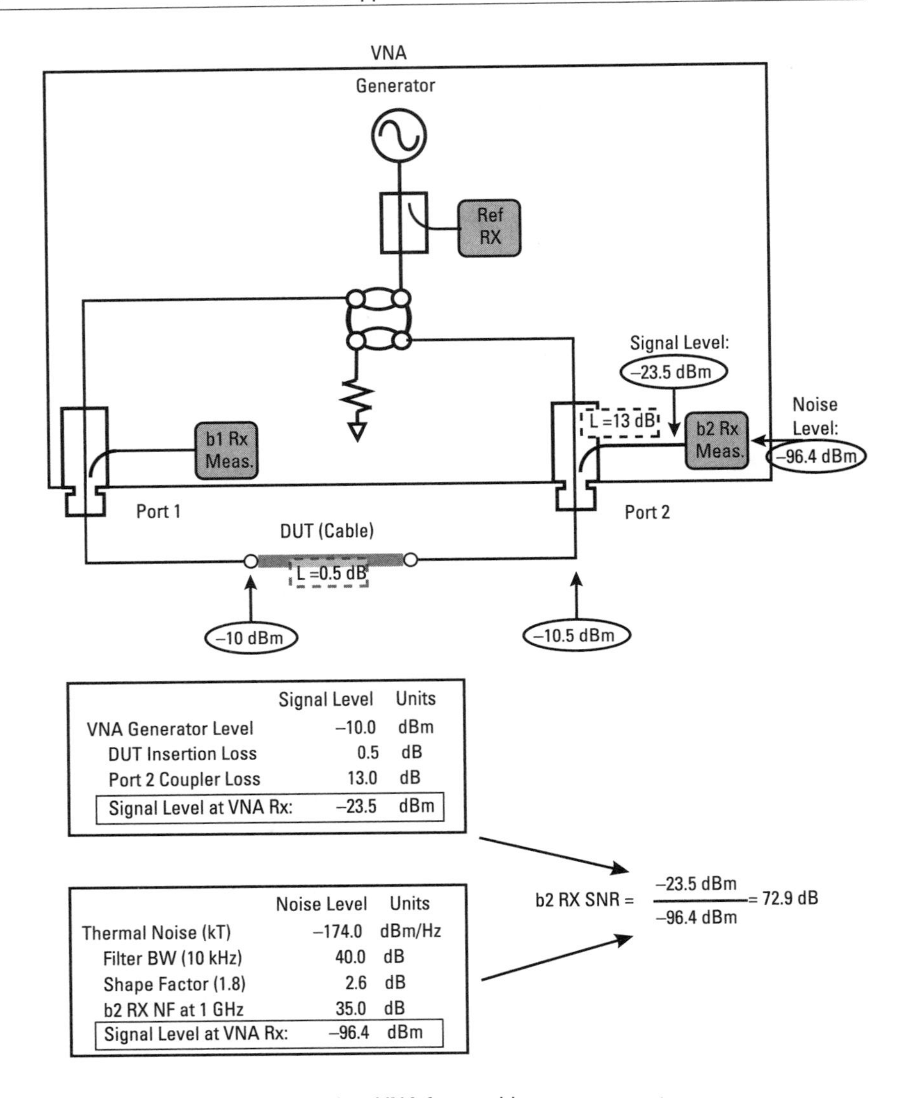

	Signal Level	Units
VNA Generator Level	–10.0	dBm
DUT Insertion Loss	0.5	dB
Port 2 Coupler Loss	13.0	dB
Signal Level at VNA Rx:	–23.5	dBm

	Noise Level	Units
Thermal Noise (kT)	–174.0	dBm/Hz
Filter BW (10 kHz)	40.0	dB
Shape Factor (1.8)	2.6	dB
b2 RX NF at 1 GHz	35.0	dB
Signal Level at VNA Rx:	–96.4	dBm

Figure 2.4 Signal and noise power in a VNA for a cable measurement.

Figure 2.5 shows both of these effects on the measurement of a bandpass filter. In the passband there is very little attenuation, so it does not really matter what values we use (within reason) for the generator power and IF bandwidth. As we make measurements past the filter's transition band into the stopband, we can see that our settings make a big difference. The top trace was obtained with a power level of −10 dBm and a 1-MHz IF BW, which roughly agrees[4]

4. For the top trace, a value of 0 dB means no gain or loss relative to the generator power of −10 dBm. Hence, a marker value of –64.3 dB means the absolute power of this marker is –74.3

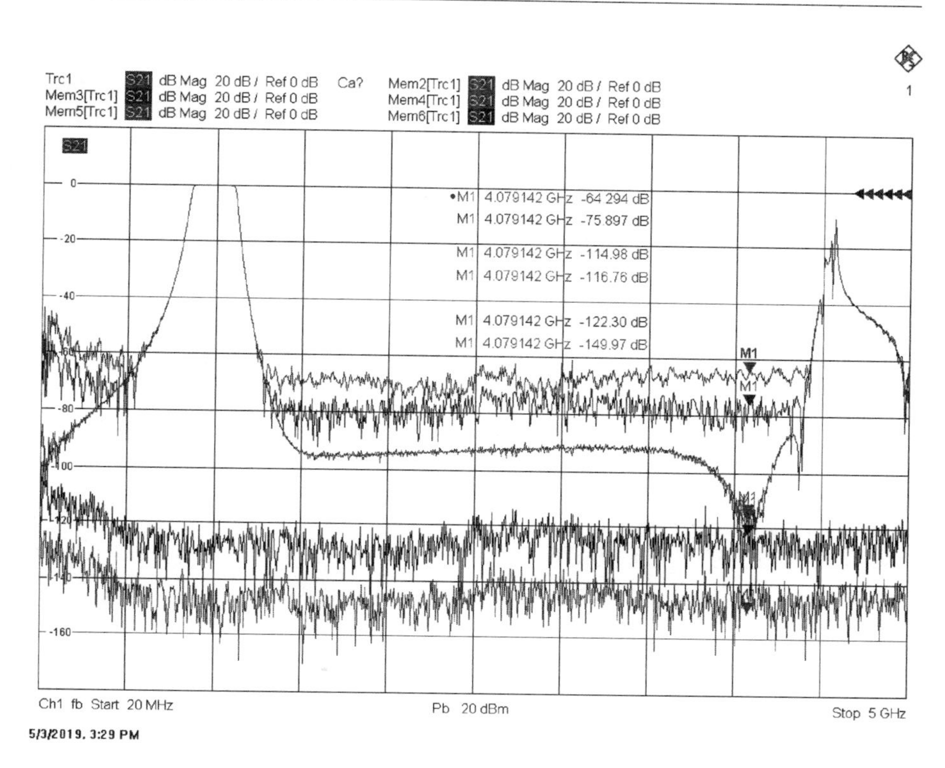

Figure 2.5 Filter measurement showing the impact of different bandwidths on stopband measurement performance (vertical scale, 20 dB/div). Top trace: 1-MHz BW, –10 dBm, 50 averages. Second trace: same as previous, but 0-dBm power. Third trace: 10-kHz BW, 0 dBm, 50 averages. Fourth trace: 10 Hz, 0 dBm, average off. (Trace 3 and 4 overlay each other.) Fifth trace: instrument dynamic range (port 1 terminated in the open standard, port 2 terminated in the match standard, 10-Hz BW, 0 dBm). Sixth trace: same as above, but VNA generator power set to maximum to achieve 150-dB dynamic range.

with noise values obtained using (2.9) and the noise level table of Figure 2.4. Notice that the vertical scale is 20 dB/division. The next trace down shows the results when we increase our signal generator power by 10 dB to 0 dBm. The noise floor decreases by 10 dB. For the third trace down, we have decreased the filter bandwidth from 1 MHz to 10 kHz, which lowers the noise floor an additional 20 dB ($=10\log_{10}(1e^6/1e^4)$), and places the noise below the signal level in the stopband. The fourth trace was obtained using an IF filter bandwidth of only 10 Hz, which represents an additional noise floor reduction of 30 dB.

---

dBm. Now subtract 60 dB for the 1-MHz IF filter and the resulting noise level is at −134.3 dBm/Hz. Finally, subtract the receiver noise figure (35 dB), and the shape factor (2.6 dB), and this back of the envelope calculation brings us close (−171.9 dBm/Hz) to the thermal noise floor (−174 dBm/Hz). This does not take into account cable losses and generator power accuracy, which can vary by several decibels.

However, the fourth trace is indistinguishable from the third trace, because the SNR in both cases is sufficient to see the actual signal in the stopband, and can even resolve the notch around 4 GHz. The fifth trace shows the VNA dynamic range for these settings (0 dBm, 10 Hz), while the sixth trace shows the maximum VNA dynamic range (150 dB) with generator power set to +20 dBm.

However, a picture is only worth a thousand words. Specifically, it does not reveal the measurement times involved for these measurements. For example, the measurement sweep associated with the top trace (1-MHz bandwidth) took only a few milliseconds, while the bottom traces (10-Hz bandwidth) took several minutes. Also, we suggested in Figure 2.4 that increasing the VNA output power was a good way to instantly improve the dynamic range (and SNR), without taking a hit on measurement time. Lower-cost or older VNAs often have restricted output power (maximum of –5 dBm or so), and all VNAs provide less power at higher frequencies. If the required dynamic range of the measurement is less than (but close to) the instrument's dynamic range, you can often make the measurement by maximizing output power and minimizing IF filter bandwidth. This might be an option for small-batch device characterization, but is unlikely to be acceptable for high-volume production testing or active bench alignment. In these latter cases, the extra performance of a modern high-end VNA (in terms of output power, receiver noise figure, and efficient digital processing) warrants the extra cost to minimize sweep (and hence measurement) time.

2.2.2 Systematic Errors

For VNAs, there are two types of systematic measurement errors: nonlinear and linear errors. We hinted at the nonlinear error type in the previous section, where we warned about overdriving the VNA receivers. Let's look at the nonlinear effects first.

2.2.2.1 Nonlinear Effects

Nonlinear effects (also called compression effects) can be observed when operating the measurement or reference receivers near their upper power limits. This is because of the electrical properties of the mixers used in the VNA receivers, which fail to faithfully reproduce the input signal level as it nears the 1-dB compression point, resulting in both amplitude and phase errors. If the signal level into both the reference and measurement receivers was the same, then the compression effects would cancel because of the ratio nature of the S-parameter measurements. However, these levels are rarely the same, due to differing signal levels between the input and output of a DUT. Furthermore, the VNA designers can adjust the input levels into the reference receivers to avoid compression and maximize dynamic range. However, the measurement receivers are at the mercy of the DUT (particularly amplifiers), which can produce power levels at the measurement receivers well in excess of the VNA's own internal signal sources. For this reason, attenuators are provided in the measurement paths of VNAs intended for active device (e.g., amplifier) measurements.

While compression effects can be quantified as a function of power level and frequency, this requires extra calculation and/or considerable memory resources, and time-consuming receiver characterization. Instead, it is much easier to simply maintain the power at the VNA receivers below compression levels using internal or external attenuators, couplers, or filters, and checking for compression (see Chapter 5). Additionally, VNA datasheets generally contain graphs of measurement uncertainty as a function of signal level for different frequency ranges. Figure 2.6 shows both compression effects and noise effects as a function of signal level and frequency for the ZNA.

2.2.2.2 Linear Effects

System error correction (SEC), also known simply as S-parameter calibration, involves mathematical compensation of the different error terms of the error network introduced in Figure 2.1(c). These error terms correspond to physical, measurable characteristics of the RF directional coupler (or reflectometer) and associcated components in the VNA test set. Most error terms can be interpreted as raw system data. After applying SEC, all system parameters should reflect an ideal VNA. In reality, some small residual error effects remain. The

Transmission measurement accuracy of the R&S®ZNA26 and R&S®ZNA43

The diagrams below show the typical accuracy of the transmission magnitude and transmission phase measurements for the R&S®ZNA26 in the frequency range from 10 MHz to 26.5 GHz and for the R&S®ZNA43 in the frequency range from 10 MHz to 43.5 GHz.

Analysis conditions: $S_{11} = S_{22} = 0$, calibration power 0 dBm, measurement power 0 dBm, high-quality semi-rigid cable. Drift effects were not considered.

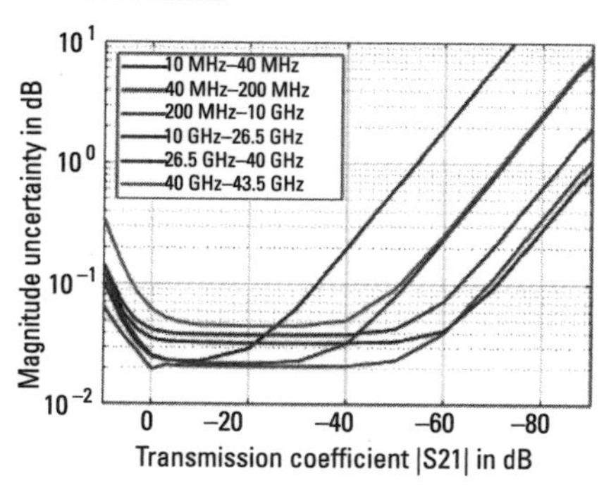

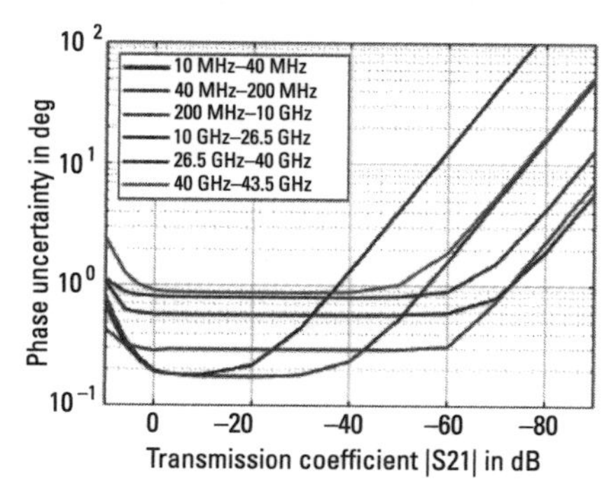

Figure 2.6 Measurement uncertainty due to compression effects (left edge) and noise (right edge) for ZNA.

resulting VNA performance is expressed in terms of effective system data. Table 2.1 provides an example of various error terms, their physical interpretation, typical raw and effective system data, and ideal system data.

The effective system data is also limited by random measurement errors such as temperature drift, noise, and repeatability. In the following sections, we discuss the different calibration standards and error-correction techniques available in modern VNAs, but largely ignore the details of the mathematic derivations. However, references are provided so that the interested reader may delve deeper into these topics.

2.3 Calibration Standards

The error-correction process relies upon calibration standards whose electrical characteristics have been carefully measured and stored as an integral part of each calibration kit. This data takes two forms: measurement-based or model-based. Measurement-based (also called "data-based") consists of an S-parameter file containing measurements over the frequency range of the calibration kit. Modeled data uses a polynomial equation whose coefficients are adjusted to fit measured S-parameter data. The modeled data approach has been around since the earliest days of the VNA when digital memory was costly and limited. By using a parametric description, the electrical behavior of a calibration standard can be calculated for any frequency from only a few parameters. This method cannot capture behavioral nuances, but it generally does a commendable job, thanks to precision machining and the simple geometry of most (coaxial) standards. However, using the actual measured S-parameters for calibration kit characterization will tend to capture behavioral nuances (most, but not all, nuances because measurements at frequencies between points covered by the calibration kit data must still be interpolated from nearest neighbor data). However, because digital memory is now cheap and plentiful, this method is

Table 2.1
Raw Versus Effective System Data

Error Term (refer to Figure 2.2(c))	System Parameter	Raw System Data	Effective System Data	Ideal System Data
e_{01}	Forward reflection tracking	<2 dB	<0.04 dB	0.0 dB
e_{00}	Forward reflectometer directivity	>29 dB	>46 dB	∞ dB
e_{11}	Forward source match	>22 dB	>39 dB	∞ dB
e_{32}	Forward transmission tracking	<2 dB	<0.06 dB	0.0 dB
e_x	Forward isolation	>130 dB	>130 dB	∞ dB
e_{22}	Forward load match	>22 dB	>44 dB	∞ dB

supplanting the modeled data approach. (Modern VNAs handle both types of kits, because there are plenty of legacy modeled-data kits still in use today.)

Regardless of the method used, the calibration standard's performance is only as good as how well it conforms to its characterization data.[5] This is why it is imperative to treat calibration kits with exceptional care. Any degradation in the physical kit means that the stored calibration data no longer faithfully represents the electrical characteristics of the physical standard. However, you will not know this by measuring the same standard that was used for the calibration, because the calibration process forces the analyzer to make the standard look like its data file. The only way to tell that something is amiss is by using an independent verification standard. This may be something as simple as calibration standards borrowed from another calibration kit, or by using an actual verification kit. Here is where having a calibration kit based on measured S-parameters is especially appealing for verification, because the S-parameters can be directly imported into the VNA from the USB stick supplied with the calibration kit. The imported data can be displayed on a memory trace, and compared with a measurement of the actual calibration standard. Formal uncertainty analysis is possible using commercially available software such as VNA Tools.[6]

Some believe that an extra standard (such as an attenuation state within an electronic calibration unit (ECAL)) can be used as a verification standard. This is only partially true, as it can only provide verification up to the connectors on the ecal unit. If a connector is damaged, this damage will propagate through every standard measured at that port, and remain invisible to the user unless an independent verification standard (from a manual calibration kit) is measured. This verification step is frequently skipped by VNA users. This leads to confusion over DUT measurement results when there are faulty calibration standards or simply a bad calibration. Figure 2.7 shows side-by-side measurement results from a bad and a good calibration kit, illustrating these principles.

The next step in our calibration journey is to take a closer look at the individual calibration standards found in VNA calibration kits. For one-port S-parameter measurements, only one-port calibration standards are required. For two (or more) port measurements, both one-port and two-port standards are required. In addition, different standards are required for 12-term and

5. The VNA capturing this data must first be calibrated and verified with standards traceable to a national bureau, like the National Institute of Standards and Technology (NIST) (in the United States) or Physikalisch-Technische Bundesanstalt (PTB) (in Germany).
6. VNA tools are available for free from the Swiss Federal Institute of Metrology Eidgenössisches Institut für Metrologie (METAS) website: www.metas.ch/vnatools. It allows the user to evaluate measurement uncertainty in compliance with the International Organization for Standardization (ISO) Guides to the expression of uncertainty in measurement (GUM). It also allows the user to extract model-based parameters for a calibration standard based on measured S-parameters.

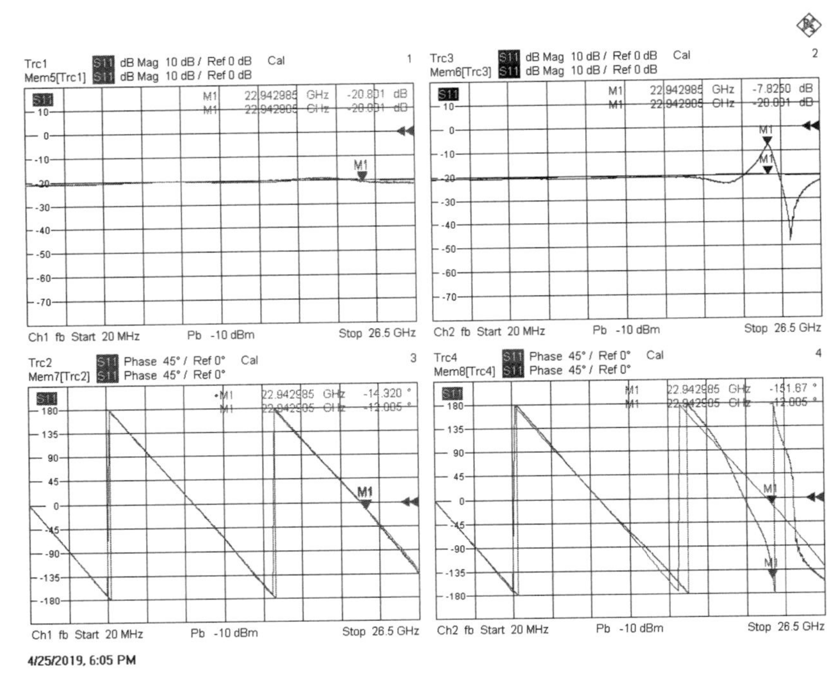

Figure 2.7 Comparing mismatch standard measurements. Standard is from a ZV-Z435 verification kit. At left, measured S11 versus S-parameter data for a good 3.5-mm calibration. (b) Measured s11 versus S-parameter data for a bad 3.5-mm calibration (damaged match standard).

7-term error-correction methods (discussed in Section 2.4). The most popular standards for both methods are presented below [6].

2.3.1 Open (O)

The cross-section of a coaxial open standard is shown in Figure 2.8(a). It consists of a metallic rod enclosed by a cylindrical metallic shell. The dimensions of the rod and shell are such that they form a short length of transmission line with 50Ω characteristic impedance. There is a gap between the end of the rod and the end of the shell. At low frequencies (below 50 MHz), the standard looks like an ideal open-circuit. As frequencies increase, fringing capacitance forms between the end of the rod and the metallic outer shell. Its frequency dependence is modeled by a third-order parametric equation:

$$C_e(f) = C_0 + C_1 f + C_2 f^2 + C_3 f^3 \tag{2.10}$$

A coaxial open standard is actually an electrical open at the end of a short length of transmission line. So as the frequency increases, the electrical phase

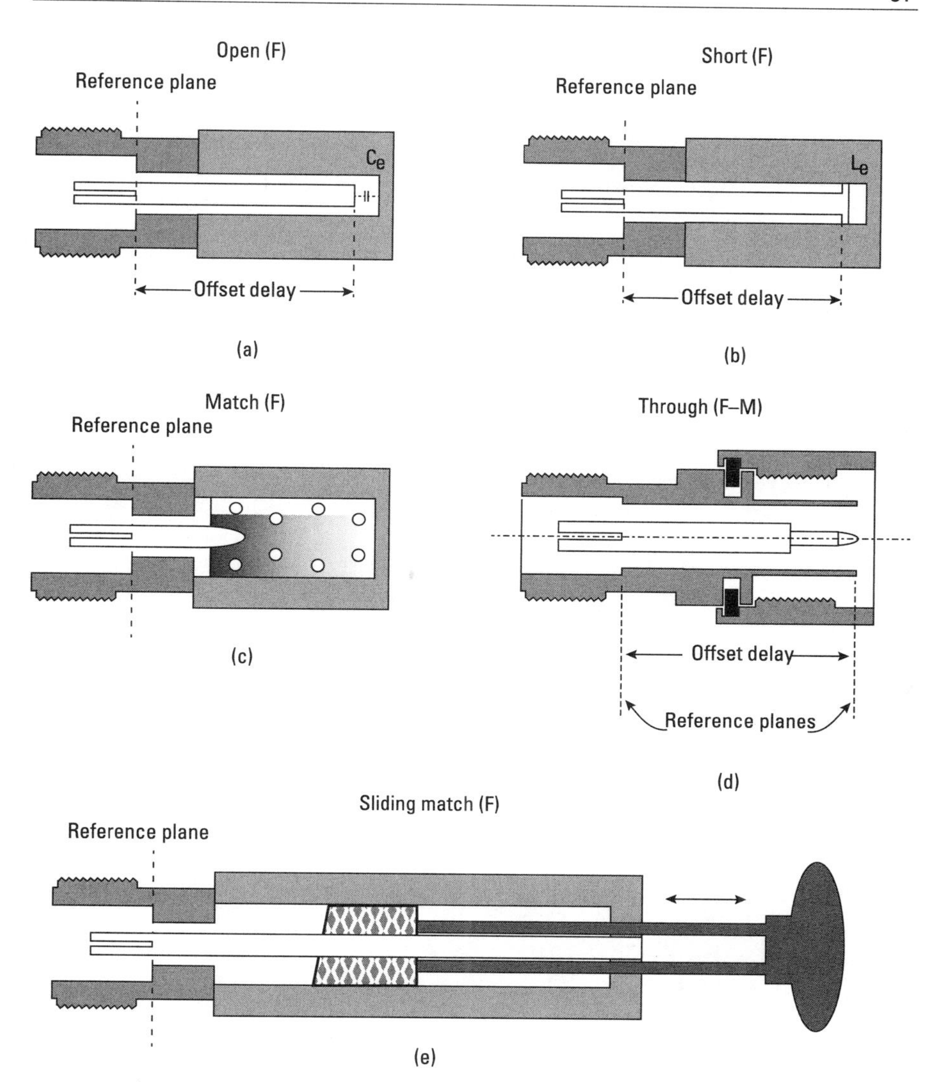

Figure 2.8 Simplified model of coaxial kit standards: (a) open, (b) short, (c) match, (d) thru, and (e) sliding match.

length (delay) associated with the line also increases. As viewed on a Smith chart, this increasing electrical phase is the reason for the clockwise rotation of the response with increasing frequency (see Figure 3.2, top left Smith chart).

To quantify this a bit, we first recognize that the reflection coefficient at any point along a transmission line is related to its terminating impedance (and associated reflection coefficient, $\Gamma(0)$), multiplied by a phase rotation:

$$\Gamma(l) = \Gamma(0)e^{-2j\beta l} \tag{2.11}$$

where the phase constant β is given by:

$$\beta = \frac{2\pi}{\lambda} \tag{2.12}$$

If the terminating impedance Z_l were a pure open (infinite impedance), the reflection coefficient would be:

$$\Gamma = \frac{Z_l - Z_o}{Z_l + Z_o} = 1 \tag{2.13}$$

However, because of the fringing capacitance, the reflection coefficient is more complicated. The impedance of a capacitor is given by:

$$Z_c = X_c = \frac{1}{j2\pi fc} \tag{2.14}$$

Inserting this into the reflection coefficient equation (2.13) yields:

$$\Gamma = \frac{Z_c - Z_o}{Z_c + Z_o} = \frac{\dfrac{1}{j2\pi fc} - Z_o}{\dfrac{1}{j2\pi fc} + Z_o} = \frac{1 - j2\pi fcZ_o}{1 + j2\pi fcZ_o} \tag{2.15}$$

Applying this result to (2.11) yields the frequency-dependent reflection coefficient for an open standard:

$$\Gamma(l) = \left[\frac{1 - j2\pi fcZ_o}{1 + j2\pi fcZ_o}\right] e^{-2j\beta l} \tag{2.16}$$

where c is C_e from (2.10)

Table 2.2 contains typical parameter values for a coaxial open standard.

2.3.2 Short (S)

The cross-section of a coaxial short standard is shown in Figure 2.8(b). It also consists of a metallic rod enclosed by a cylindrical metallic shell. The dimensions of the rod and shell are such that they form a short length of transmis-

Table 2.2
Parameter Values for a Coaxial Open Standard

Offset Length	Phase Uncertainty		Polynomial Coefficients for the Fringing Capacitance			
	0 to 8 GHz	8 to 26.5 GHz	C0	C1	C2	C3
5.0 mm	<0.5	<2.5	13.6348 fF	–0.2164 fF/GHz	0.0189 fF/GHz^2	–0.00028 fF/GHz^3

sion line with 50Ω characteristic impedance. The rod is electrically connected to the shell at the end, forming a short-circuit. At low frequencies (below 50 MHz), the standard looks like an ideal short-circuit. As frequency increases, the behavior of the electrical short is dominated by parasitic inductance similar to how the open is affected by fringing capacitance (see Figure 3.2, top right Smith chart). This frequency-dependent parasitic inductance is modeled by a third-order parametric equation:

$$L_e(f) = L_0 + L_1 f + L_2 f^2 + L_3 f^3 \tag{2.17}$$

An ideal short has a terminating impedance of 0Ω. Inserting this value into (2.13) yields a reflection coefficient of −1. The parasitic impedance results in a more complicated expression. The impedance of an inductor is given by:

$$Z_L = X_L = j2\pi fL \tag{2.18}$$

Inserting this expression into the equation for reflection coefficient (2.13) yields:

$$\Gamma = \frac{Z_L - Z_o}{Z_L + Z_o} = \frac{j2\pi fL - Z_o}{j2\pi fL + Z_o} \tag{2.19}$$

Inserting this into (2.11) yields the frequency-dependent reflection coefficient for a short standard:

$$\Gamma(l) = \left[\frac{j2\pi fL - Z_o}{j2\pi fL + Z_o}\right] e^{-2j\beta l} \tag{2.20}$$

where the parasitic inductance L is L_e from (2.17).

Table 2.3 contains typical parameter values for a coaxial short standard.

Table 2.3
Parameter Values for a Coaxial Short Standard

Offset Length	Phase Uncertainty		Polynomial Coefficients for the Parasitic Inductance			
	0 to 8 GHz	8 to 26.5 GHz	L0	L1	L2	L3
5.0 mm	<0.5	<2.0	0 pH	0 pH/GHz	0 pH/GHz^2	0 pH/GHz^3

2.3.3 Match (M)

A match (or load) is a precision impedance standard with a value designed to mimic the system impedance. It is designed to provide this impedance over as broad a frequency range as possible, using advanced manufacturing techniques such as laser trimming. The match standard has the same basic form factor/packaging concept as the open and short standards, but with a resistive termination (Figure 2.8(c)). In the distant past, models for the match standard simply assumed an ideal 50Ω impedance ($\Gamma = 0$) at all frequencies. This is a reasonable assumption up to perhaps 6 GHz (for a well-designed standard), but match performance degrades with increasing frequency. As a result, a calibration kit based on the ideal match model leads to reduced measurement accuracy at microwave frequencies. The preferred modern-day approach is to use a calibration kit employing measured return-loss values (S-parameters) for each standard over the frequency range of the kit. One other issue affecting match standards is the power-handling capability. Although the open and the short are more forgiving (since they provide complete reflection), the match standard must be able to absorb all incoming power. Not surprisingly, match standards are more frequently damaged by excessive power than the other standards.

Table 2.4 provides typical performance values for a coaxial match standard.

2.3.4 Sliding Match (Sliding Load)

The sliding load combines an accurate airline with a movable ferrite cylinder (Figure 2.8(e)). The airline has dimensions that are aligned with the connector's pin and shield diameters to minimize impedance discontinuities/reflections and achieve a characteristic impedance of 50Ω. The ferrite material absorbs most of the RF energy, beginning at around 2 GHz. The measured return loss may be as low as 20 dB, which is not great for a termination. By moving the position of

Table 2.4
Performance Values for a Coaxial Match Standard

DC Resistance	Return Loss		Maximum Power
	0 to 4 GHz	4 to 26.5 GHz	
$50\,\omega \pm 0.5\,\omega$	>40 dB	>30 dB	0.5W (27 dBm)

the ferrite material along the airline, the phase angle of its reflection coefficient changes. As seen in Figure 2.9, it takes three unique points around the reflection coefficient circle to determine the center. Notice that each pair of unique points is connected by a line segment. The perpendicular bisector of this line segment is then drawn on the Smith chart. We know that the center of the reflection coefficient circle lies somewhere along this perpendicular bisector, but we need another independent measurement point to generate a second perpendicular bisector. Once we have two perpendicular bisectors, we can determine the center of the circle by finding the point where they meet. This value, Γ represents the reflection coefficient of the airline. (The airline impedance can be calculated from (2.13) by rearranging the terms to solve for Z_l.)

In practice, the ferrite is placed at one position along the airline and a frequency sweep is performed. This process is repeated for other ferrite positions.

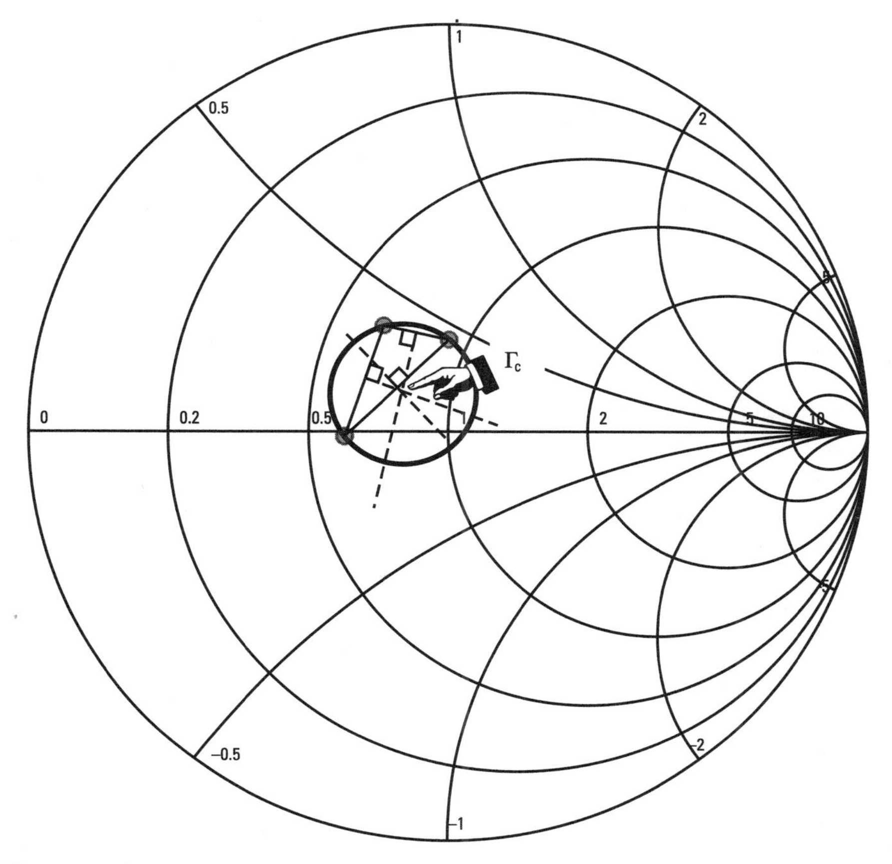

Figure 2.9 Determining the reflection coefficient of a sliding match airline using a 3-point geometrical method. By drawing line segments between pairs of points and then constructing a perpendicular bisector for each line segment, the center of the circle, Γ can be determined.

However, as frequency increases, the reflection coefficient phase angle associated with each ferrite position will change. Phase angles associated with longer airline positions will move more quickly around the Smith chart than shorter airline positions. At some frequencies, the reflection coefficients of the longer airline will overtake the shorter airline, resulting in two measurement points on top of each other. Thus, it is common practice to use 6 different ferrite positions to ensure that there will always be a sufficient number (3) of unique measurement points to determine the center of the airline's reflection coefficient circle.

Another interesting aspect of the sliding match is that the center pin is movable. This allows its position to be adjusted relative to the mating connector. Generally, a small gap is required to ensure that the pin is not jammed into the mating receptacle, resulting in connector damage. A gauge allows this gap to be very accurately established, because there is a trade-off between gap size and reflection coefficient magnitude. For example, a gap of 0.15 mm in an *N* connector yields a parasitic reflection coefficient of 33 dB at 18 GHz. If the gap is halved (0.067 mm), the parasitic reflection coefficient improves to 40 dB. This parasitic reflection impacts the airline reflection coefficient that is determined from the geometrical method described above.

What about measurements below 2 GHz? Since the ferrite material becomes decreasingly effective at absorbing RF energy below 2 GHz, a sliding match is often complemented by a fixed match for this lower frequency range.

2.3.5 Thru (T)

The cross-section of a coaxial thru standard is shown in Figure 2.8(d). It is a two-port standard consisting of a short airline section (metallic rod enclosed by a cylindrical metallic shell, often supported by dielectric spacers, or simply held in place by the connector assembly at each end). The dimensions of the rod and shell are such that they form a short length of transmission line with 50Ω characteristic impedance. The electrical length and loss are known and included in the calibration kit. If the test cables use the same connector types but with opposite genders, this is called an insertable connection. The connectors can be directly mated to form a zero length thru with 0-dB loss and 0 ps of electrical delay. Care must be taken when using older calibration kits that contain no thru standard because the calibration kit characterization data assumes an insertable thru connection. However, the most common situation today is for a DUT to have the same connector type and gender at the input and output (e.g., SMA female connectors). This noninsertable situation calls for a noninsertable thru standard, most often a SMA female-female "bullet." However, this bullet standard has both electrical delay and loss. If you calibrate with this standard using characterization data that assumes zero loss and zero phase, you will end up incorporating this error into your DUT measurements. Essentially, the measured loss of your DUT will be reduced by the loss of the thru standard. (Also, the measured electrical length of your DUT will also be shorter than its actual electrical length by an amount equal to the electrical length of the thru.) This is an easy mistake to make, and while the error introduced at lower frequencies (perhaps up to UHF, for example, 500 MHz) is usually negligible, the size of the amplitude and phase error grows with frequency. In a day and age where engineers must eke out every tenth of a decibel of performance from their wireless devices in the 3 to 6-GHz range, making this calibration error can cause measurement discrepancies (e.g., between manufacturer and customer) if either fails to account for the thru standard's actual loss.

2.3.6 Reflect (R)

As opposed to the open standard of Section 2.3.1, the one-port reflect standard has considerably relaxed requirements. Specifically, the reflection coefficient must be greater than zero, and its phase must be known to +/–90° (e.g., in the top half of the Smith chart (inductive) or bottom half (capacitive)). This is great news when building your own standards, since precise information about its electrical length and loss is not necessary. The reflect standard is used in some of the 7-term error-correction routines, since these allow partially known standards. The only caveat is that the 7-term method assumes that the reflect standard is identical at both ports ($s_{11} = s_{22}$) and if there is any phase delay that extends the phase response beyond 0° to +90° or 0° to –90°, this delay must be accounted for.

2.3.7 Line (L)

The line standard is a two-port standard that is used in conjunction with a thru standard for the TRL 7-term error correction method. The line standard establishes the measurement impedance and is matched as closely as possible to the desired characteristic impedance. The length of the line standard must be longer that the length of the thru standard. In practice, the line's electrical length is established as the difference between the line and thru lengths, and this electrical phase length varies from 20° at the lowest operating frequency to 160° at the highest frequency of interest. (This design range provides conservative margin from 0° and multiples of 180°, which cause calculation singularities.)

2.3.8 Symmetrical Network (N)

The symmetrical network is a two-port standard with similar partially known requirements as the reflect standard. The reflection coefficients of the ports must be identical ($S_{11} = S_{22}$) and known to behave "more like an open" or "more like a short." The transmission coefficient (S_{21}, S_{12}) is not used, so it can have values ranging anywhere from 0 dB to $-\infty$ dB.

2.3.9 Attenuator Standard (A)

The attenuator standard is a two-port standard that has an insertion loss ranging from 10 to 55 dB (the exact value does not need to be known), but must be well matched at both ports (high return loss).

2.3.10 Unknown Thru (U)

The unknown thru is a two-port standard with unknown loss and length characteristics. However, it must be a reciprocal device ($S_{21} = S_{12}$), and its delay time/electrical phase must be known approximately, as the calibration procedure delivers two possible solutions. However, it is usually pretty easy to decide on the appropriate solution based on an assessment of the device's physical length.

2.4 Calibration Techniques

The modern VNA supports a wide variety of calibration techniques to support a large number of different applications. The various methods are differentiated according to their calibration kit requirements and resulting measurement performance. By calibration kit requirements, we refer to the particular calibration standards needed for the calibration, and the degree to which the electrical characteristics of the standards must be known. Some techniques, like the TOSM method, require four fully characterized standards and a total of eight measurements, whereas the TNA method requires only three standards (one of which is only partially characterized), and only six measurements. While the former method might be preferable where a suitable commercial calibration kit is available, the latter would be much more attractive for a symmetrical test fixture with largely unknown electrical characteristics.

In this section, we will consider different methods based on user complexity (number of standards and connect/disconnect cycles). In general, as the complexity increases, so does the error-correction performance. Best multiport system error correction is achieved with 7- to 12-term error correction methods.

2.4.1 Normalization

In normalization, a single calibration standard is used to flatten a response. This can be very useful for "go/no-go" testing, for example, to tell whether a component (or cable) is short-circuited or open-circuited. Before applying normalization, the calibration standard is connected and the aggregate vector combination of all (unknown) error terms is measured. During the normalization process, this measured response is compared with the calibration kit data for the standard, and correction values (for amplitude and phase) are applied to force the displayed measurement to mimic the calibration kit data. After normalization, these same amplitude and phase correction values are applied to every subsequent measurement. Normalization is better than no calibration at all, but will generally exhibit much more ripple on DUT results than a more comprehensive calibration technique.

2.4.1.1 One-Port Normalization

One-port normalization involves reflection normalization (S_{11} or S_{22}) with either an open (Refl. Norm Open) or short (Refl. Norm Short) standard. The appropriate open or short standard is attached to the end of the desired VNA test port cable, and a single measurement sweep is performed. While the resulting response looks impressive (mirroring the characterization data for the standard), only the aggregate of the different internal error contributors (test port match, reflection tracking, and directivity) is corrected. Because the DUT reflection characteristics will affect these individual error terms differently than the highly reflective calibration kit standard, the DUT measurement will show more ripple.

2.4.1.2 Two-Port Normalization

In two-port normalization, a two-port thru standard is used for either normalization of the response in the forward direction (Trans Norm) or both forward and reverse direction (Trans Norm Both). Like the one-port method, the objective is to force S_{21} (and/or S_{12}) to the amplitude and phase values of a known thru standard. Again, only the aggregate effect of the 12 (or more) error terms is corrected with a single amplitude and phase value at each frequency point. Any deviation in electrical length, loss, or match of the DUT from the normalization thru standard will affect the individual error contributors differently, resulting in ripple. However, this can be a quick and easy way to determine if a transmission path is damaged (open) or excessively attenuated [7].

2.4.2 Full Single-Port Correction (OSM)

In a full single-port correction (Refl OSM), three standards are used to generate three independent equations to solve for the three unknown error terms associated with the single-port error model. This requires three completely characterized calibration standards (open, short, and match).[7] The signal flow graph [8]

7. Other values are also possible, but these three are conceptually easy to construct and characterize. They also result in values approaching the reflection coefficient extremes ($\Gamma = 1, -1$)

for a one-port measurement is presented in Figure 2(d). The equation resulting from this flow graph is presented in (2.21), where the DUT's measured reflection coefficient (M_{DUT}) consists of the true reflection coefficient (Γ_{DUT}) masked by the influences of undesired error terms e_{11}, e_{10}, and e_{11}. (We use the terms M and Γ rather than Γ^m and Γ^a to simplify the equations, and to emphasize the difference between the measured value M (affected by errors) and the true value Γ.)

$$M_{DUT} = e_{00} + \frac{e_{10}e_{01}\Gamma_{DUT}}{\left(1 - e_{11}\Gamma_{DUT}\right)} \tag{2.21}$$

Going forward, since the e_{10} and e_{01} terms always appear together as a product ($e_{10}e_{01}$), we make the simplifying assumption that $e_{01} = 1$, and lump the "out and back" error effects into e_{10}.

To solve for the unknown error terms in (2.21), the calibration kit standards are connected one at a time, and the VNA performs a measurement sweep to collect raw S-parameter data for each standard. The corresponding equation for each standard is presented below.

$$\text{Open: } M_O = e_{00} + \frac{e_{10}\Gamma_O}{\left(1 - e_{11}\Gamma_O\right)} \tag{2.22}$$

$$\text{Short: } M_S = e_{00} + \frac{e_{10}\Gamma_S}{\left(1 - e_{11}\Gamma_S\right)} \tag{2.23}$$

$$\text{Match: } M_M = e_{00} + \frac{e_{10}\Gamma_M}{\left(1 - e_{11}\Gamma_M\right)} \tag{2.24}$$

The next step is to solve these three equations for the three unknowns e_{00}, e_{11}, and e_{10}. All other information is known: the actual reflection coefficients (Γ_O Γ_S and Γ_M) from the calibration kit data, and the measured reflection coefficients (M_O M_S M_M) from the VNA raw measurements. Starting with (2.24) for the match standard, we note that the match termination is equal to the system impedance by definition, and therefore the actual match standard has no reflections ($\Gamma_M = 0$) leading to:

---

and also the center of the Smith chart, ($\Gamma = 0$), respectively.

$$M_M = e_{00} \tag{2.25}$$

We will use this information while rearranging (2.22) and (2.23) to solve for e_{10}:

$$\frac{(M_O - M_M)(1 - e_{11}\Gamma_O)}{\Gamma_O} = e_{10} \tag{2.26}$$

$$\frac{(M_S - M_M)(1 - e_{11}\Gamma_S)}{\Gamma_S} = e_{10} \tag{2.27}$$

Setting (2.26) equal to (2.27) and solving for e_{11}, we eventually obtain:

$$e_{11} = \frac{\Gamma_S(M_O - M_M) - \Gamma_O(M_S - M_M)}{\Gamma_S\Gamma_O(M_O - M_S)} \tag{2.28}$$

Similarly, we can solve (2.22) and (2.23) for e_{11}:

$$\frac{(M_O - M_M) - e_{10}\Gamma_O}{(M_O - M_M)\Gamma_O} = e_{11} \tag{2.29}$$

$$\frac{(M_S - M_M) - e_{10}\Gamma_S}{(M_S - M_M)\Gamma_S} = e_{11} \tag{2.30}$$

Equating (2.29) to (2.30) and solving for e_{10} eventually yields:

$$e_{10} = \frac{(M_S - M_M)(M_O - M_M)(\Gamma_O - \Gamma_S)}{(M_O - M_S)\Gamma_O\Gamma_S} \tag{2.31}$$

Equipped with the calculated error terms e_{00}, e_{10}, and e_{11}, we can rearrange (2.21) to determine the true reflection coefficient of any DUT, by peeling off these error terms from the measured reflection coefficient:

$$\Gamma_{DUT} = \frac{M_{DUT} - e_{oo}}{e_{10} + e_{11}(M_{DUT} - e_{oo})} \tag{2.32}$$

At the completion of the one-port OSM calibration, the VNA calculates the error correction coefficients e_{00}, e_{11}, and e_{10} using (2.25), (2.28), and (2.31), respectively, for each frequency point and stores them in memory. Then, during a measurement sweep, the VNA applies (2.32) to each frequency point to obtain an error-corrected trace of the DUT's true reflection coefficient.

2.4.3 One-Path, Two-Port Correction

One-path, two-port correction combines a full OSM calibration at port 1 with transmission normalization between port 1 and port 2. It is a very useful method for situations involving unidirectional transmission (from port 1 to port 2), such as when only the gain and input match of an amplifier are of interest. Typically, a high-value attenuator is placed at the output of the amplifier (to protect port 2 of the VNA from damage). With a conventional two-port measurement technique (Sections 2.4.4 to 2.4.5), if the attenuator is considered a part of the DUT (during measurement), the measured DUT gain will be reduced by the loss of the attenuator, and port 2 of the VNA will measure only the return loss of the attenuator, not the amplifier. If the attenuator is considered part of the VNA (during calibration), this high attenuation will greatly reduce the raw directivity of the port-2 reflectometer, making the resulting error correction very susceptible to phase changes caused by slight test cable movement. More significantly, the high attenuation will reduce the SNR of any port-2 VNA measurements, and this will induce noise into the calibration. (Since a conventional two-port calibration requires measurement sweeps in both the forward and reverse direction, the effect of this port-2 noise and low SNR will propagate throughout the calibration, causing potentially noisy results in all four S-parameters.)

With one-path, two-port, a full OSM calibration is performed at port 1 (where there is good SNR), and a simple normalization is performed between port 1 and port 2 (a total of four connections/four sweeps). When a high-quality attenuator is applied to VNA port 2, it provides a nearly reflection-free load for both the normalization standard and for the amplifier. This is important, because uncorrected reflections cause ripple. Although VNA port 2 is uncorrected (mathematically), the attenuator provides a nearly reflection-free match (physically), resulting in a ripple-free normalization.

2.4.4 Seven-Term Error Correction

Seven-term error correction methods take advantage of modern VNA architectures, which have dedicated measurement and reference receivers for each port. As a result, four receivers collect data during a two-port measurement. The error model/flow diagram appears in Figure 2.2(c). There is a complete two-port error model at each VNA port. These models contain error terms that are consistent for both a forward measurement and a reverse measurement. The terms of the error models can be related back to physical characteristics of the analyzer as shown in Table 2.5 [9]. Additionally, since S-parameters rely on ratio measurements, one of the thru terms can be normalized to a value of 1.0. Referring to Table 2.5, if we normalize e_{10} to 1, then once e_{32} is determined, it can be used to solve for e_{23} in the reverse reflection tracking term. Similarly, e_{01} is determined from the forward reflection tracking term. This leaves only 7 truly unique error terms that must be solved.

Seven unknowns imply that we need a total of 7 independent equations. As it turns out, each thru measurement provides four equations (two in the forward direction, two in the reverse direction). This leaves only three additional equations, which can be provided by one-port measurements. The equations and measurement standards required by the different 7-term methods are shown in Table 2.6 and discussed below.

Table 2.5
Mapping the Error Terms of the 7-Term Error Model to Physical Parameters

Raw System Data	Forward Measurement	Reverse Measurement
Directivity	e_{00}	e_{33}
Reflection tracking	$e_{10} \cdot e_{01}$	$e_{23} \cdot e_{32}$
Source match	e_{11}	e_{22}
Transmission tracking	$e_{10} \cdot e_{32}$	$e_{01} \cdot e_{23}$
Load match	e_{22}	e_{11}

Table 2.6
Number of Equations Associated with Various 7-Term Error Correction Techniques

	TOM	TRM	TRL	TNA	UOSM
Two-Port Standards					
Thru (T)	4	4	4	4	—
Line (L)	—	—	2	—	—
Network (N)	—	—	—	1	—
Attenuator (A)	—	—	—	2	—
Unknown Thru (U)	—	—	—	—	1
One-port standards					
Open (O)	2	—	—	—	2
Short (S)	—	—	—	—	2
Match (M)	2	2	—	—	2
Reflect (R)	—	1	1	—	—
Total Number of Equations	8	7	7	7	7

2.4.4.1 TOM

The TOM method is the only three-term technique that uses fully characterized standards [10]. It results in a total of eight equations: two for the open and two for the match standard (at each port), and four for the thru standard. As a result, there are actually more equations than unknowns, meaning that the technique is overdetermined. This allows the extra equation to serve as a plausibility check. Hence, if there is a calibration issue (wrong standard(s) applied, broken standard(s), poor operator technique), this will be flagged via the implicit plausibility check at the end of the calibration.

2.4.4.2 TRM

The TRM is very similar to the TOM method, but the open standard is replaced by a reflect standard [11]. (Ideally, the same reflect standard is used at every port.) Since the reflect standard is only partially characterized, it provides only one equation ($s_{11} = s_{22}$). The match still provides two equations and the thru provides four equations, resulting in the necessary total of seven independent equations. The TRM method is a candidate for probe-station calibrations in frequency ranges where the two probes can be considered to have identical probe characteristics (when lifted in the air).

2.4.4.3 TRL

The TRL method is very popular for higher-frequency measurements because the technique completely eliminates the difficult-to-build microwave or millimeter-wave match standard [12]. Instead, it generally relies on a coaxial airline for the line standard with characteristic impedance that is determined by parameters traceable to known physical measurement standards (the diameters of the conductors). The thru standard in this method is always assumed to have a length and loss of zero.[8] Figure 2.10 shows an example of a microstrip calibration substrate for TRL. Here, the reference plane is assumed by the TRL method to be exactly at the midpoint of the thru line. This establishes the reference plane for all other standards. In this example, the reflect standard extends beyond the midpoint, so we must add the associated phase delay to the calibration kit definition. Also, the electrical delay of the line standard is the difference between the length of the line and the thru standards. As mentioned in Section 2.3.7, the electrical length of the line standard ranges between 20° and 160°, and multiple lines can be used to increase the frequency range of the TRL method beyond the 8:1 limitation imposed by a single line. As with all the three-letter methods of this section, the thru measurement provides four equations, the line provides two equations, and the reflect standard provides one equation, for the necessary total of seven independent equations.

---

8. The LRL (line reflect line) standard accommodates a thru standard with a length > 0.

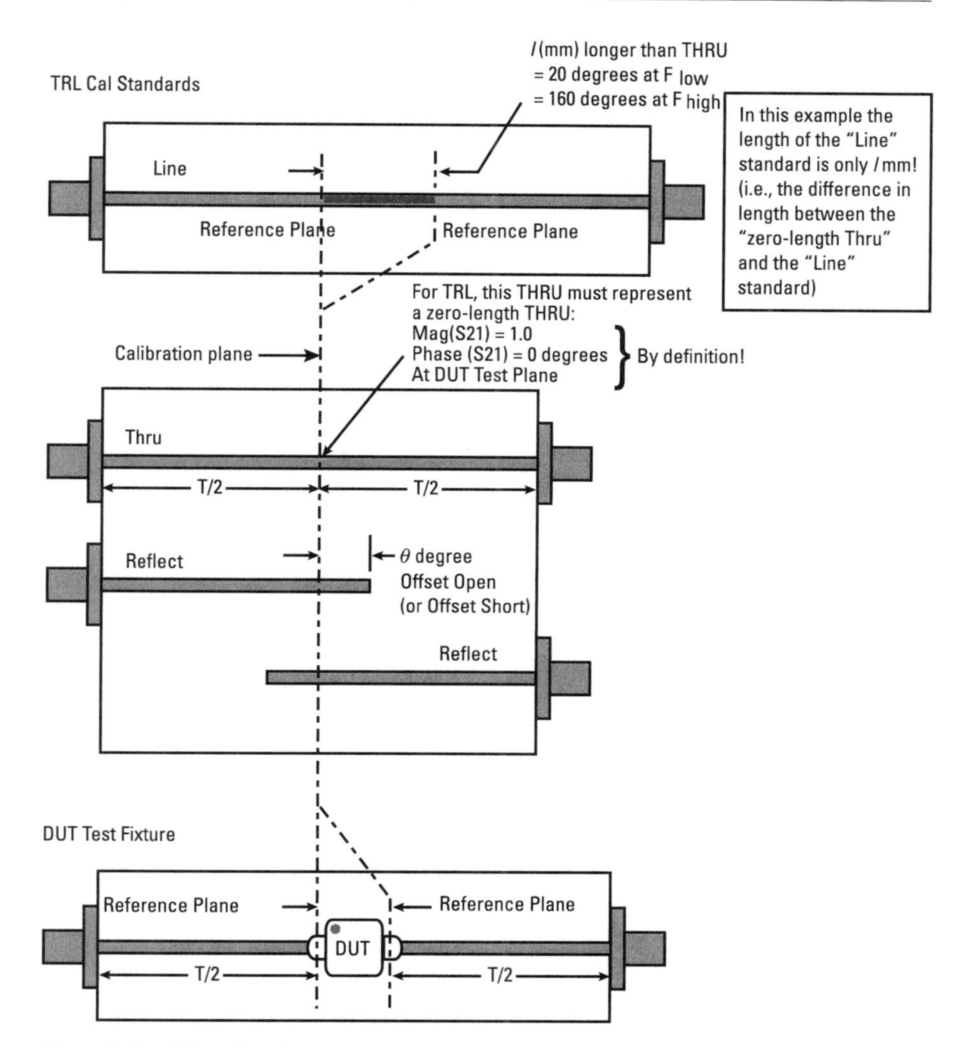

Figure 2.10 TRL calibration substrate example.

2.4.4.4 TNA

The TNA method is unique in that it allows the reflect standard to be replaced by a symmetrical network (measurement fixture) [7]. An attenuator with excellent match but arbitrary attenuation value (see Section 2.3.9) replaces the line standard. The TNA technique requires very minimal knowledge concerning the electrical characteristics of the standards. As with the reflect standard, the network standard provides one equation, the attenuator provides two equations, and the thru provides four equations for the required total of seven independent equations.

2.4.4.5 UOSM

The UOSM method is particularly useful among calibration methods because it allows the electrical characteristics of an unknown thru to be entirely determined during the calibration process [13]. The only a priori requirement is that the unknown thru must be reciprocal ($S_{21} = S_{12}$).

Consider the test setup in Figure 2.2(c). Here, the DUT (unknown thru) provides one equation (reciprocity condition, $s_{21} = s_{12}$), and the fully characterized O, S, and M standards each provide two equations for a total of seven independent equations. We first encountered the necessary mathematics for this method in the section on one-port OSM calibration, where we could determine e_{00}, e_{11}, and $e_{10}e_{01}$ using one-port OSM standards on port 1, and e_{22}, e_{33}, and $e_{23}e_{32}$ using OSM standards on port 2. Using some fairly lofty matrix calculations [14], the actual value of S_{21} (and by reciprocity, S_{12}) can be determined down to a sign (+/–) ambiguity. The user requires only a rough knowledge of the unknown thru's S_{21} phase shift (<180°) to select the correct sign.

The UOSM method is very useful for determining the electrical characteristics of coaxial adapters, such as a male 3.5 mm to female *N* adapter, or a 1.85-mm male to WR19 (40–60 GHz) waveguide adapter. To characterize an adapter, you generally need manual calibration kits containing one-port standards (OSM) for each side of the adapter. The adapter itself serves as the unknown thru. If the adapter is the last standard measured during the calibration process, then at the completion of the UOSM calibration the VNA displays the loss and phase of the adapter. The two-port S-parameters of the adapter can then be exported to a file and recalled for future applications requiring adapter de-embedding.

2.4.5 12-Term Error Correction

A two-port VNA architecture that uses only three receivers (two for measuring reflected waves and one for a common reference) seems a reasonably economical approach. Since the S-parameter definition allows only a single port to be excited at a time, why bother with dedicated reference receivers for each port (Figure 2.11)?

While this argument makes economic sense, a price must be paid when it comes to calibrating such an instrument. A closer examination reveals that the single reference receiver samples the outgoing wave prior to the switch, regardless of the measurement direction. This makes the switch part of the test set, and the error model must account for its impact on measurements. From an error model standpoint, two separate submodels are required to account for error terms in the forward and reverse directions (Figure 2.12) [15, 16].

Notice that each submodel has seven error terms (including cross-talk). They can be mapped to physical parameters similar to what was done for

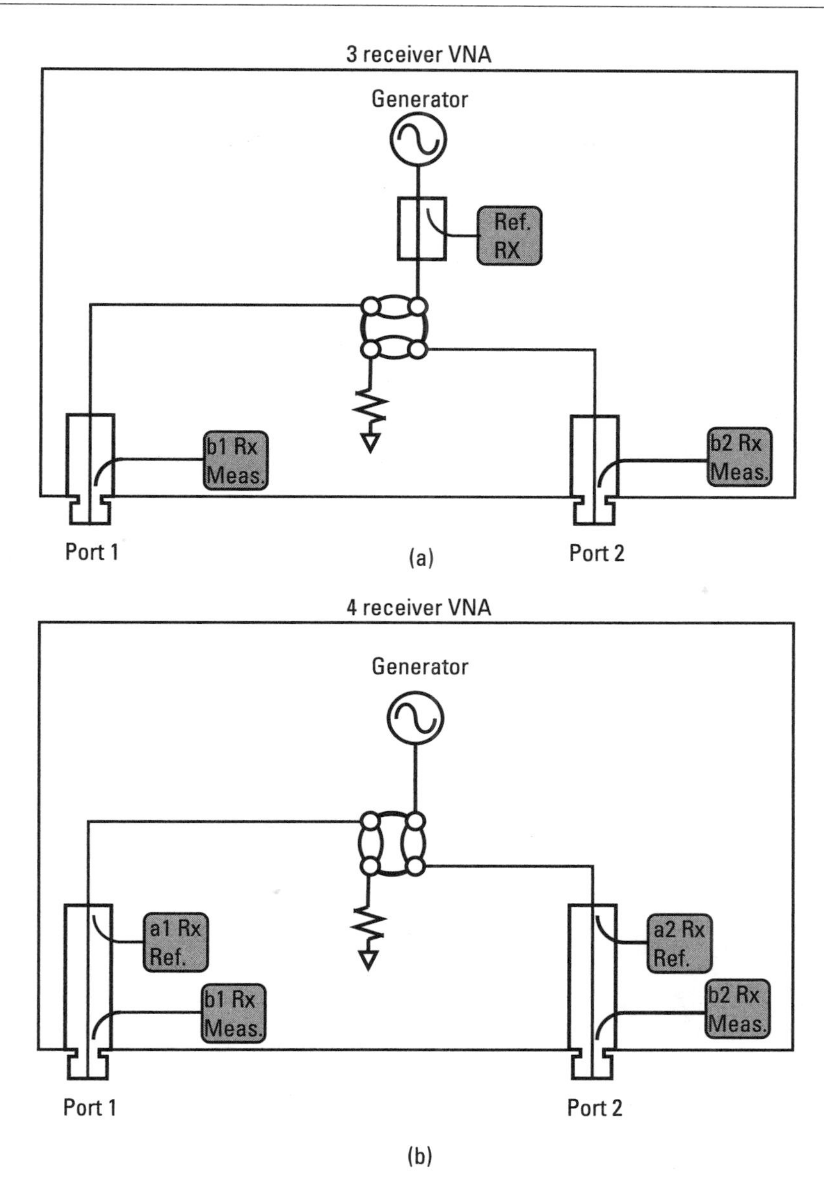

Figure 2.11 Architecture comparison between a 3- and 4-receiver VNA.

seven-term error correction in Table 2.5. The resulting mapping is shown in Table 2.7.

Just as with the seven-term error model, we can treat one of the transmission quantities as equal to 1 (either e_{10}, e_{01}, e_{23}, e_{32} or e'_{01}, e'_{23}, e'_{32}, or e'_{10}). If we set $e_{10} = 1$, then we have 6 unknowns in the first column and 8 in the second column. Notice that e'_{32} only appears once, in the reflection tracking term.

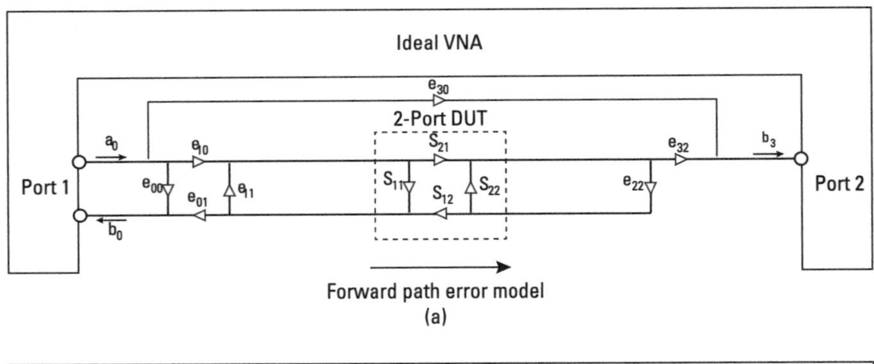

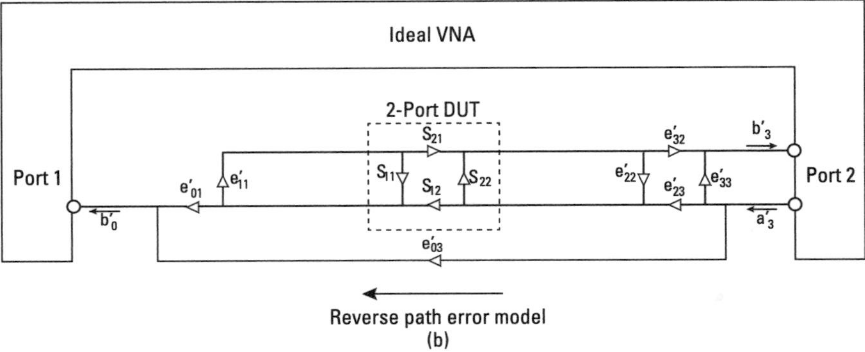

Figure 2.12 The 12-term error submodels for a three-receiver VNA

Table 2.7
Mapping the Error Terms of the 12-Term Error Model to Physical Parameters

Raw System Data	Forward Measurement	Reverse Measurement
Directivity	e_{00}	e'_{33}
Reflection tracking	$e_{01} \cdot e_{10}$	$e'_{32} \cdot e'_{23}$
Source match	e_{11}	e'_{22}
Transmission tracking	$e_{32} \cdot e_{10}$	$e'_{01} \cdot e'_{23}$
Load match	e_{22}	e'_{11}
Crosstalk	e_{30}	e'_{03}

Similarly, e'_{01} only appears once, in the transmission tracking term. This means that we do not have to solve down to the level of the individual error terms e'_{32} or e'_{01}, but merely solve for the composite $e'_{32}e'_{23}$ and $e'_{01}e'_{23}$. This effectively leaves us with only 6 unknowns in the second column, for a total of 12 error terms. Furthermore, if we assume no crosstalk ($e_{30} = e'_{03} = 0$), then we are left with only 10 unknowns.

2.4.5.1 TOSM

Fortunately, the TOSM provides 10 separate equations: enough to solve for the 10 unknowns in the three-receiver error model. All standard must be fully characterized. As in the case of the seven-term error correction techniques, the fully characterized thru standard provides a total of four equations, and each one-port standard (OSM) provides a total of two independent equations (one for each VNA port), as shown in Table 2.8.

Generally, crosstalk error correction is provided as an additional (optional) step. By terminating both ports with match standards, we force $s_{21} = 0$ and $s_{12} = 0$ (two equations for the two unknowns e_x and e'_x). Crosstalk is sometimes used when making measurements on a probe station, where the probes are in proximity. The disadvantage of crosstalk correction is that it is very susceptible to position. If a probe is moved only a small distance away from the crosstalk calibration position, the crosstalk correction vectors may reinforce, rather than cancel, the crosstalk error, making the crosstalk problem worse (over some portion(s) of the measurement frequency range). In most modern VNAs, the crosstalk level is below the noise floor and may be safely ignored.

Table 2.9 summarizes the different calibration techniques discussed in this chapter and suggests possible uses for each.

2.5 Power Calibration

S-parameters are used to examine the ratio of different receiver quantities. Complex rather than scalar calculations are performed, and error-correction is applied. The resulting amplitude values are frequently displayed in decibel units. However, there are instances where absolute rather than relative ampli-

Table 2.8
Number of Equations Associated with the TOSM Method

	TOSM
Two-port standards	
Thru (T)	4
One-port standards	
Open (O)	2
Short (S)	2
Match (M)	2
Total number of equations	10

Table 2.9
Calibration Technique Summary

		Calibration Technique						
Application	**Normalization**	**OSM**	**TOM**	**TRM**	**TRL**	**TNA**	**UOSM**	**TOSM**
Two-port gain/loss measurements	Limited (e.g., Go/NoGo)		✓	✓	✓	✓	✓	✓
Test fixture de-embedding		✓			✓	✓[1]		
Calibration with implicit verification			✓					
Calibration with partially known standards			✓	✓	✓	✓	✓	
Applications requiring standards with different genders			✓	✓[2]	✓[2]	✓[2]	✓	✓
Applications requiring different connector types							✓	
Calibration supporting sliding match for microwave applications		✓	✓	✓			✓	✓
On-wafer measurements				✓	✓	✓	✓	

1. Band-limited; possible singularities if design frequency range is exceeded.
2. Requires standards that produce symmetrical reflections.

tude values are desired. For example, a user may want to accurately specify the stimulus power going into his or her DUT. Perhaps harmonic or intermodulation characteristics are required, but S-parameters cannot be used because they are only defined for identical frequencies at all ports. These applications warrant power calibration. Modern VNAs accommodate scalar power calibration using external power meters or power sensors. Generally, power calibration takes two forms: source power calibration and receiver power calibration. These are discussed below, and step-by-step power calibration procedures are provided in subsequent chapters. Table 2.10 presents a number of applications and their associated calibration requirements.

2.5.1 Source Power Calibration

Source power calibration is used to adjust the absolute level of the VNA generator to a specific value at the calibration plane. This is ideally the same plane used for S-parameter calibration, but certain applications (like high-power measurements or probe station calibration) may require that power calibration be performed at a different reference plane. In these instances, network embedding or

Table 2.10
Calibrations Required to Support Various Applications

Two-Port Application	Vector System Error Correction (Also Known as S-Parameter Calibration)	Power Calibration
S-parameter measurements on passive or active (linear)		
Devices	Required	Not needed
S-parameter measurements on		
Nonlinear DUTs	Required	Generator flatness (port 1)
Amplifier hot-s_{22} measurements	Required	Generator flatness (port 1)
Amplifier harmonics	N/A	General flatness (port 1), receiver (port 2)
Amplifier two-tone IMD	N/A	General flatness (for both generators), receiver (port 2)
Mixer or frequency converting measurements	Optional	General flatness (RF port, LO port), receiver (IF port)
1-dB compression point (based on gain compression (s_{21}))	Required	General flatness (port 1), receiver (port 2)

de-embedding is used to align the power calibration plane with the S-parameter calibration plane.

In practice, source power calibration begins by connecting a power sensor to the end of the port 1 test cable and setting the VNA to a reasonably high power level. (We want a good SNR at the power sensor: 0 or −10 dBm are typical default levels for power calibration.) A frequency sweep is performed, during which a correction table is established for the generator's reference receiver. The power sensor transfers measured power values to the reference receiver for every point in the frequency sweep. This gives the VNA receiver the same absolute power measuring accuracy as the power sensor. After this initial sweep, source power calibration is considered complete, even if the generator still has a wide variation in output level (+/–2 dB). Additional source power iterations may be performed to reduce this variation to a user-defined tolerance. Fortunately, the (rather slow) power sensor measurements are no longer needed after the first measurement sweep. By using only the power-corrected reference receiver for subsequent sweeps, the generator leveling process can proceed quickly.[9]

2.5.2 Receiver Power Cal

Receiver power calibration is a separate step performed after source power calibration. As mentioned above, source power calibration establishes absolute

9. For leveling the source power, 10 iterations are typically performed, with a user-defined power tolerance of 0.1 dB.

power measurement accuracy for the VNA port 1 reference receiver. This means that the port 1 reference receiver knows exactly how much power is at the end of the port 1 test cable. By connecting the port 1 and port 2 test cables together with a through adapter, the receiver power calibration process transfers this absolute power level information to the port 2 measurement receiver using its own power correction table. Measurements can now be made and the analyzer will know exactly how much power is supplied to the DUT (at port 1) and the amount of power entering the DUT (at port 2).

The source power calibration process may then be repeated at port 2, and the receiver power calibration repeated at port 1 to support bidirectional VNA power measurements.

2.5.3 SMARTerCal

SMARTerCal is a calibration procedure used in R&S VNAs, which combine a full n-port S-parameter calibration (for two or more ports) with a scalar power calibration at a single port. It is denoted by a "p" in front of the familiar S-parameter calibration types (e.g., PTOSM, PTRL). This type of calibration utilizes the calculated S-parameter error terms to provide the most accurate power measurements possible. The mathematics are straightforward and based on S-parameter definitions. For example,

$$S_{11} = \frac{b_1}{a_1} |_{a_2=0} \tag{2.33}$$

We start by performing a power calibration with a power sensor at the end of the port 1 test cable. This yields scalar wave quantity a_1 with a particular amplitude and phase (relative to the generator) at each frequency point. By mathematically applying a phase rotation to each frequency point, we establish a reference vector $|a_1| < 0$. Then we can determine wave quantity b_1 from the reference value of a_1 by using the definition of S-parameter S_{11}:

$$b_1 = S_{11} a_1 \tag{2.34}$$

Also, since

$$S_{21} = \frac{b_2}{a_1} \tag{2.35}$$

we can calculate b_2 from a_1:

$$b_2 = S_{21} a_1 \tag{2.36}$$

To find a_2, we note that:

$$S_{22} = \frac{b_2}{a_2} \tag{2.37}$$

So,

$$a_2 = \frac{b_2}{S_{22}} \tag{2.38}$$

Using (2.36) to substitute for b_2 yields:

$$a_2 = \frac{S_{21}}{S_{22}} a_1 \tag{2.39}$$

So all four wave quantities can be calculated from the known value of a_1 at each frequency point. In addition, because the generator is measured during each new sweep, any changes to the a_1 amplitude or phase (due to power changes within the VNA or device mismatch) can easily be propagated to the other receivers using (2.34) through (2.39).

It is important to note that SMARTerCal calibrates the reference and measurement receivers but does not provide a flatness calibration for the generator(s). This can be accomplished with a subsequent power calibration step, where the user is prompted to attach the actual DUT. This allows the source flatness routine to remove ripple that would otherwise result because of a poorly matched device.

2.5.4 Automatic Level Control (ALC)

Alternatively, ALC can be used to provide precise power levels to a DUT regardless of device mismatch. With ALC, a digital loop monitors and adjusts the generator power to attain a desired target level at the DUT input. When the generator settles to the desired value (within some tolerance window, e.g., 0.05 dB), the loop ceases adjustment and the VNA makes a measurement (Figure 2.13(a)). The frequency (in a frequency sweep) or power (in a power sweep) is then incremented to the next step value, and the process repeats. Modern VNAs have very sophisticated ALC circuitry, often with dedicated receivers and IF filtering bandwidths that may be adjusted separately from the VNA's main

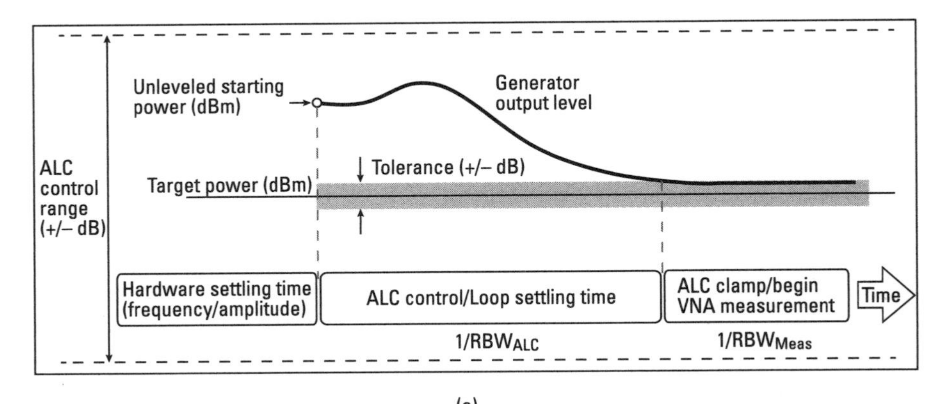

(a)

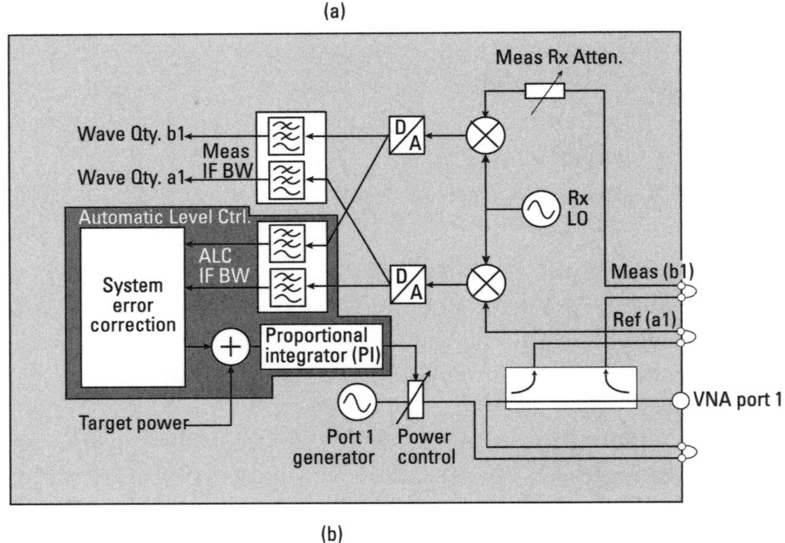

(b)

Figure 2.13 ALC loop settling time diagram (a) and architectural details (b) revealing separate, dedicated IF filters for ALC operation.

reference and measurement receivers (Figure 2.13(b)). These advanced features allow great flexibility for optimizing loop settling time performance.

References

[1] "AN 95-1 S-Parameter Techniques for Faster, More Accurate Network Design," 5952-1130, Agilent Technologies, 1967, p. 3.

[2] Pozar, D., *Microwave Engineering*, Reading, MA: Addison-Wesley, 1990, pp. 76–79.

[3] Hiebel, M., *Fundamentals of Vector Network Analysis*, Munich, Germany: Rohde & Schwarz, 2007, pp. 80–88.

[4] Sklar, B., *Digital Communications Fundamentals and Applications*, Englewood Cliffs, NJ: Prentice Hall, 1988, p. 202.

[5] Paech, A., et al., "Noise Figure Measurement without a Noise Source on a Vector Network Analyzer," App Note 1EZ61_2E, Rohde & Schwarz, October 27, 2010, p. 13.

[6] Hiebel, M., *Fundamentals of Vector Network Analysis*, Munich, Germany; Rohde & Schwarz, 2007, pp. 89–99.

[7] "AN 1287-3: Applying Error Correction to Network Analyzer Measurements," 5965-7709E, Agilent Technologies, 2002, pp. 10–11.

[8] Pozar, D., *Microwave Engineering*, Reading, MA: Addison-Wesley, 1990, pp. 245–250.

[9] Janjusevic, N., Application Note AN-49-016, "UVNA-63 Application Note Error Correction," Mini-Circuits, 2019, pp. 11–12.

[10] Eul, H., and B. Schiek, "A Generalized Theory and New Calibration Procedures for Network Analyzer Self-Calibration," *IEEE Transactions on Microwave Theory and Techniques*, Vol. 39, No. 4, April 1991, pp. 724–731.

[11] Eul, H., and B. Schiek, "Thru-Match-Reflect: One Result of a Rigorous Theory for De-Embedding and Network Analyzer Calibration," *Proc. 18th European Microwave Conference*, (Stockholm), 1988, pp. 909–914.

[12] Engen, G., and C. Hoer, "'Thru-Reflect-Line': An Improved Technique for Calibrating the Dual Six-Port Automatic Network Analyzer," *IEEE Transactions Microwave Theory and Techniques*, Vol. MTT-27, No. 12, December 1979, pp. 987–992.

[13] Basu, S., and L. Hayden, "An SOLR Calibration for Accurate Measurement of Orthogonal On-Wafer DUTS," *IEEE MTT-S Digest*, 1997, TH2D-3, pp. 1335–1338.

[14] Ferrero, A., and U. Pisani, "Two-Port Network Analyzer Calibration Using an Unknown 'Thru'," *IEEE Microwave and Guided Wave Letters*, Vol. 2, No. 12, December 1992, pp. 505–507.

[15] "AN 1287-3, Applying Error Correction to Network Analyzer Measurements," 5965-7709E, Agilent Technologies, 2002, p. 6.

[16] Janjusevic, N., Application Note AN-49-016, "UVNA-63 Application Note Error Correction," Mini-Circuits, 2019, pp. 7–11.

Selected Bibliography

"Agilent Electronic vs Mechanical Calibration Kits: Calibration Methods and Accuracy," 5988-9477EN, Agilent Technologies, 2003.

"AN 1291-1B 10 Hints for Making Better Network Analyzer Measurements," 5965-8166E, Agilent Technologies, 2001.

App Note 11410-00270 Rev A, "What Is Your Measurement Accuracy?" Anritsu, Morgan Hill, CA, 2001.

Bednorz, T., "Measurement Uncertainties for Vector Network Analysis," App Note 1EZ29-1E, Rohde & Schwarz, May 29, 1998.

Donecker, B., "Determining the Measurement Accuracy of the HP 8510 Microwave Network Analyzer," *RF & Microwave Measurement Symposium,* Hewlett-Packard, March 1985.

Janjusevic, N., and J. Langner, Application Note AN-49-017, "UVNA-63 Application Note Calibration Standards and the SOLT Method," Mini-Circuits, 2019.

Marks, R., "Multi-Line Calibration for MMIC Measurement," *36th ARFTG Conference Digest,* Vol. 18, 1990, pp. 47–56.

Ostwald O., "Measurement Accuracy of the ZVK Vector Network Analyzer," App Note 1EZ48-0E, Rohde & Schwarz, January 24, 2001.

Rytting, D., "An Analysis of Vector Measurement Accuracy Enhancement Techniques," Hewlett-Packard, RF & MW Symposium, 1980.

"Understanding the Fundamental Principles of Vector Network Analysis," 5965-7707E, Agilent Technologies, 2012.

Zeier, M., et al., "Contemporary Evaluation of Measurement Uncertainties in Vector Network Analysis," *Technisches Messen,* Vol. 84, No. 5, 2017, pp. 348–358.

Zeier, M., et al., "Establishing Traceability for the Measurement of Scattering Parameters in Coaxial Line Systems," *Metrologia,* Vol. 55 No. 10, 2017, pp. 1088/1681–7575/aaa21c.

3

Passive and Active One-Port Device Measurements

3.1 Passive One-Port Devices

Passive one-port devices include antennas, terminations, resonators, and two-port devices with one of the ports terminated into a specific impedance (open, short, and/or matched conditions are popular). Examples include fundamental RF components such as resistors, capacitors, and inductors. Another important one-port application involves phase matching of cables, where only one end of the cable is equipped with an RF connector, while the other end is trimmed to obtain a particular electrical length.

In this chapter, we begin by focusing on applications involving passive one-port devices and later extend the discussion to consider simultaneous measurement of multiple one-port devices. Key specifications for this class of components include frequency-domain reflection measurements (return loss, S_{11}, voltage standing wave radio (VSWR), reflection coefficient magnitude/phase and impedance/admittance). For many passive devices, measurements may be carried out using default or preset VNA settings. However, as with any measurement challenge, optimal performance lurks in the details. Users must ensure a sufficient number of points across the device's frequency response to satisfy measurement requirements, and use appropriate VNA power and bandwidth settings to ensure sufficient dynamic range for the task at hand. Table 3.1 provides an overview of the one-port topics covered in this chapter.

Table 3.1
Example Applications and Measurements for Passive and Active One-Port Devices

Example Application	Category	Typical Measurements	Section Number
Antenna matching	Passive	Return loss/VSWR	3.1
Characterize terminated two-port devices	Passive	Return loss/impedance	3.2
Measurements of RF components (RLC)	Passive	Return loss/impedance	3.1, 3.2
Phase-Matching of Cables	Passive	Electrical/mechanical length	3.3
Characterize oscillator, synthesizer, Signal generator	Active	Return loss/impedance	3.5

3.1.1 Steps for Setting Up a Single-Port Measurement

The following are the steps for setting up a single-port measurement:

- *Step 1:* Determine the appropriate interface between the VNA and the DUT. If a cable is required, use only a high-quality phase-stable test cable with appropriate connectors (and sexes) to provide this interface. If possible, avoid RF adapters, which only increase reflections and can cause problems with maintaining tight connections throughout the calibration and measurement process. In some instances, it might be convenient to dispense with the test cable altogether and use a high-quality connector saver right at the VNA port. Other applications (such as testing of discrete packaged components) will require a test fixture and appropriate calibration and de-embedding techniques to move the calibration plane close to the DUT.
- *Step 2:* Start from an instrument preset. Configure the start frequency, stop frequency, and number of points (or equivalently, frequency step size) appropriately for the measurement at hand. Special considerations for oscillator measurements are discussed in Section 3.5.
- *Step 3:* As mentioned in the introduction to this chapter, many one-port applications (device development, prototyping, characterization, and bench tuning) are well suited to VNA default settings. Chief among these are default power levels (–10 dBm to 0 dBm) and default intermediate frequency (IF) bandwidths (1 kHz to 10 kHz). For production measurements where speed is important, the user may elect to sacrifice dynamic range for faster sweep speed by selecting a wider IF bandwidth

(up to 1 MHz depending on your willingness to tolerate trace noise). If you are using a microwave VNA (capable of measurements to 24 GHz or higher), and the particular application requires frequency sweeps beginning below 500 MHz, it is a good idea to use the dynamic bandwidth at a lower-frequency mode. This mode compensates for the loss of coupling effectiveness in microwave directional couplers at lower frequencies by automatically reducing the bandwidth to reduce noise and maintain good SNR. For some instruments/manufacturers, this mode is active by default. In R&S VNAs, this mode may be activated by the following key sequence: "Channel" > "Power BW AVG" > "Meas Bandwidth" > "Fine Adjust…" (see Figure 3.1).

- *Step 4:* Perform a single-port open-short-match (OSM) calibration (SEC) using a calibration kit appropriate for the DUT connector style and sex (e.g., a 3.5-mm calibration kit for SMA or 3.5-mm connectors). It is good practice to save the calibration and/or the test setup after completing this step, in case of subsequent operator mistake (like accidentally pressing preset).
- *Step 5:* After completing the calibration, one should take the time to verify it by installing an open or a short standard from a different calibration kit or from a verification kit[1]. Configure the instrument to measure S_{11} in trace format "Smith" (Smith chart). Does the trace look as expected for a short or open standard? (See Figure 3.2.) The distance that the phase travels around the perimeter of the Smith chart will be greater for wide frequency sweeps and physically long calibration standards. Traces that wrap multiple times around the Smith chart are not uncommon. (See Appendix 3A for an explanation.) If the Smith chart display

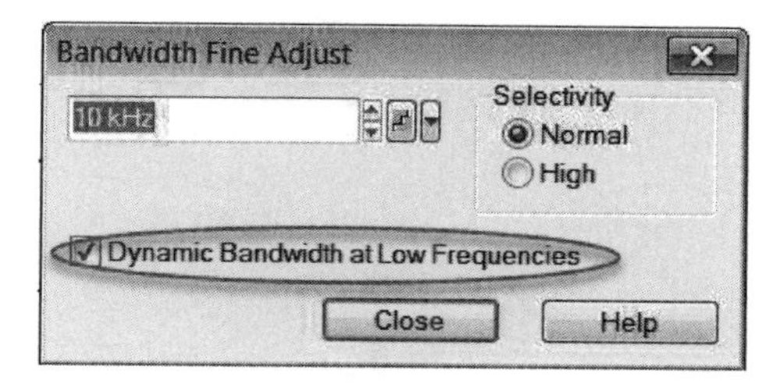

Figure 3.1 Dialog for selecting "Dynamic BW at Low Frequencies."

1. Calibration verification, measurement uncertainty and traceability are important and extensive metrology topics. The interested reader is referred to the references for a deeper dive. Our objective here is to provide the practicing engineer or technician with a procedure for weeding out gross setup problems that would otherwise result in inaccurate measurements or wasted time trying to uncover the source of measurement results that do not make sense.

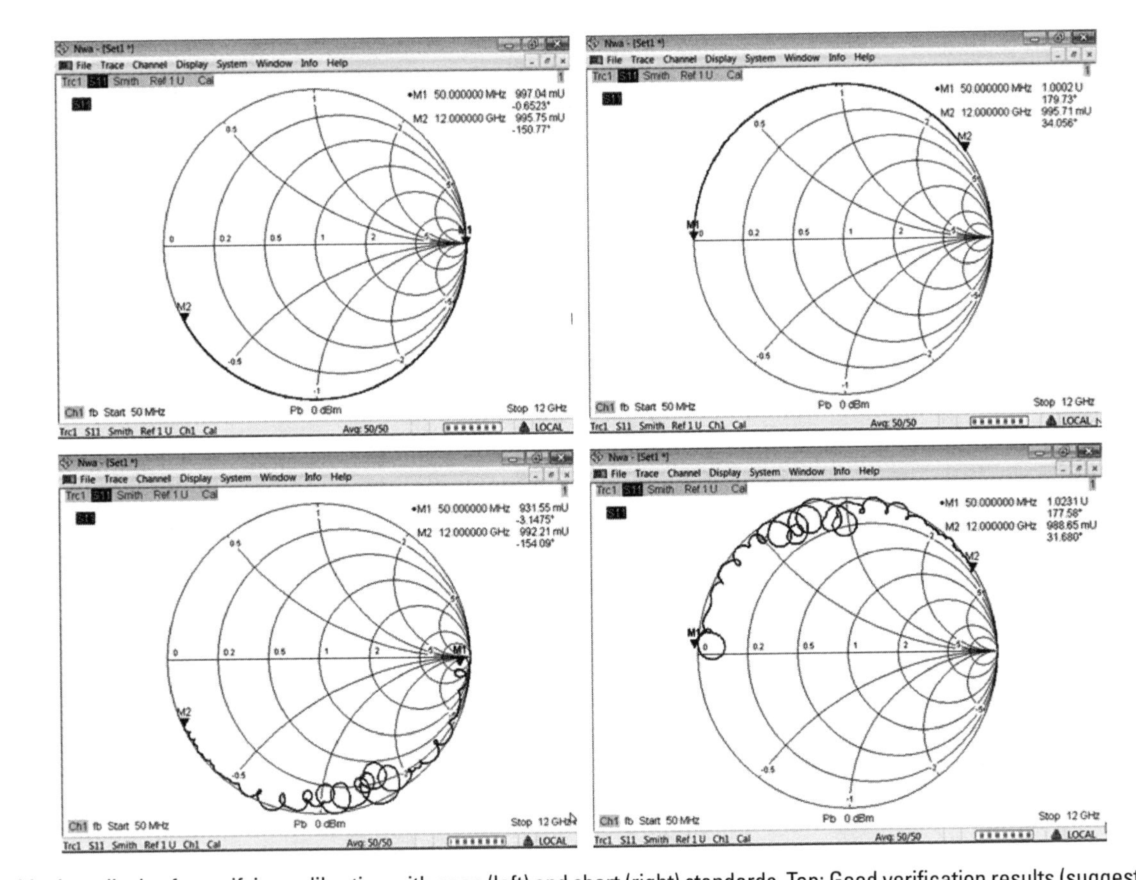

Figure 3.2 Smith chart display for verifying calibration with open (left) and short (right) standards. Top: Good verification results (suggests a good calibration). Bottom: Bad verification results (suggests a problem with the verification or calibration).

looks reasonable, move the cable around and see if the response remains stable. If either the magnitude or phase of the display changes, you must determine the faulty component(s) in the setup or you will make poor/invalid measurements. Ensure that the VNA cable is not worn or damaged and that the connector pins or receptacles are not dirty, damaged, or recessed. (Check the latter with a precision connector gauge.[2])

- *Step 6:* If the calibration is valid, connect the DUT and configure the trace format and scale for the desired measurement. For a one-port device, the starting point will be an S_{11} measurement. The appropriate formats are shown in Table 3.2.

3.1.2 Calibration for Multiple Single-Port Devices

Since we are measuring multiple single-port devices, we can calibrate each port using a single port calibration (OSM). In this example, we will assume a four-port analyzer.

For a Rohde & Schwarz (R&S) model ZVA, under the R&S ZVA Channel group, choose "Cal" > "Start Cal" > ("- more –") > "Other…". Select the desired test cable connector type and gender, and select the appropriate calibration kit. Then press "Next" and click "Add Calibration..." For the "Type" pull-down menu, select "Full One-Port" and click the check boxes for all four ports, as shown below. Then press "OK" to exit this dialog and "Next" to begin the calibration procedure, which requires 12 measurements (three OSM standards for each port).

3.1.3 Port Configuration for Multiple Simultaneous Single-Port Measurements

With calibration completed, we need to configure the instrument to apply source power to all four sources simultaneously and make simultaneous measurements (represented by S_{11}, S_{22}, S_{33}, and S_{44}) rather than conventional sequential measurements. This configuration process is not available for every VNA model, and where it is available, it is unique to a particular VNA manufacturer and model. In this section we consider the R&S ZVA.

Table 3.2
Common One-Port Measurements and VNA Formats

Desired Measurement	VNA Format
Return loss	dB Mag
VSWR	SWR
Reflection coefficient	Lin Mag, Polar, or Smith (for Smith chart)

2. Refer to Section 2.2.1.2 for a discussion of this topic.

Under the R&S ZVA Channel group, choose "Mode" > "Port Config," and click the "Balanced Ports and Port Groups" on the dialog (Figure 3.3).

On the next dialog, click the "Port Groups" tab and "Add Group" button. In the "Port Group" box, use the pulldown menu to change the "LastPort" to 1. Then, a new entry will appear. Change the "LastPort" to 2. Repeat for ports 3 and 4. At the end, the dialog should appear as shown in Figure 3.3. Click "OK" to exit this dialog and "OK" to exit the "Port Configuration" dialog. The instrument will now make simultaneous measurements at the four ports.

To restore conventional VNA sweeping, either preset the instrument or click on the "Channel" group, choose "Mode" > "Port Config," and click the "Balanced Ports and Port Groups" in the dialog. Then, click the "Predefined Configs" tab, and select the first one "(A) 4x Single." Click "OK" to exit this dialog and "OK" to exit the Port Configuration dialog.

3.2 Impedance Measurements

The impedance of any one-port device connected to a VNA may be easily calculated from the reflection coefficient [1], which is a fundamental VNA measurement capability. The relationship is as follows:

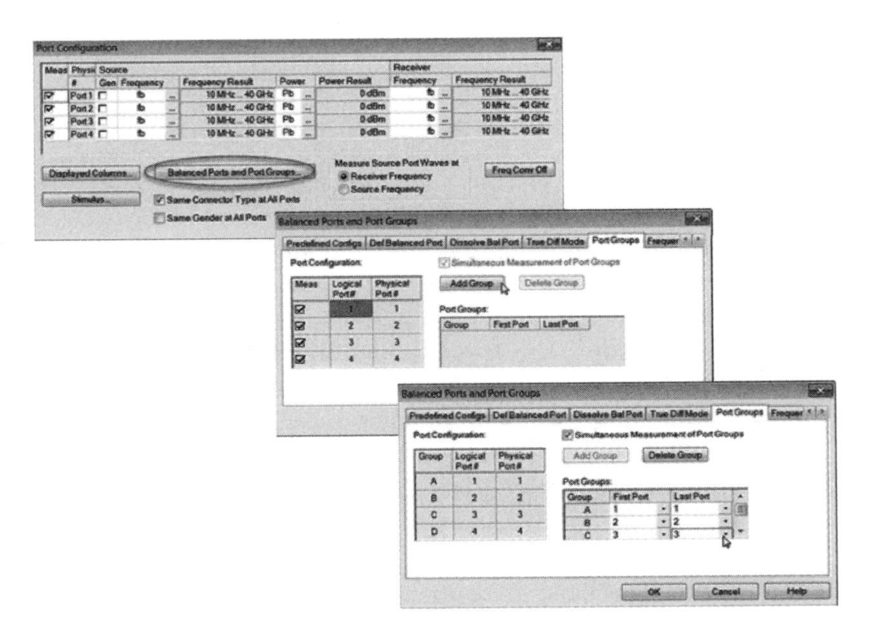

Figure 3.3 Dialogs for creating simultaneous one-port measurements.

$$Z_x = Z_o \left(\frac{1+\Gamma}{1-\Gamma} \right) \tag{3.1}$$

where Z_x = unknown impedance, Z_o = characteristic impedance (generally 50Ω), and Γ = reflection coefficient (generally a complex value).

There are two ways to make impedance measurements on a VNA. One method uses dedicated traces to display the real and imaginary impedance values. The other method employs markers configured to display impedance values on an otherwise unrelated trace (like dB Mag (S_{11})). We will look at both.

3.2.1 Impedance Traces

The following are the steps for creating impedance traces:

- *Step 1:* To measure the impedance of a DUT connected to VNA port 1, select the "Z ← S11" measurement. (This is usually under the menu containing different S-parameters, ratios, or wave quantities.) Selecting this measurement may automatically set the format to "Lin Mag" (which is the magnitude of the complex impedance, in ohms).
- *Step 2:* To measure the real or imaginary component of impedance, change the trace format to "Real" or "Imag," respectively.
- *Step 3:* Add a second trace, if desired, and change the trace format as required to measure both real and imaginary components concurrently. Using this same approach, you could have several traces measuring impedance components, and another trace set to simultaneously display reflection coefficient in Smith chart format (preferably in a different display area).

3.2.2 Impedance Markers

The following are the steps for impedance markers:

- *Step 1:* To measure the impedance of a DUT connected to VNA port 1 with a marker, select an "S11" measurement.
- *Step 2:* Markers have different default settings depending on the measurement format. For example, a marker on a Smith chart (format "Smith") will provide impedance (R + JX) by default, whereas a marker on a "Db Mag" trace presents the decibel value of the magnitude of the reflection coefficient at the marker position. To change this default operation, select "Marker" > "Marker Properties" and look for the "Marker

Format" On a ZVA or ZNB, change the setting from "Default" to "R + jX" format to display the impedance values.

3.3 Phase and Electrical Length Measurements

One popular single-port application for a VNA is to measure the electrical or mechanical length of a cable. This is important when phase matching two or more cables. There are two popular ways to accomplish this (without resorting to time-domain reflectometry methods): calibration offset or group delay measurements.

3.3.1 Calibration Offset Method

Modern VNAs have provisions for moving the calibration plane either closer to or further away from the end of the test cable. This is useful for cable length measurements. We attach the unknown cable to the end of our VNA test cable and use a calibration offset function to move the calibration plane out to the end of the unknown cable. We directly read the amount of electrical delay incurred in this process (in units of time). The instrument may also determine the cable's mechanical length (in millimeters) if supplied with the cable's velocity factor or permittivity, E_r.

- *Step 1:* Ensure that the VNA port of interest has a valid one-port (OSM) calibration for the frequency range of interest.
- *Step 2:* Connect the unknown cable to VNA port 1 and measure "dB Mag" of S_{11} to get a sense for the cable's effective frequency operating range. Leave the far end of the unknown cable unterminated. For example, Figure 3.4 shows magnitude responses for two unknown transmission lines. The black trace shows the magnitude response of an unterminated 2.92 mm thru adapter from a manual calibration kit, which shows very little ripple or loss up to 40 GHz. The gray trace shows the magnitude response of an inexpensive SMA-terminated cable swept from 10 MHz to 40 GHz. The frequency response of this SMA cable reveals a resonance at approximately 18 GHz (Marker 2) and significant ripple even below this frequency. As a result, the user should select a measurement frequency range that covers only the practical operating range of the cable (in this case 10 MHz to 8 GHz (Marker 1)) for subsequent electrical length measurements. Perform a new calibration if warranted (e.g., to obtain measurements at specific frequency points without interpolating between calibration points).

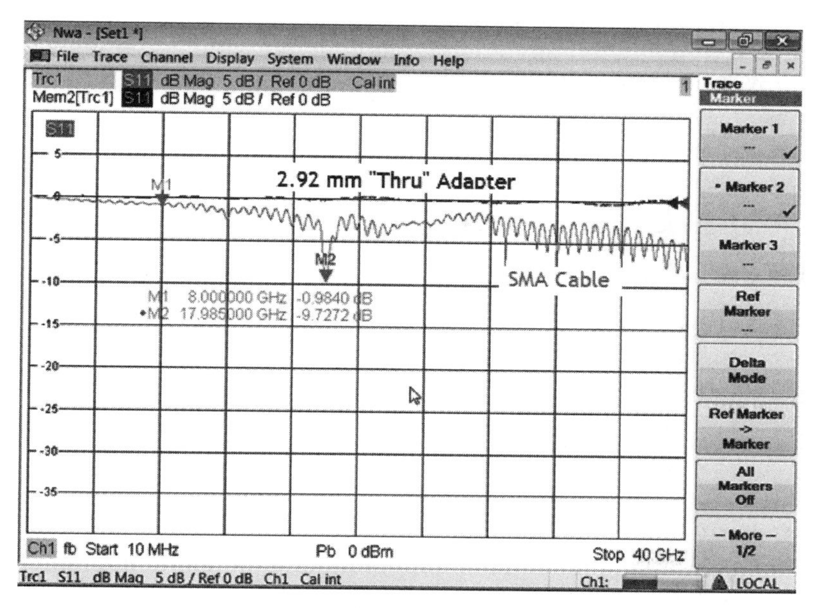

Figure 3.4 Magnitude responses of a 2.92-mm thru adapter and an inexpensive SMA cable.

- *Step 3:* Change the trace format to "Phase." Locate the calibration offset menu. (For the ZVA, this is "Channel" Group > "Offsets.") You should see many +180° to −180° phase wraps across the displayed frequency range. (These wraps will appear denser (closer together) for a longer versus a shorter cable.) (See Figure 3.5(a).)
- *Step 4:* Press the "Autolength" button. This will automatically flatten the trace by effectively moving the calibration plane out to the end of the unknown cable. (See Figure 3.5(b).)
- *Step 5:* Press the "Delay" softkey. This displays a table with the calculated least-squares fit value of the delay that provides the flattest (closest to zero-degree) phase result for the unknown cable connected to port 1. This is the electrical length of your cable.
- *Step 6:* This is optional for mechanical length. Press the "Mechanical Length" softkey and enter either the cable velocity factor or effective permittivity. This will yield the mechanical length of the unknown cable.

3.3.2 Group Delay Method

Group delay is an important VNA function that calculates the propagation delay of a wave through a device. Group delay (τ_g) is a real quantity calculated

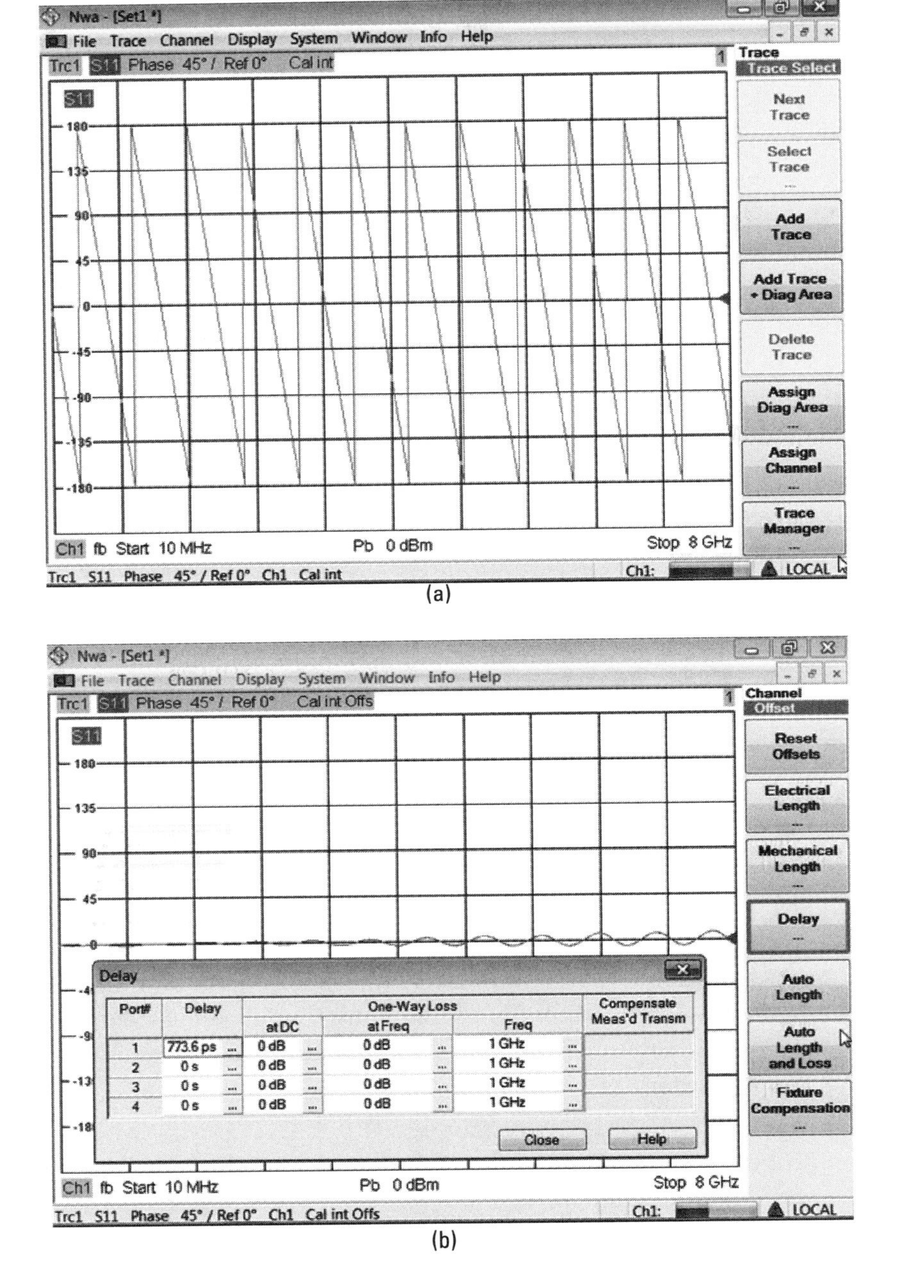

Figure 3.5 S11 phase response of a cable (a) before and (b) after applying the Autolength function.

from the change in a device's phase response across a given frequency interval. With reference to Figure 3.6(a), the mathematical definition of group delay is:

$$\tau_g = \frac{-1}{360^\circ}\left(\frac{\Delta\varphi}{\Delta f}\right) \tag{3.2}$$

where $\Delta\varphi$ = change of phase in degrees, Δf= change of frequency in hertz, and τ_g= group delay in seconds.

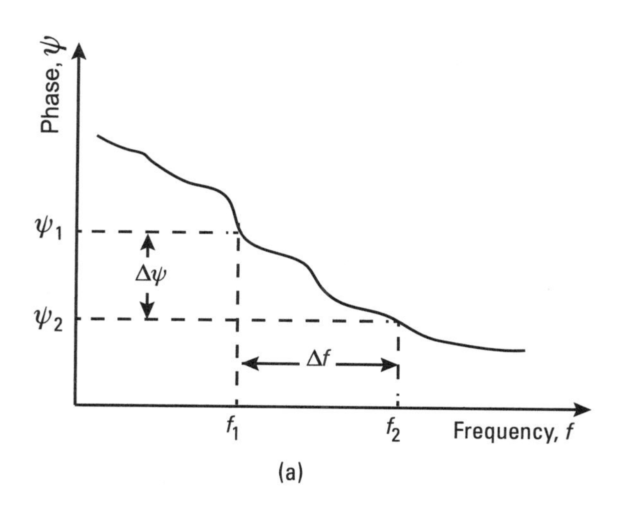

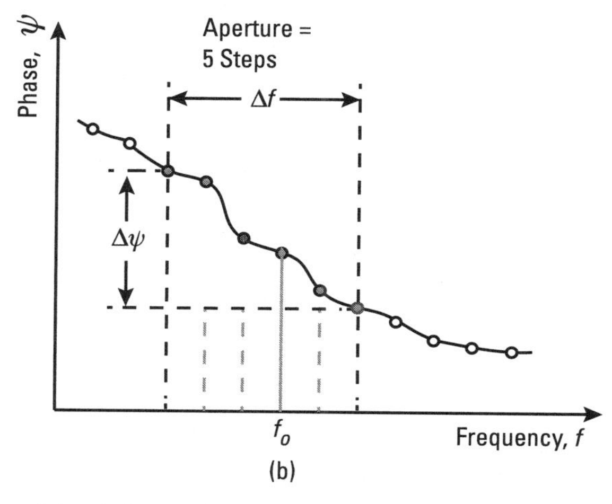

Figure 3.6 Group delay definition. (a) Frequency and phase definition for group delay. (b) VNA implementation using a frequency aperture composed of consecutive points in a frequency sweep.

Nondispersive DUTs have a linear phase response which produces a constant delay (i.e., the ratio of phase change for a given frequency change is constant).

Group delay measurements are typically performed on two-port devices, but can be applied to one-port cable measurements as well, provided the user recognizes that the displayed group delay value is twice the actual cable length. (The signal launched into the cable propagates to the end of the cable, where it fully reflects and travels back through the cable.) In this application the VNA calculates round-trip rather than one-way delay. The group delay "Aperture" setting is very important. It defines the number of measurement points (with total frequency spacing Δf) over which the phase change will be calculated. For example, in Figure 3.6(b), the aperture is set to 5, which represents a set of five consecutive frequency points. The measurement is defined at f_o, but the phase values that are used for the group delay calculation start at three frequency steps before f_o to two frequency steps after. These two end-points are used to calculate a linear estimate of the phase change across this frequency interval, which, in turn, is used to calculate the group delay at point f_o. The analyzer then advances to the next frequency point in the frequency sweep and repeats the group delay calculation. It repeats this process to the end of the VNA sweep.

In the following example, the VNA performs a linear frequency sweep from 10 MHz to 8 GHz, using 5-MHz steps (1,599 measurement points). Figure 3.7 shows four different delay measurements for the same cable. The upper

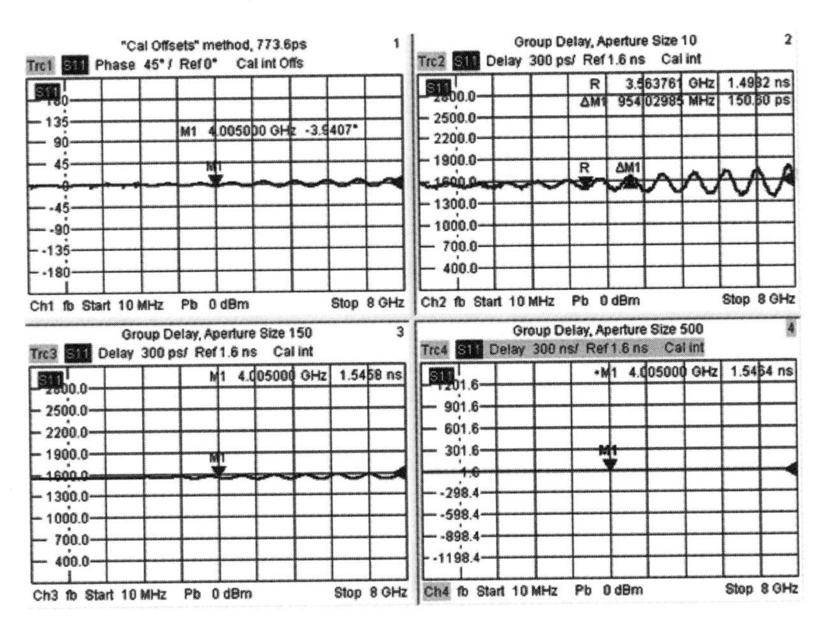

Figure 3.7 Group delay example.

left window shows the delay as measured using the previous Cal Offset method, which yields a delay value of 773.6 ps. The other three windows measure the delay using the group delay method for the same cable, but with three different apertures applied (upper right: 10, lower left: 150, lower right: 500). This reveals some interesting characteristics of the group delay calculation, which we can summarize as follows:

- As the aperture increases, the resulting wider frequency interval results in greater smoothing of the calculated group delay value (at the expense of fine detail in the phase response).
- As the aperture decreases, the resulting narrow frequency interval results in better phase resolution and consequently better group delay resolution, at the expense of greater trace noise and/or ripple.

The trace in the lower right window with aperture of 500 gives a round-trip delay value of 1.546 ns, or an equivalent one-way cable electrical length of 773 ps, which is in close agreement with the Cal Offset method (upper left window).

3.4 Measurement Uncertainty for Passive One-Port Devices

Suppose that after following the setup procedure described in Section 3.1, you measure a device and find that the return loss is 10 dB. How confident are you in this measurement? (How confident should you be?) Or what is your measurement uncertainty?

A VNA, even with properly performed SEC, does not yield error-free measurements. In this section, we explore the largest uncertainty contributors for one-port reflection measurements (directivity) and source match and quantify these uncertainties through several example calculations. One of the most interesting takeaways from this exercise is that you will find the uncertainties vary depending on the reflection characteristics of the DUT itself.

3.4.1 Directivity

Figure 3.8(a) shows a conventional signal flow through a directional coupler. Typically, a three-port directional coupler is driven by a generator at port 1 and terminated by a load at port 2. The coupled port (port 3) provides a convenient, small sample of the signal level incident at port 1. In this example, the signal flow from port 1 to port 2 is the forward direction, while the signal flow from port 2 to port 1 is the reverse direction.

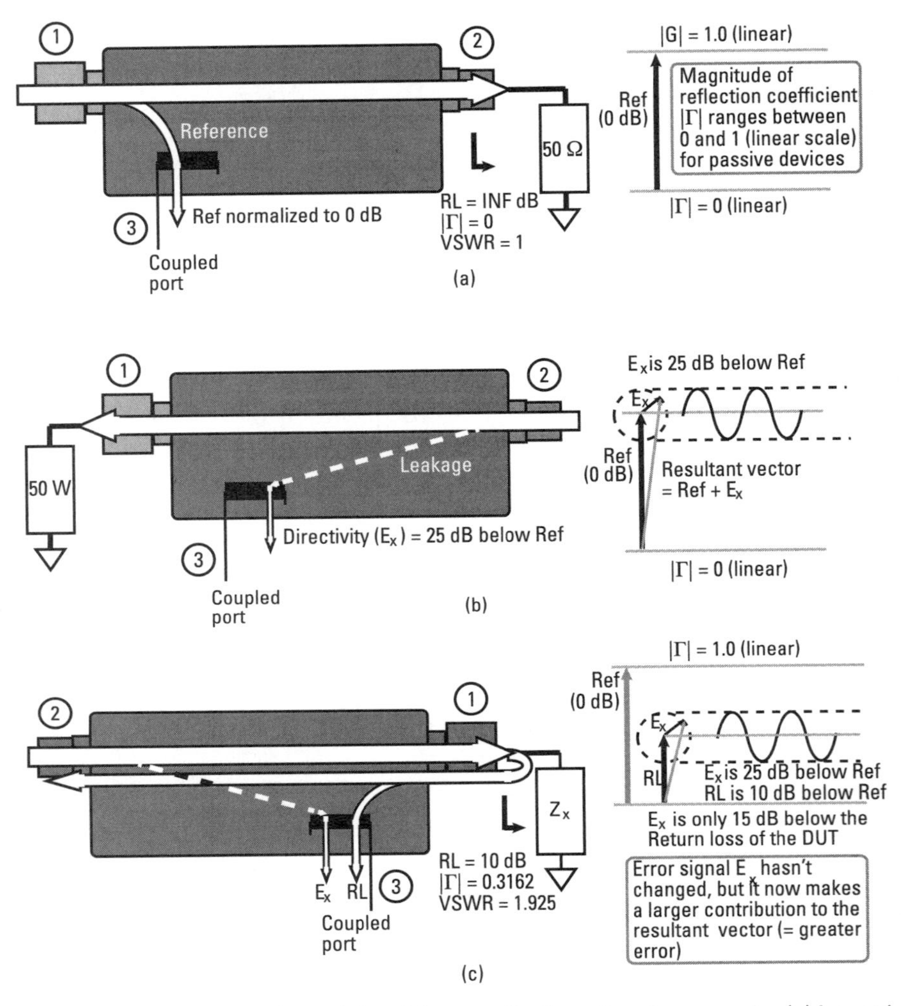

Figure 3.8 Directional coupler and directivity error. Configurations for measuring (a) forward coupling coefficient and (b) reverse leakage. Example (c) shows effects of limited coupler directivity on a typical return loss (RL) measurement.

To determine the directivity of the directional coupler, we connect a signal generator to port 1 and terminate port 2 in a 50Ω load. This termination has the same impedance as the system characteristic impedance (usually 50Ω), ensuring that all energy flowing into the load is dissipated (no reflections). We then measure the signal level at the coupled port (port 3) and record this reference value. All subsequent measurements will be expressed as ratios relative to this reference value (in decibels). Next, the generator is applied to port 2 and the 50Ω load is applied to port 1 (Figure 3.8(b)), and again the signal level at the coupled port is measured. In an ideal directional coupler, there should be no signal detected. However, a real coupler suffers leakage due to factors like limited isolation from the signal generator, as well as connector mismatch, deviation from ideal geometry and nonideal internal terminations. When this leakage is expressed relative to the reference value, it is called the coupler directivity.

To use a directional coupler to measure the return loss of a one-port device, it is flipped end-for-end as shown in Figure 3.8(c). In this arrangement, the coupled port measures reflections from a device under test. To calibrate this setup, we terminate the coupler's test port (port 1) in a short (or open) circuit so that all incident power reflects back into the coupled port. This provides our reference value, because a short (or open) circuit condition produces the largest possible reflection condition for a passive device.

Next, we remove the short-circuit and apply a termination. In this case, all incident power is absorbed (in the ideal case), and there should be no energy reflected from the termination back into the coupler. However, because the directional coupler is not ideal, there is a small amount of leakage due to the previously mentioned sources. By measuring this leakage value and comparing it to the reference signal, we determine the coupler directivity, which is given by the difference (in decibels) between the leakage and the reference value. If we draw this on a phasor diagram (shown at the right in Figure 3.8), we represent the (normalized) reference signal as a vector of length 1 and the error as a vector of length $|E_x|$. Depending on the relative phase of the error signal compared to the reference signal, the signals can add constructively or destructively, yielding values that range from $1 + |E_x|$ to $1 - |E_x|$. Consider an example: a signal generator applies a reference value of 0 dBm to port 2. A short-circuit is applied at port 1, and a spectrum analyzer measures –5 dBm at port 3. Next, we replace the short-circuit by a 50Ω termination and the spectrum analyzer measures –30 dBm at port 3. The directivity is the difference between these values: –5 – (–30) = 25 dB. Next, we attach a test device to port 1. Assuming for a moment that this device has a (known) return loss of 10 dB, the difference between this device's return loss (10 dB) and the coupler directivity (an unchanging leakage value 25 dB below the reference value) is 25 dB – 10 dB = 15 dB. The phasor diagram in Figure 3.8(c) shows the result. In this instance, the error vector $|E_x|$ is only 15 dB lower than the measured return loss value (10 dB). To calculate the associated measurement uncertainty, we must convert this number into an equivalent linear reflection coefficient: 10^(–15/20) = 0.1778. Thus, the error can range from 1 +/– 0.1778. Converting to decibels, we get: $20\text{Log}_{10}(1.1778) = 1.42$ dB and $20\text{Log}_{10}(0.8221) = -1.70$ dB.

These represent deviations from the true measurement value (10 dB). However, since we are measuring return loss, the signs need to be switched. To understand why, consider the phasor diagram, which is actually displaying linear reflection coefficient (Γ), which can vary from 0 to 1. A smaller reflection coefficient (Γ approaching 0) corresponds to a higher return loss, while a larger reflection coefficient (Γ approaching 1) corresponds to a lower return loss. The equation quantifying this relationship is given by:

$$RL = -20\log10(|\Gamma|) \tag{3.3}$$

Hence, from the phasor diagram, a longer vector indicates a higher reflection coefficient, which corresponds to a lower return loss, so we flip the signs of the error values for return loss. The coupler will report the DUT as having a return loss somewhere between 10 – 1.42 and 10 + 1.7 dB or 8.58 dB to 11.7 dB. That is a measurement uncertainty range of over 3 dB. If we repeat this experiment with a better coupler having 40-dB directivity, we will obtain measurements ranging from 10.28 to 9.73 dB, reducing the measurement uncertainty to less than +/–0.3 dB. We can summarize our directivity findings as follows. Directivity error is a function of both the return loss of the device and the directivity of the coupler. As a rule of thumb, using a coupler with directivity 30 dB better than the return loss of the DUT ensures return loss uncertainty of less than +/–0.3 dB.

To simplify these calculations, Table 3.3 can quickly determine the resultant measurement uncertainty. The first column, labeled $|x|$, represents the ratio of the length of the error vector to the reference vector in decibels. For example, if a coupler has directivity of 42 dB and the DUT has a return loss of 19 dB, the difference, 23 dB, represents the magnitude of the error vector relative to the DUT's return loss. Entering the table at 23 dB gives an error range of +0.58 dB to –0.64 dB. Since we are working with return loss, we flip the signs and obtain the expected measurement range of 18.42 dB to 19.64 dB.

System error correction in a VNA uses calibration standards with known amplitude and phase characteristics to mathematically remove errors. By applying vectors of equal amplitude and opposite phase at each frequency point, the error terms can be greatly reduced. In the case of directivity, once the VNA has been calibrated, the amplitude and phase of the isolation path leakage are mathematically removed from the VNA's directional coupler, yielding a much better effective directivity. These corrections apply to other VNA sources of error as well, leading to something called "effective system data" that is often shown on VNA specification sheets (Table 3.4). The resulting effective directivity greatly helps reduce measurement uncertainty, but care must still be taken when measuring DUTs with high return loss (>25 dB, generally). Typically, data sheets also show that effective system data is strongly dependent upon the quality of the calibration kit. Effective system data also varies across frequency, generally decreasing (sometimes quite substantially) with increasing frequency.

This data is valid between +18°C and +28°C, provided the temperature has not varied by more than 1°C since calibration. Frequency points, measurement bandwidth, and sweep time have to be identical for measurement and

Table 3.3
Uncertainty Error Chart

	Measurement Uncertainty					
	$\lvert x \rvert$	$1+\lvert x \rvert$	$1-\lvert x \rvert$	$1+\lvert x \rvert$	$1-\lvert x \rvert$	$\Delta\varphi$
(lin.)	(lin.)	(lin.)	(dB)	(dB)	degrees	
0 dB	1.0000	2.0000	0.0000	6.0206 dB	$-\infty$	90.00
1 dB	0.8913	1.8913	0.1087	5.5350 dB	–9.2715 dB	63.03
2 dB	0.7943	1.7943	0.2057	5.0780 dB	–13.7365 dB	52.59
3 dB	0.7079	1.7079	0.2921	4.6495 dB	–10.6907 dB	45.07
4 dB	0.6310	1.6310	0.3690	4.2489 dB	–8.6585 dB	39.12
5 dB	0.5623	1.5623	0.4377	3.8755 dB	–7.1773 dB	34.22
6 dB	0.5012	1.5012	0.4988	3.5287 dB	–6.0412 dB	30.08
7 dB	0.4467	1.4467	0.5533	3.2075 dB	–5.1405 dB	26.53
8 dB	0.3981	1.3981	0.6019	2.9108 dB	–4.4096 dB	23.46
9 dB	0.3548	1.3548	0.6452	2.6376 dB	–3.8063 dB	20.78
10 dB	0.3162	1.3162	0.6838	2.3866 dB	–3.3018 dB	18.43
11 dB	0.2818	1.2818	0.7182	2.1567 dB	–2.8756 dB	16.37
12 dB	0.2512	1.2512	0.7488	1.9465 dB	–2.5126 dB	14.55
13 dB	0.2239	1.2239	0.7761	1.7547 dB	–2.2013 dB	12.94
14 dB	0.1995	1.1995	0.8005	1.5802 dB	–1.9331 dB	11.51
15 dB	0.1778	1.1778	0.8222	1.4216 dB	–1.7007 dB	10.24
16 dB	0.1585	1.1585	0.8415	1.2778 dB	–1.4988 dB	9.12
17 dB	0.1413	1.1413	0.8587	1.1476 dB	–1.3227 dB	8.12
18 dB	0.1259	1.1259	0.8741	1.0299 dB	–1.1687 dB	7.23
19 dB	0.1122	1.1122	0.8878	0.9237 dB	–1.0337 dB	6.44
20 dB	0.1000	1.1000	0.9000	0.8279 dB	–0.9151 dB	5.74
21 dB	0.0891	1.0891	0.9109	0.7416 dB	–0.8108 dB	5.11
22 dB	0.0794	1.0794	0.9206	0.6639 dB	–0.7189 dB	4.56
23 dB	0.0708	1.0708	0.9292	0.5941 dB	–0.6378 dB	4.06
24 dB	0.0631	1.0631	0.9369	0.5314 dB	–0.5661 dB	3.62
25 dB	0.0562	1.0562	0.9438	0.4752 dB	–0.5027 dB	3.22
26 dB	0.0501	1.0501	0.9499	0.4248 dB	–0.4466 dB	2.87
27 dB	0.0447	1.0447	0.9553	0.3796 dB	–0.3969 dB	2.56
28 dB	0.0398	1.0398	0.9602	0.3391 dB	–0.3529 dB	2.28
29 dB	0.0355	1.0355	0.9645	0.3028 dB	–0.3138 dB	2.03
30 dB	0.0316	1.0316	0.9684	0.2704 dB	–0.2791 dB	1.81
31 dB	0.0282	1.0282	0.9718	0.2414 dB	–0.2483 dB	1.62
32 dB	0.0251	1.0251	0.9749	0.2155 dB	–0.2210 dB	1.44
33 dB	0.0224	1.0224	0.9776	0.1923 dB	–0.1967 dB	1.28
34 dB	0.0200	1.0200	0.9800	0.1716 dB	–0.1751 dB	1.14
35 dB	0.0178	1.0178	0.9822	0.1531 dB	–0.1558 dB	1.02
36 dB	0.0158	1.0158	0.9842	0.1366 dB	–0.1388 dB	0.91
37 dB	0.0141	1.0141	0.9859	0.1218 dB	–0.1236 dB	0.81

Table 3.3 (continued)

	Measurement Uncertainty					
	$\lvert x \rvert$	$1+\lvert x \rvert$	$1-\lvert x \rvert$	$1+\lvert x \rvert$	$1-\lvert x \rvert$	$\Delta\varphi$
39 dB	0.0112	1.0112	0.9888	0.0969 dB	–0.0980 dB	0.64
40 dB	0.0100	1.0100	0.9900	0.0864 dB	–0.0873 dB	0.57
41 dB	0.0089	1.0089	0.9911	0.0771 dB	–0.0778 dB	0.51
42 dB	0.0079	1.0079	0.9921	0.0687 dB	–0.0693 dB	0.46
43 dB	0.0071	1.0071	0.9929	0.0613 dB	–0.0617 dB	0.41
44 dB	0.0063	1.0063	0.9937	0.0546 dB	–0.0550 dB	0.36
45 dB	0.0056	1.0056	0.9944	0.0487 dB	–0.0490 dB	0.32
46 dB	0.0050	1.0050	0.9950	0.0434 dB	–0.0436 dB	0.29
47 dB	0.0045	1.0045	0.9955	0.0387 dB	–0.0389 dB	0.26
48 dB	0.0040	1.0040	0.9960	0.0345 dB	–0.0346 dB	0.23
49 dB	0.0035	1.0035	0.9965	0.0308 dB	–0.0309 dB	0.20
50 dB	0.0032	1.0032	0.9968	0.0274 dB	–0.0275 dB	0.18

Table 3.4

Effective System Data for a ZNB Network Analyzer

R&S®ZNB4 and R&S®ZNB8		
Calibrated using R&S ZV-Z270	9 kHz to 100 kHz	100 kHz to 4.5 GHz
Directivity	>46 dB	>45 dB
Source match	>41 dB	>40 dB
Load match	>44 dB	>45 dB
Reflection tracking	<0.02 dB	<0.02 dB
Transmission tracking	<0.028 dB	<0.018 dB
R&S®ZNB20		
Calibrated using R&S ZV-Z235	100 kHz to 10 GHz	10 GHz to 20 GHz
Directivity	>46 dB	>41 dB
Source match	>43 dB	>38 dB
Load match	>44 dB	>40 dB
Reflection tracking	<0.05 dB	<0.05 dB
Transmission tracking	<0.025 dB	<0.035 dB
R&S®ZNB40		
Calibrated using R&S ZV-Z229	100 kHz to 4 GHz	4 GHz to 20 GHz
Directivity	>42 dB	>38 dB
Source match	>38 dB	>36 dB
Load match	>40 dB	>38 dB
Reflection tracking	<0.05 dB	<0.05 dB
Transmission tracking	<0.02 dB	<0.03 dB

calibration (no interpolation allowed). The data is based on a measurement bandwidth of 10 Hz.

Finally, the effective data is only as good as the stability of the measurement system, consisting of the VNA, test cables, adapters, and operator experience/technique (things like torquing of connections, cleaning connectors, minimizing cable movement). Remember, SEC is obtained at a single snapshot in time (during calibration), with cables in a particular physical orientation. After calibration, if the cables are moved, the reflections that were cancelled out during the calibration process may no longer cancel (depending on the quality of the phase stable cables), leading to deterioration of effective system data, and higher measurement uncertainty (often indicated by significant ripple in the measurement traces).

Since a VNA is also used for measuring phase, it is worth taking a moment to consider phase uncertainty. We use the same basic phasor diagram as Figure 3.8 to set the stage. Here, we can see that the largest phase error (θ) will result when the resultant vector (equal to the vector sum of the reference and error vector) is exactly tangent to the circle (Figure 3.9). Since the length of the reference vector is in all cases normalized to 1, we can determine the angle from simple trigonometry.

Since

$$\sin(\theta) = \left(|E_x|/1\right) \tag{3.4}$$

then

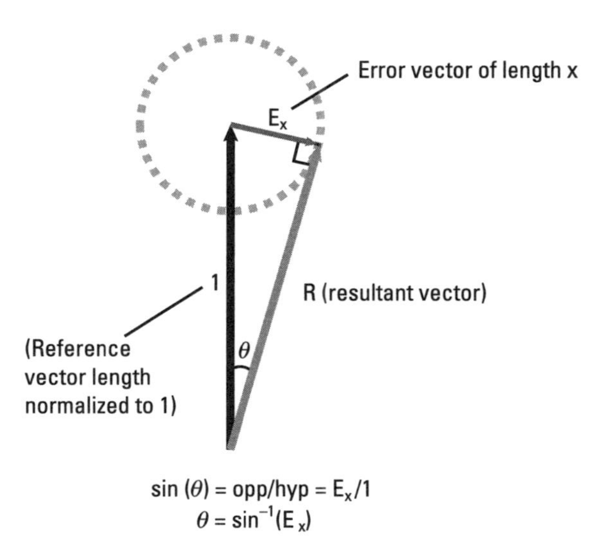

Figure 3.9 Phasor diagram for calculating phase uncertainty.

$$\Theta = \sin^{-1}\left(\left|E_x\right|\right) \tag{3.5}$$

So if $|E_x| = 30$ dB, $|E_x|$ in linear terms is 0.0316 and $\sin^{-1}(0.0316) = 1.81°$. Thus the uncertainty of any phase measurement will be +/−1.81°.

3.4.2 Source Match

Source match is the second potential contributor to measurement uncertainty in one-port measurements. Figure 3.10 shows the same directional coupler as before. This time, total reflection from the open or short termination encounters a second reflection at the coupler's test port due to its nonideal source match. This reflection propagates back to the open/short termination, re-reflects, and enters the directional coupler's receiver at a value 18 dB below the total reflection. (For completeness, a portion of this signal also reflects back

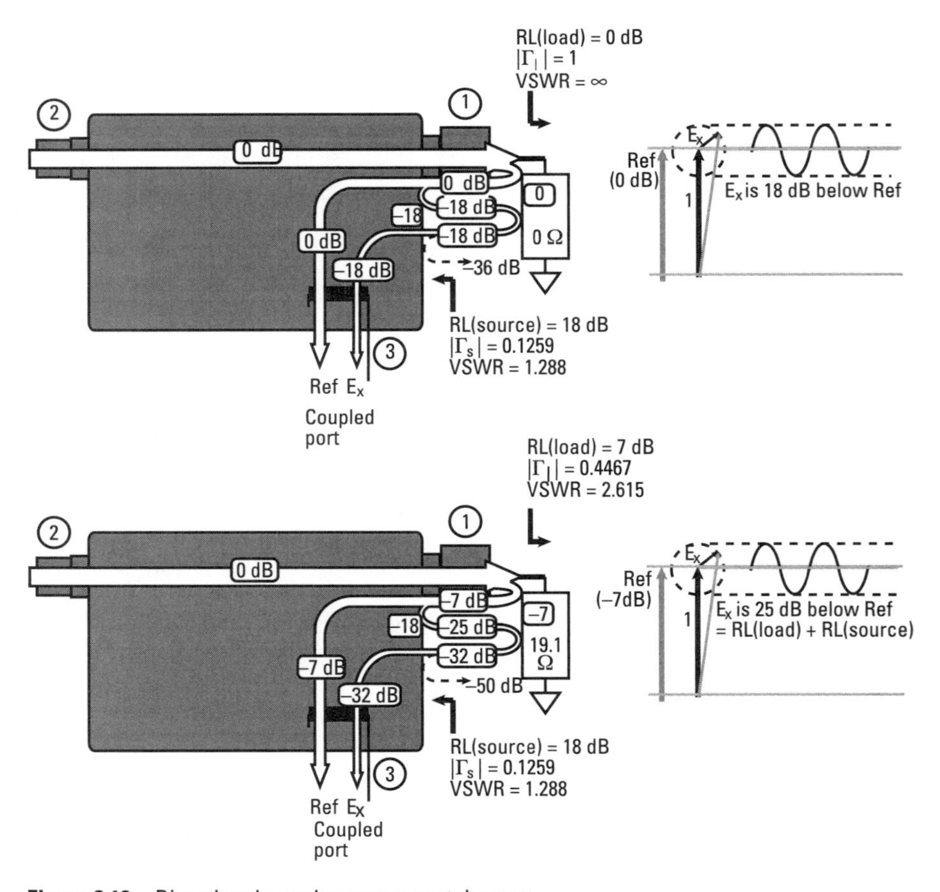

Figure 3.10 Directional coupler source match error.

towards the termination, but the amplitude is so low (–36 dB) compared to the original reflection that its impact (and further reflections) can be ignored.) The net result is an error signal that combines with the original (total) reflection in the receiver. Depending on the phase of this error signal, the measured return loss can vary from –1.03 dB to +1.17 dB (Table 3.3). An ideal short or open termination has a return loss equal to 0 dB, but because of a source match error, the measured value can range between –1.03 dB to 1.17 dB. Passive devices have only positive return loss values, so a measurement of –1.03 dB implies a reflection coefficient greater than 1. An uninformed observer might conclude the passive termination is sending back more power than it is receiving, which is an obvious fallacy.

Fortunately, source match contributions taper off rather quickly with increasing DUT return loss. As an example, substitute a device with 7-dB return loss. Now, the source match error vector is 18 dB + 7 dB = 25 dB below the initial reflection. This produces an error range of –0.48 dB to +0.5 dB. So, the influence of the source match has gone from over +/–1 dB for an open/short termination to +/–0.5 dB for a DUT with 7-dB return loss.

3.5 Active One-Port DUTs: Oscillators, VCOs, and Signal Generators

The classic VNA architecture has been optimized for measuring S-parameters of primarily passive components without frequency conversion. In this classic application, only two internal frequency synthesizers are used. One provides RF stimulus to each port, and a second feeds a common LO signal to all VNA receivers. The LO signal mixes with the RF stimulus at each REF and MEAS receiver, producing a common IF frequency with a phase that depends on the RF energy reflected from (or transmitted through) a DUT.

This configuration permits an extremely simple heterodyne receiver architecture providing broad frequency coverage with minimal hardware. This simple architecture is important for reducing drift and other error contributions. However, several unintended consequences come into play when introducing external signals into a VNA from test devices such as oscillators, VCOs, or signal generators. These unintended consequences include spurious signals, heterodyne images, and receiver compression/overload.

3.5.1 Spurious Signals

When external signals combine with internally generated LO and RF signals in a VNA receiver, spurious signals often result. Figure 3.11(a) presents the spectrum of a 279.2-MHz oscillator as measured with a spectrum analyzer. This particular oscillator's spectrum is not very clean (compare with Figure 3.13).

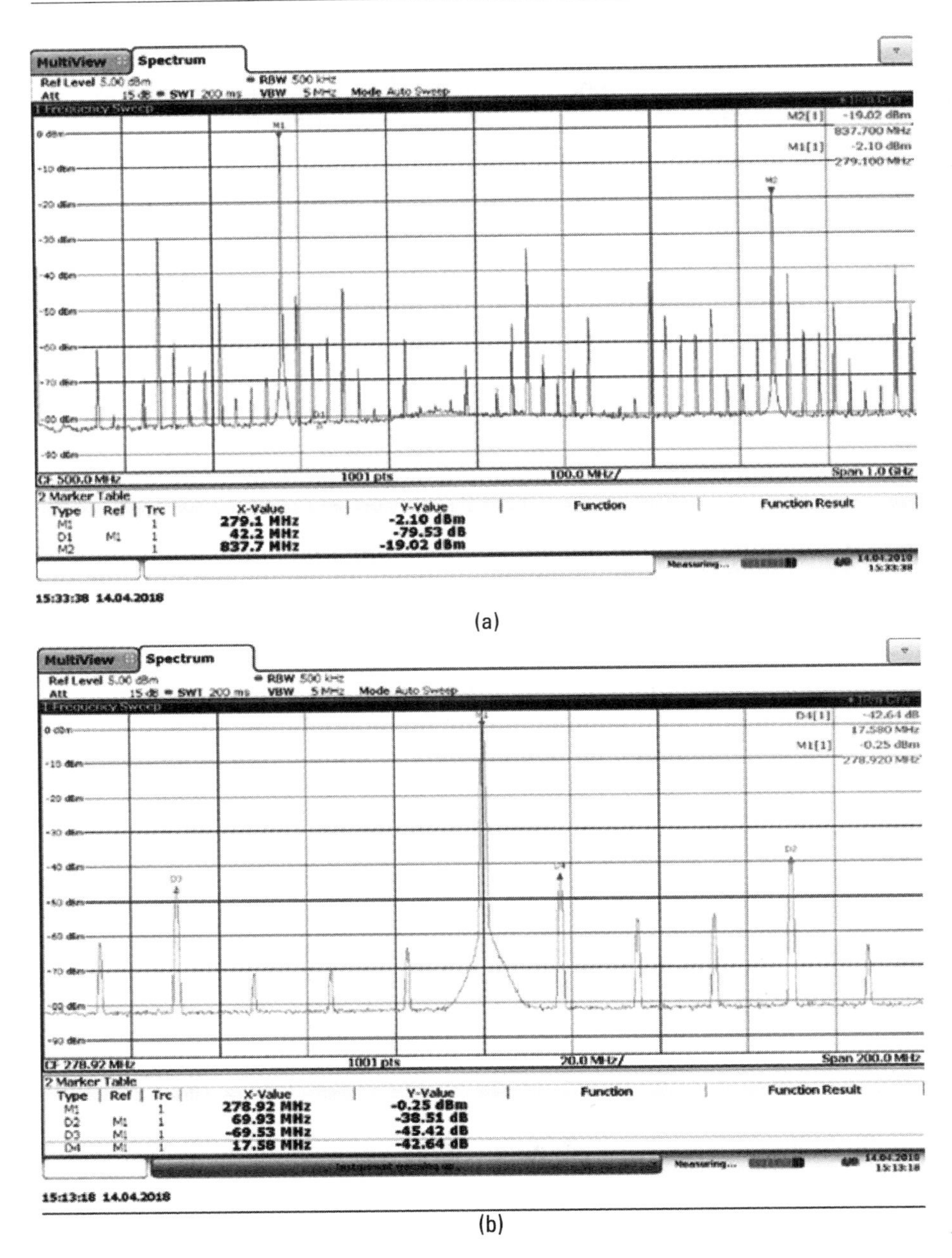

Figure 3.11 (a) Wideband spectrum of 279.2-MHz oscillator. (b) Narrowband spectrum examined around fundamental frequency.

There are a number of nonharmonically related spurious signals. Figure 3.11(b) shows the spectrum zoomed in around the fundamental frequency, with plenty of nearby spurious signals at fairly low (<40 dBc) signal levels.

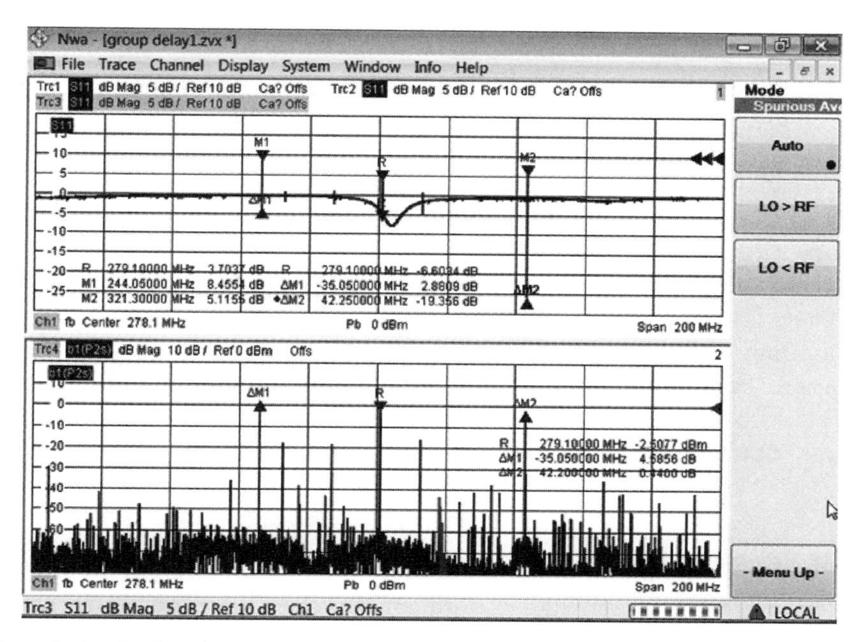

Figure 3.12 Oscillator return loss and power spectrum as measured by a VNA.

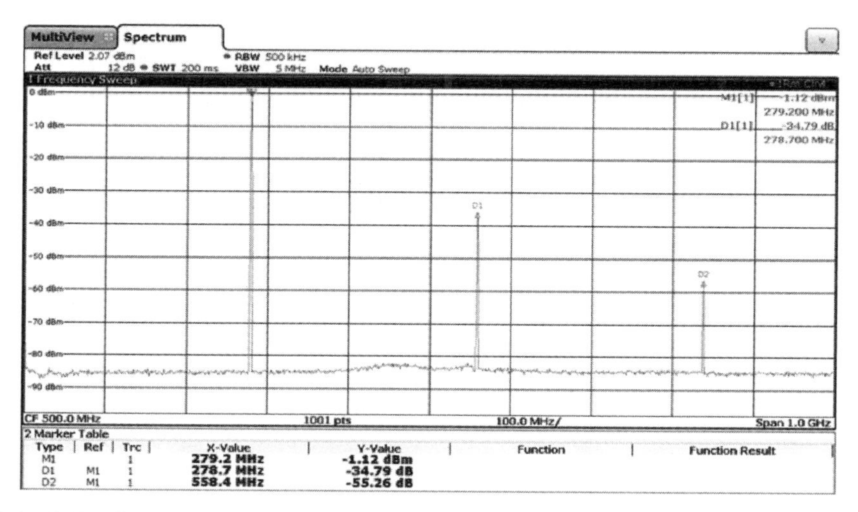

Figure 3.13 Spectrum analyzer view of a clean signal generator.

If we connect this oscillator to a VNA to measure its return loss, we obtain the result shown in Figure 3.12. Here, the top half of the display shows the S_{11} dB Mag measurement, while the bottom half shows the amplitude response seen by the b_1 measurement receiver (in dBm). In this example, we see the oscillator component at 279.2 MHz (center of the screen) leaking through the S_{11} measurement, along with two image frequencies (described in Section 3.5.2), and some other smaller spurious signals.

Since the oscillator is not phase-locked to the VNA, the phase of all these signals randomly adds to or subtracts from the magnitude of the VNA's internal RF and LO generator signals at these frequencies, causing spikes. (The display in Figure 3.12 was obtained by displaying "Max Hold" and "Min Hold" of the S_{11} traces simultaneously to emphasize the impact of these interfering signals.) Of equal interest is the spectrum recorded by the b_1 measurement receiver at the bottom of Figure 3.12. You can see many more spurious responses here than were captured by the spectrum analyzer. This is caused by the mixing of the oscillator's spurious signals with the VNA's internal LO generator. (In the bottom display, the VNA's internal RF generator has been switched off during the b_2 receiver sweep.)

3.5.2 Heterodyne Images

The classic heterodyne receiver uses a hardware-based mixer to convert an RF signal to a lower, fixed IF by using a swept local oscillator (LO) tuned to f_{rf} +/− f_{if}. Figure 6.1(a) shows the concept. Assume the receiver IF is 30 MHz. This means the difference between the LO and the RF stimulus signal must be 30 MHz. As an example, assume that a VNA steps its RF from 100 to 500 MHz in 100-MHz increments. If the VNA is designed to use a LO frequency that is higher than the RF (LO > RF), the corresponding receiver LO synthesizer steps would be 130, 230, …, 530 MHz. At each of these measurement points, the VNA's measurement receiver (essentially a mixer) produces a difference frequency of 30 MHz. Similarly, if the VNA uses LO < RF, the receiver LO synthesizer steps would be 70, 170, …, 470 MHz, which also produces an IF signal at 30 MHz at each frequency step. For either case, the VNA knows exactly which LO frequency and RF it is generating at any given time to produce the IF frequency at the receivers.

Image problems emerge when we introduce an external signal into the VNA. Consider the same example, but this time use a 10-MHz frequency step (100, 110, ..., 490, 500 MHz). Suppose that we connect an external oscillator with 200 MHz output at port 1 of this VNA. The VNA begins stepping its internal RF generator (100, 110, 120, ..., 500 MHz…) and if the receiver mode is LO > RF, the LO generator produces corresponding steps at 130, 140, 150, …, 530 MHz…). When the internal RF generator reaches 140 MHz, the corresponding LO generator frequency is 170 MHz, and the mixing process produces a difference signal at the measurement receiver IF (30 MHz). However, there is another IF signal produced by mixing between the 200-MHz fixed oscillator attached to the port and the LO generator (170 MHz). Furthermore, these two 30-MHz IF signals may combine at any phase, resulting in a changing value displayed at 140 MHz on the screen (the RF).

The analyzer continues stepping its RF and LO frequency. When the internal RF generator reaches 200 MHz, the LO reaches 230 MHz, and the VNA's measurement receiver again sees two IF signals: one due to mixing of its internal LO generator signal with the internal RF generator and the other due to mixing of its internal LO generator signal with the external 200-MHz oscillator. The result is again a combination of these two 30-MHz IF responses displayed on the VNA screen at 200 MHz. Because the oscillator and the RF generator are not likely phase-locked, the amplitudes of these displayed signals will vary from sweep to sweep. Notice that these two responses (the image response at 140 MHz and the real response at 200 MHz) are separated by twice the IF (2×30 MHz).

Figure 3.13 presents the spectrum of a clean signal generator (R&S SMW 200A) as displayed on a spectrum analyzer. The fundamental tone is at 279.2 MHz, while the second and third harmonics are approximately –35 and –55 dBc, and there are no noticeable spurious signals.

If we connect this signal generator to a VNA and measure its return loss, we obtain the result shown in Figure 3.14. Here, the top half of the screen shows the S_{11} dB Mag measurement, while the bottom half shows the amplitude response seen by the b_1 measurement receiver (in dBm). In this example, we see the oscillator component at 279.2 MHz in the center of the screen, as well as two image frequencies: a lower image at twice the IF below the true signal (–34.55 MHz) and another image at twice the IF above the true signal (+42.55 MHz). This leads to two immediate conclusions:

- In this default operating mode (Spurious Avoidance = "Auto"), the instrument is changing from LO < RF to LO > RF during the sweep.
- The actual IF is changing during the sweep (from 34.55/2 = 17.275 MHz to 42.55/2 = 21.275 MHz).

The instrument can be forced to use either LO < RF or LO > RF via the spurious avoidance mode. Results for this signal generator measurement are presented in Figures 3.15 and 3.16, respectively.

Why provide this spurious avoidance capability? There are some measurements where one mode does a better job at eliminating spurious products. In this particular case, the LO > RF mode produces fewer spurious signals than the LO < RF mode (or Auto mode).[3]

3. In the Chapter 9, we discuss methods for using these two modes for producing an image-free response with a VNA.

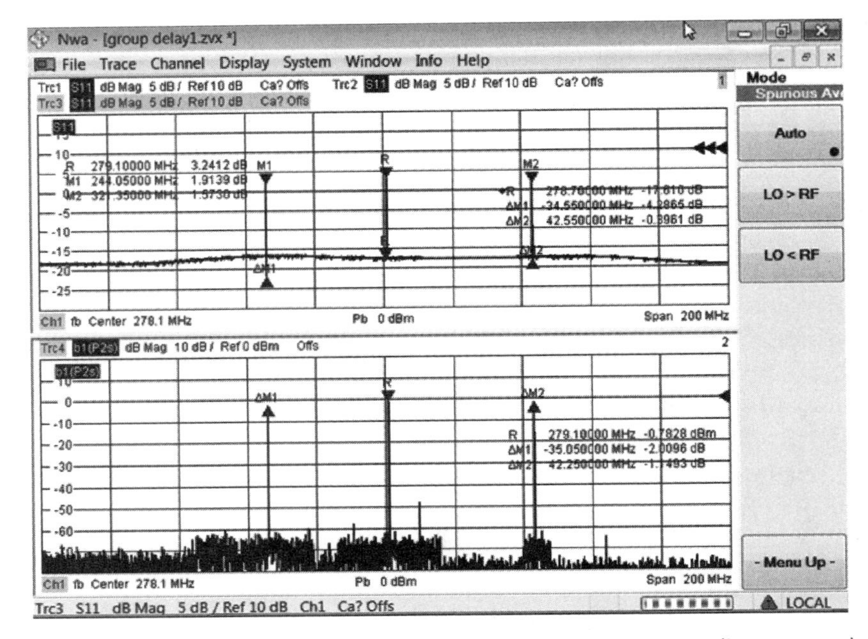

Figure 3.14 Signal generator return loss (top trace) and power spectrum (bottom trace) as measured by a VNA. Notice multiple image frequencies.

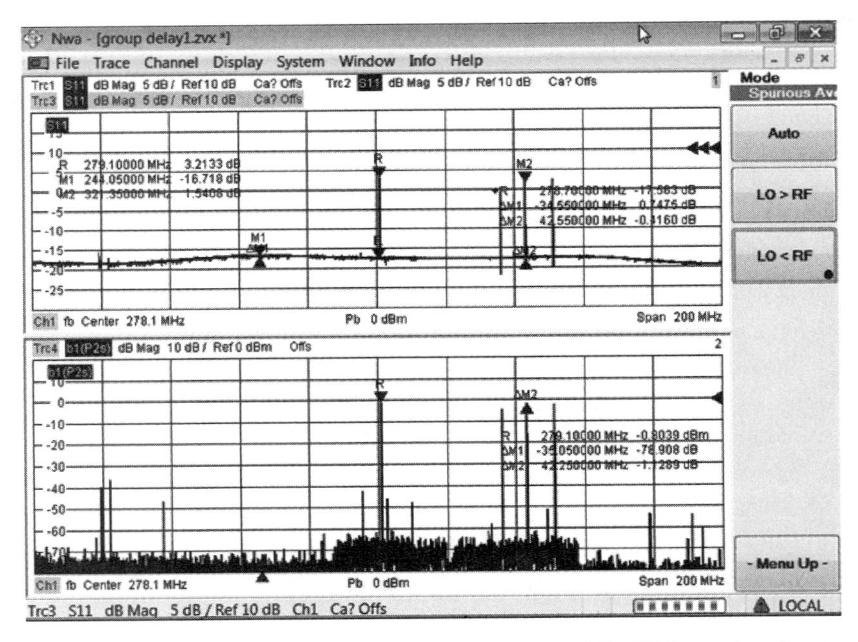

Figure 3.15 Measurement results with spur avoidance set to LO < RF. Return loss (top trace) and power spectrum (bottom trace).

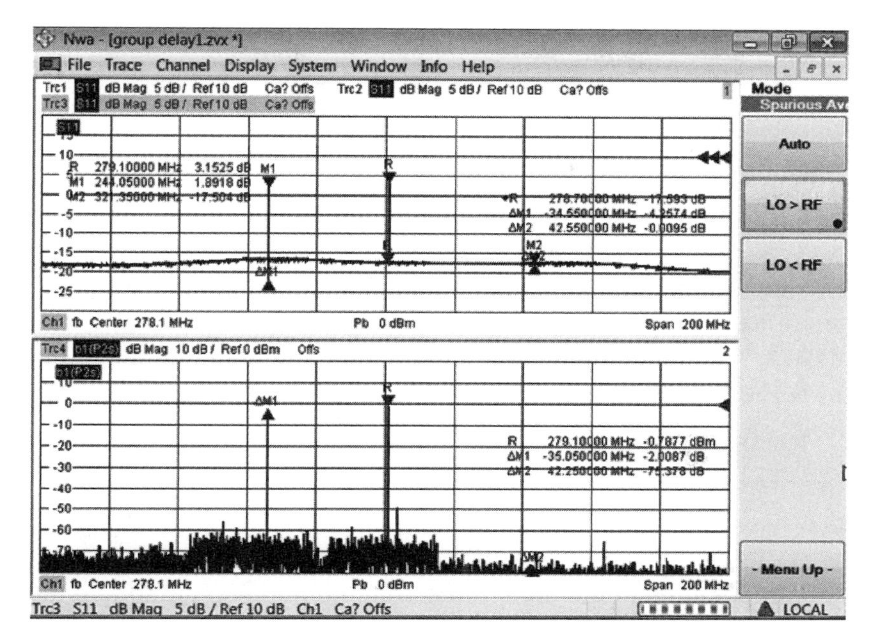

Figure 3.16 Measurement results with spur avoidance set to LO > RF. Return loss (top trace and power spectrum (bottom trace).

Besides the spurious avoidance mode, spurs and images can often be tamed by a judicious selection of measurement points and filter BW. Since a VNA (unlike a spectrum analyzer) makes measurements at discrete frequencies, choosing frequency points in a segmented sweep or a linear sweep that avoid known image frequencies can result in a clean response. However, there is one more consequence associated with making measurements on active one-port devices that we have not yet considered: receiver compression/overload.

3.5.3 Receiver Compression/Overload

When measuring an active one-port device, care must be taken to avoid overloading the VNA receiver. Even though the oscillator under test may operate at only a single frequency, the VNA receiver/mixer has no input filtering, so it sees the full power of this signal at all times. If the amplitude of this signal is high enough to drive the receiver into compression, it can compromise measurement results across the entire frequency measurement range. Figure 3.17 shows a real-world example. Here, the DUT is the same signal generator used in Figure 3.13. We measure S_{11} (dB Mag format) across a wide frequency range. Spurs and image responses are avoided by choosing larger frequency steps and a narrower IF BW. However, the receivers are still impacted by the full brunt of the generator's signal at 279.2 MHz, even though it does not appear on the screen. Thus, we need to verify that this signal is not causing receiver compression. In

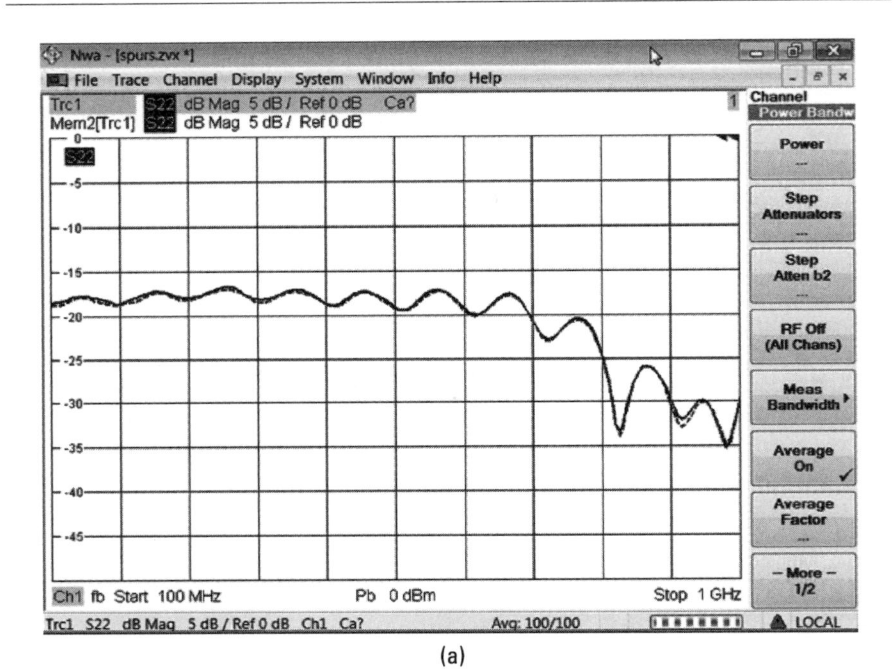

(a)

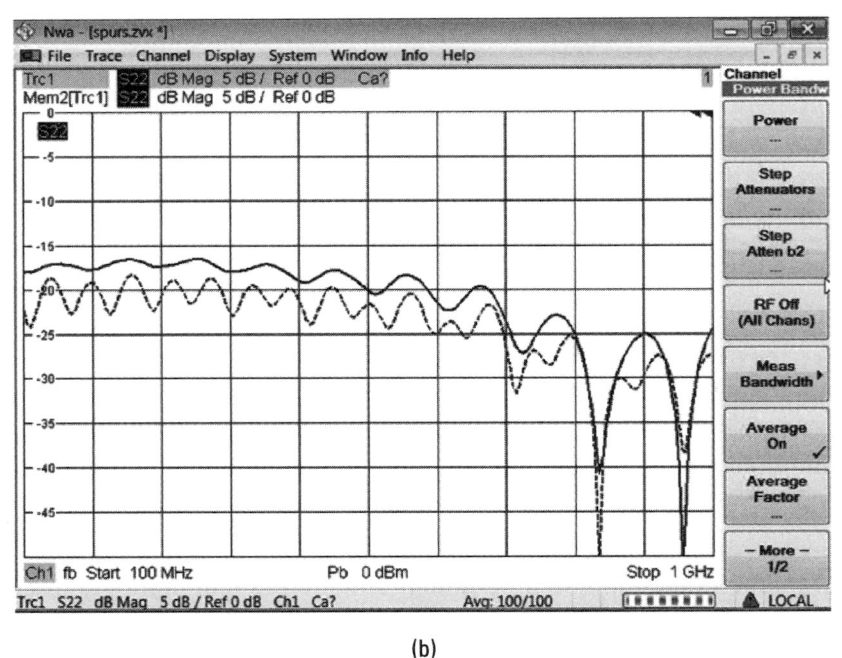

(b)

Figure 3.17 (a) S11 response of a signal generator (output level +9 dBm). No receiver compression observed. (b) S11 response of a signal generator (output level +20 dBm). Receiver compression is compromising the S11 measurement.

Figure 3.17(a), the signal generator is applying +9 dBm to the VNA port. To make sure the receiver is not being compressed, we increase the mechanical attenuator connected between the directional coupler and the MEAS receiver in 5-dB steps. This screenshot shows the results for both 0 dB and 30 dB receiver attenuation. There is virtually no difference in the response, so we can be sure that the receiver is not undergoing any compression at the default 0-dB setting.[4]

In Figure 3.17(b), the signal level is increased to +20 dBm. Again, measurements are made with 0 dB (dotted trace) and 30 dB (solid trace) attenuation in front of the b_2 receiver. There is a significant difference in the S_{11} dB Mag response across the entire frequency range. Because the b_2 measurement receiver is operating in compression with default attenuation (0 dB), it shows an artificially low S_{11} value. The key takeaway is that when measuring an active one-port device, even if you cannot directly see its response on the screen, it is nevertheless present and may be compressing the VNA receiver. Always check for overload.

Reference

[1] Pozar, D., *Microwave Engineering*, Reading, MA: Addison-Wesley, 1990, p. 83.

Selected Bibliography

Ginley, R., "Confidence in VNA Measurements," *IEEE Microwave Magazine*, Vol. 8, No. 4, August 2007, pp. 54–58.

Ridler, N., et al., "Measurement Uncertainty, Traceability, and the GUM," *IEEE Microwave Magazine*, Vol. 8, No. 4, August 2007, pp. 44–53.

Rumiantsev, A., and N. Ridler, "VNA Calibration," *IEEE Microwave Magazine*, Vol. 9, No. 3, June 2008, pp. 86–99.

Appendix 3A: Phase Wrap Resulting from an Offset Open or Offset Short Calibration Kit Standard

Conventional wisdom suggests that after performing a calibration (also known as SEC) with a manual calibration kit, if you reinstall the open or the short standard, you should see a nice tight dot at the right edge or left edge of a Smith

4. Because 30 dB is a significant amount of attenuation, the SNR of the signal entering the b_2 receiver was drastically reduced, leading to significant trace noise. To compensate, Trc1 was measured with 100 averages to clean up the noise (average out its effects) and restore a clean trace.

chart, respectively. Many engineers and technicians are surprised to see the dot smeared out into an arc. For high-frequency sweeps, that arc can sometimes wrap entirely around the edge of the Smith chart, several times. What is wrong?

The issue is in the nature of the open or short provided by the standard. It would be very difficult to provide a precise open or short-circuit condition right at the end of a coaxial test cable. Instead the true open or short condition is established at a distance from the connector's test plane, connected via a short, but precisely known length of (usually air dielectric) transmission line. Figure 3A.1 shows a cross-section view of a female offset short calibration kit standard, and the equivalent circuit diagram.

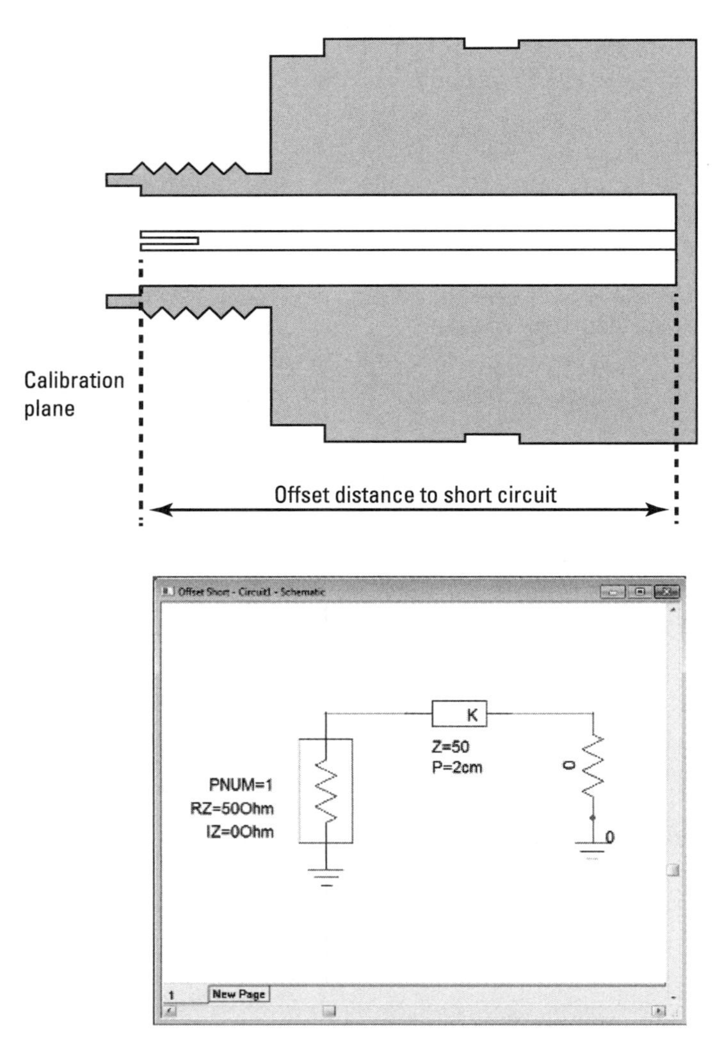

Figure 3A.1 Cross-section of a female offset short calibration standard.

At low frequencies, this transmission line has a nearly negligible impact on the short-circuit condition, but as the frequency extends up into microwave frequencies, this short length of transmission line begins to have a significant effect on the impedance. For example, engineers often utilize a quarter-wave section of transmission line as a handy impedance inverter. Consider that a 2-cm-long transmission line appears as quarter-wave inverter for an 8-cm wavelength (3.747 GHz). Thus, a 2-cm-long offset-open standard would be transformed into a short circuit at this frequency.

The mathematics of this process is straightforward. Begin with the formula for determining the input impedance (Z_{in}) of a lossless line (Z_o) terminated in a known impedance (Z_L):

$$Z_{in} = Z_o \frac{(Z_L + JZ_o \tan\theta)}{(Z_o + JZ_L \tan\theta)} \tag{3A.1}$$

For the case of a short-circuit termination ($Z_L = 0$), (3A.1) becomes:

$$Z_{in} = Z_o \frac{(JZ_o \tan\theta)}{(Z_o)} \tag{3A.2}$$

Simplifying further, we get:

$$Z_{in} = JZ_o \tan\theta \tag{3A.3}$$

Here, θ is the fractional length of the transmission line. For example, assume that we are performing the measurement at 1 GHz. The corresponding free-space wavelength is:

$$\Lambda = c/f \approx 3.0e10\left(\frac{cm}{s}\right)\Big/1.0e9s^{-1} = 30\ cm \tag{3A.4}$$

Using an offset short with offset length of 2 cm, the fractional wavelength is expressed as:

$$\lambda_{frac} = 2cm/300cm = 0.0666\lambda \tag{3A.5}$$

In terms of degrees, the effect on electrical length will be doubled because of the 2-way travel (out and back) of the wave:

$$\Theta = 2\lambda_{frac} x \frac{360°}{1\lambda} = 47.88° \; cm \tag{3A.6}$$

Applying this value to (3A.3), we obtain:

$$Z_{in} = JZ_o \tan\theta = j(50)\tan(47.88) = j55.3 \; ohms \tag{3A.7}$$

We can normalize this impedance and plot it directly on a Smith chart, or we can crank through the math that converts this impedance to an equivalent reflection coefficient. The advantage of this latter approach is that it gives us insight into the wrap effect, where this short circuit is swept through an arc associated with the offset length.

Reflection coefficient (Γ) is calculated as a function of impedance using this equation:

$$\Gamma = \frac{(Z_x - Z_0)}{(Z_x - Z_o)} \tag{3A.8}$$

Z_x is the impedance we want to convert (in this case, Z_{in} from (3A.3)). Let's substitute this value into (3A.8), and do a bit more simplifying:

$$\Gamma = \frac{(JZ_o \tan\theta - Z_o)}{(JZ_o \tan\theta + Z_o)} = \frac{Z_o(J\tan\theta - 1)}{Z_o(J\tan\theta + 1)} = \frac{(-1 + J\tan\theta)}{(1 + J\tan\theta)} \tag{3A.9}$$

For small values of θ, tan(θ) << 1, so Γ= –1, or 1 < 180° (left edge of the Smith chart, along the real axis).

Taking the same approach for the open standard, we obtain:

$$\Gamma = \frac{(-JZ_o \cot\theta - Z_o)}{(-JZ_o \cot\theta - Z_o)} = \frac{(-1 - J\cot\theta)}{(1 - J\cot\theta)} \tag{3A.10}$$

Tabulated values for offset open and short standards are provided in Table 3A.1 and displayed graphically in Figure 3A.2.

For small values of θ, cot(θ) >> 1, so Γ= 1 (or 1 <0°) (right edge of the Smith chart, along the real axis.)

Below, a MATLAB script was created using these equations to show the phase wrap versus Frequency for an offset short and offset open on both a polar plot and a Smith chart. (Basically, both the polar plot and Smith charts are used to plot Γ. Only the grid lines (and their interpretation) are different.

Table 3A.1
Tabulated Γ for Offset Open and Short Standards Versus Frequency

Results for Offset Short				
Frequency (MHz)	**Fractional λ**	**Impedance**	**Reflection Coefficient**	
		R + jX	**Mag(Γ)**	**<(Γ)**
50.0	0.00334	0.00 + j1.05	1.00	177.60°
150.0	0.01001	0.00 + j3.15	1.00	172.80°
250.0	0.01668	0.00 + j5.26	1.00	167.99°
350.0	0.02335	0.00 + j7.39	1.00	163.19°
450.0	0.03002	0.00 + j9.54	1.00	158.39°
550.0	0.03669	0.00 + j11.74	1.00	153.58°
650.0	0.04336	0.00 + j13.97	1.00	148.78°
750.0	0.05003	0.00 + j16.26	1.00	143.98°
850.0	0.05670	0.00 + j18.61	1.00	139.17°
950.0	0.06338	0.00 + j21.03	1.00	134.37°
Results for Offset Open				
Frequency	**Fractional**	**Impedance**	**Reflection Coefficient**	
MHz	Λ	**R + jX**	**Mag(Γ)**	**<(Γ)**
50.0	0.00334	0.00 – j2,385.38	1.00	–2.40°
150.0	0.01001	0.00 – j794.20	1.00	–7.20°
250.0	0.01668	0.00 – j475.40	1.00	–12.01°
350.0	0.02335	0.00 – j338.37	1.00	–16.81°
450.0	0.03002	0.00 – j261.93	1.00	–21.61°
550.0	0.03669	0.00 – j213.03	1.00	–26.42°
650.0	0.04336	0.00 – j178.95	1.00	–31.22°
750.0	0.05003	0.00 – j153.77	1.00	–36.02°
850.0	0.05670	0.00 – j134.35	1.00	–40.83°
950.0	0.06338	0.00 – j118.86	1.00	–45.63°

For completeness, it is worth mentioning that there are actually a few instances where the user will see a single dot at the left edge or right edge of the Smith chart after performing a calibration and reconnecting a calibration standard:

1. When calibrating with waveguide, in a conventional waveguide calibration, the calibration plane is established right at the waveguide opening, so the offset distance is zero (see Chapter 9).

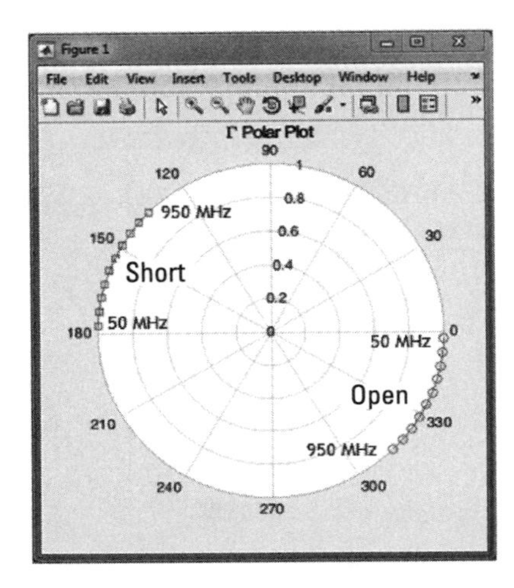

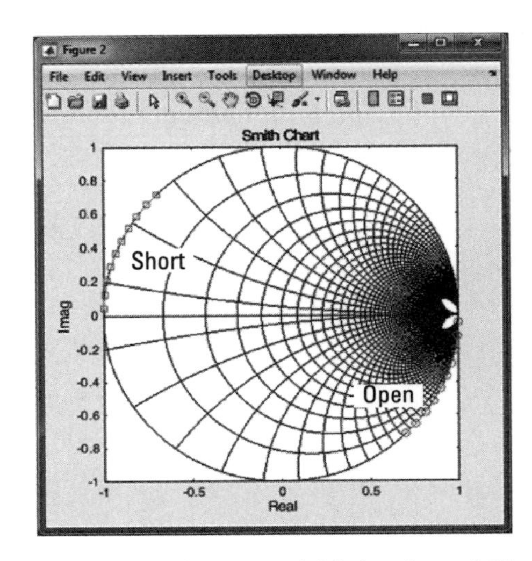

Figure 3A.2 Γ for offset open and short standards versus frequency. (a) Polar plot and (b) Smith chart.

2. When applying a port extension, the port extension function was discussed in Section 3.3 and may be used to move the calibration plane from the coaxial connector to the location of the actual open-circuit or short-circuit in the calibration kit standard. By eliminating this offset distance, the Smith chart will (ideally) show a single dot at the left or right edge. In modern network analyzers, if a port extension is active and a calibration is started, the instrument will temporarily disable the port extension until the completion of the calibration.
3. When using an incorrect calibration kit definition, if a generic calibration kit file is selected (where the open and short standards are defined with zero offset length), then at the completion of the calibration process you will again see a single dot at the left and right edge of the Smith chart for a short or open standard, respectively. However, the measurements will be in error, because the instrument has failed to take the precise physical length offset(s) into account. Any vector reflection measurements will be affected by twice the offset length (out-and-back), leading to significant impedance errors.

Appendix 3A.2: MATLAB Script for Generating Data for Offset Open and Short Standards Versus Frequency

```
% calkit_offset.m
```

```
    % script displays complex Gamma (Reflection Coeffi-
cient) as a function of
    % frequency and offset length of a Cal Kit (OPEN or
SHORT) standard
    % Result is displayed on Polar Plot and Smith Chart
    % Author: Greg Bonaguide
    % Date: 08/12/2018
    clc
            % clear Command Window
    try
    if exist('f1','var')
    close (f1)
            % close figure
    end %if
    catch
    end %try
    try
    if exist('f2','var')
    close (f2)
            % close figure
    end %if
    catch
    end %try
    clear all;
            % delete all variables
    Zo = 50;
            % characteristic impedance
    offset_length = 2.0;
      % offset length (in cm)
    c = 2.998e10;
      % speed of light (in cm/s)
    f = (50:100:1000)' ;
      % freq sweep array (in MHz)
    if f(1) > 0
    else
    f(1) = 0.00001;
    end %if
    lambda = c./(f*1e6);
      % wavelength (in cm)
    frac_wavelength = offset_length./lambda;          %
fractional wavelength of offset
    x = frac_wavelength.*360.0;
% converted to degrees
    Zin_short = i.*Zo*tan(x./180*pi());               %
calculate complex input
    % impedance(SHORT)
    Zin_open = -i.*Zo*cot(x./180*pi());               %
calculate complex input
```

```
    % impedance(OPEN)
    Gamma_s = (Zin_short - Zo)./(Zin_short + Zo);      %
calculate complex Gamma(SHORT)
    Gamma_o = (Zin_open - Zo) ./ (Zin_open + Zo);      %
calculate complex Gamma(OPEN)
    f1 = figure;
      % create first figure
    polarplot(angle(Gamma_s),   abs(Gamma_s),'r-s');   %
plot offset SHORT as a polar plot
    title('\Gamma Polar Plot');
    set(gcf,'Position',[1350,600,400,400]);            %
position it on my PC screen
    hold on % add a second trace
    polarplot(angle(Gamma_o),   abs(Gamma_o),'b-o');   %
plot offset OPEN as a polar plot
    LineWidth = 1.5;
    rlim([0 1]);
      % set outer ring magnitude
    % of polar plot to 1.0
    f2 = figure;
      % create a second figure
    gb_smithchart;
            % it will be a Smith Chart
    set(gcf,'Position',[1350,75,459,425]);
% position it on my PC screen
    plot(real(Gamma_s),imag(Gamma_s),'r-s');           %
plot offset SHORT on Smith Chart
    hold on;
            % add a second trace
    plot(real(Gamma_o), imag(Gamma_o),'b-o');          %
plot offset OPEN on Smith Chart
    %set(gca,'LineWidth',1.0);
      % set Line Width
    str = sprintf('Results for Offset SHORT');         %
plot table for offset SHORT
    disp (str);
    str = sprintf('Freq Fractional Impedance Reflection
Coefficient');
    disp (str);
    str = sprintf('MHz %c R + jX Mag(%c) <(%c)',char
    (955),char(915),char(915));
    disp (str);
    for n = 1:1:size(f,1)
    str = sprintf('%6.1f %7.5f %5.2f + j%-8.2f %4.2f
%6.2f%c'...
     ,f(n),frac_wavelength(n),real(Zin_
short(n)),imag(Zin_short(n)),...
```

```
    abs(Gamma_s(n)),angle(Gamma_s(n))*180/
pi(),char(176));
    disp (str);
    end
    str = sprintf('\n\nResults for Offset OPEN');      %
plot table for offset OPEN
    disp (str);
    str = sprintf('Freq Fractional Impedance Reflection
Coefficient');
    disp (str);
    str = sprintf('MHz %c R + jX Mag(%c) <(%c)',char
    (955),char(915),char(915));
    disp (str);
    for n = 1:1:size(f,1)
    str = sprintf('%6.1f %7.5f %5.2f + j%-8.2f %4.2f
%6.2f%c'...
    ,f(n),frac_wavelength(n),real(Zin_
open(n)),imag(Zin_open(n)),...
    abs(Gamma_o(n)),angle(Gamma_o(n))*180/
pi(),char(176));
    disp (str);
    end
```

4

Passive Two-Port Device Measurements

4.1 Passive Two-Port Devices

Passive two-port devices constitute a complex constellation of different RF functions, with similarly broad measurement requirements. However, most of these devices share a common need for insertion loss and return loss measurements. With this in mind, we begin this chapter by establishing a baseline approach for the vast majority of two-port measurements and work our way into the unique requirements of more specialized and challenging applications.

Key specifications for this class of components include frequency-domain reflection measurements (similar to those covered in Chapter 3), as well as transmission measurements (insertion loss, two-port group delay, and phase response). However, some two-port applications place additional burdens on dynamic range, or involve measurements of more than two ports. Isolators and circulators are nonreciprocal and thus require bidirectional measurements. RF couplers are often tested using two-port analyzers even though they may have three or four ports.

System error correction (SEC) for a two-port measurement is more complex, involving correction of anywhere from seven and sixteen error terms (depending on calibration method) rather than just the three error terms associated with a one-port measurement. To correct all four S-parameters, the VNA must sweep in both the forward and reverse direction. Even if a user needs to measure only the input match (S_{11}) of a two-port device connected between port 1 and port 2 of a VNA, it must still perform both a forward and a reverse sweep. (See

Chapter 2 for details.) Table 4.1 provides an overview of the passive two-port topics covered in this chapter.

4.1.1 Steps for Establishing a Two-Port Measurement Baseline

- *Step 1:* Determine the appropriate interface between the VNA and the device under test (DUT). If test cables are required, use only high-quality phase-stable test cables with appropriate connectors (and genders). If possible, install any required RF adapters before performing system error correction. Otherwise, these adapters will only increase reflections and measurement ripple. In some instances, it may be advantageous to dispense with one of the test cables and use a high-quality connector saver right at a VNA port. Other applications (such as testing of discrete packaged components) require test fixturing and appropriate calibration and de-embedding techniques to move the calibration plane close to the DUT.
- *Step 2:* Start from an instrument preset. Configure the start frequency, stop frequency, and number of points (or equivalently, frequency step size) for the measurement at hand.
- *Step 3:* Set receiver bandwidth and source level. Test devices such as cables and delay lines are well suited to VNA default settings. Chief among these are default power levels (−10 dBm to 0 dBm) and default IF bandwidths (1 kHz to 10 kHz). For production measurements where speed is important, the user may elect to sacrifice dynamic range for

Table 4.1
Example Devices and Unique Measurement Requirements for Passive Two-Port Devices

Device/ Application	Unique Measurement Requirements (Beyond Inserstion (IL) and Return Loss (RL)	Section
Cable	Phase matching/group delay, TDR	4.1.2
Delay line	Group delay, phase	4.1.2.1
Attenuator	Accurate (high) loss (requires high-dynamic range)	4.1.2
Filter	Passband, stopband, and transition band loss, phase/group delay (requires high dynamic range)	4.1.3
Power combiners/ splitters	Path comparison (loss, phase), multiple two-port measurements	4.1.4.1
Directional couplers/ bridges	Forward versus reverse loss between multiple ports	4.1.4.3
Circulators/isolators	Forward versus reverse loss between multiple ports	4.1.4.2
Switches	Thru versus isolated path loss, transient switching (amplifier and phase)	4.1.5

faster sweep speed by selecting a wider measurement bandwidth (up to 1 MHz depending on trace noise tolerance). If you are using a microwave VNA (capable of measurements to 24 GHz or higher), and the particular application requires measurements beginning below 500 MHz, it is a good idea to use the Dynamic Bandwidth at Lower Frequencies mode (see Figure 3.1).

- *Step 4:* Perform a two-port calibration (SEC) using a calibration kit that matches the DUT connector type and gender. The most popular two-port manual calibration type is TOSM (thru, open, short, match), also known as SOLT (short, open, load, thru). Follow the SEC instructions for your VNA, and then save the calibration and/or the test setup after completing all required SEC steps.
- *Step 5:* After completing the calibration, one should take the time to verify it. See Section 3.1.1, Step 5 for discussion.
- *Step 6:* If the calibration is valid, connect a DUT. Configure the trace format and scale for the desired measurement. For two-port passive device measurements, place S_{11} and S_{22} in a single display area. Similarly, place S_{21} and S_{12} together in a second display area. Whenever multiple traces occupy a single display area, it is desirable to couple the trace scaling factors so that changing one changes the other(s) automatically. This is especially helpful when measuring passive reciprocal devices, where S_{21} and S_{12} should be identical. If you do not couple the scaling factors and change the scale of the S_{21} trace, you might forget and later question why this trace appears different from the S_{12} trace.

Coupling two or more scaling factors is quite easy on an R&S VNA. From the Trace group, select "Trace Config" > "Traces" > "Trace Manager" to open the Trace Manager dialog. As seen in Figure 4.1, if you want to link Trc3 to Trc1, follow along the row for Trc3 until you get to the pull-down in the last column (Scale). Click on this, and select Trc1 to couple the scaling of Trc3 to Trc1. Then click "Close." Alternatively, if you wanted to couple all four traces

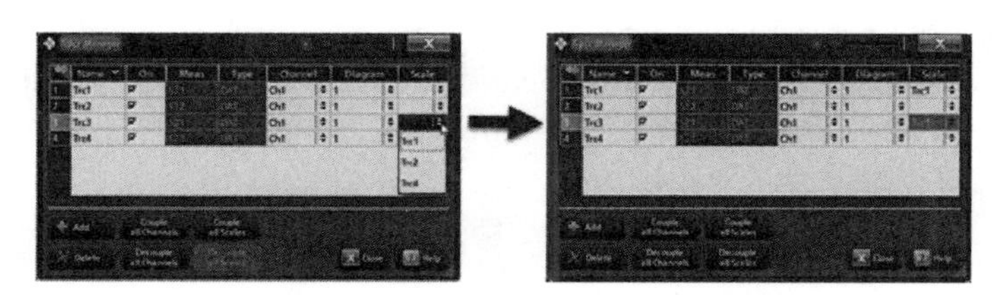

Figure 4.1 Dialog to couple trace scaling factors. Here, Trc3 is coupled to Trc1.

together, simple click the button "Couple all Scales" in this dialog. Traces may by uncoupled in a similar manner.

4.1.2 Cable Measurements

Two-port cable measurements are conceptually easy to make on a VNA using the two-port baseline procedure outlined in Section 4.2. However, cable measurements are also unforgiving when it comes to VNA test cable quality, cable adapter usage, and operator technique. The inherent nature of most cables (low insertion loss, combined with generally high return loss) compounds the problem, as illustrated in Figure 4.2. Multiple significant transmission and reflection paths lead to a resultant vector with many moving parts. Note that many of these error vector components contain a transmission loss (TL) factor. Fortunately, if the two-port device is lossy (e.g., non-negligible transmission loss), the TL will tend to attenuate these reflections, especially when they pass through the DUT multiple times.

However, in a cable with low transmission loss, these reflections are attenuated only slightly and therefore tend to show up as a complicated ripple pattern on the S_{21} and S_{12} measurements. Under ideal conditions (good SEC and good VNA cables), the VNA source and load return losses will be very low, so there will be very little ripple. However, if the test cables lack phase stability, then any cable movement will upset the condition for perfect cancellation of error vectors, resulting in considerably larger reflections and noticeable ripples in the S_{12} and S_{21} responses (Figure 4.3). Similarly, an operator who does not minimize cable movement, neglects proper connector care, or lacks experience making repeatable connections will likely see ripples in his measurements, even if his or her VNA cables have good phase stability.

At any particular frequency, the SEC should ideally cancel the error vectors, leading to S-parameter measurements that accurately reflect the true DUT characteristics. Realistically, we can get close to complete cancellation. Any residual error will be exposed as a small ripple in the DUT response. When the DUT is a device with extremely low loss (like a short cable or adapter), it is possible for the S_{21} (S_{12}) response to rise above 0 dB due to these small residual errors, suggesting the cable has gain (Figure 4.4). In these cases, there is nothing wrong with the analyzer. The best one can do is to enforce best practices: use known-good phase-stable VNA cables, perform careful SEC, and apply operator skill, patience, and attention to detail to obtain the best possible measurements.

4.1.2.1 Delay Lines

Measuring delay lines requires the same care and attention to detail as conventional cable measurements. Depending on the amount of delay required, delay-

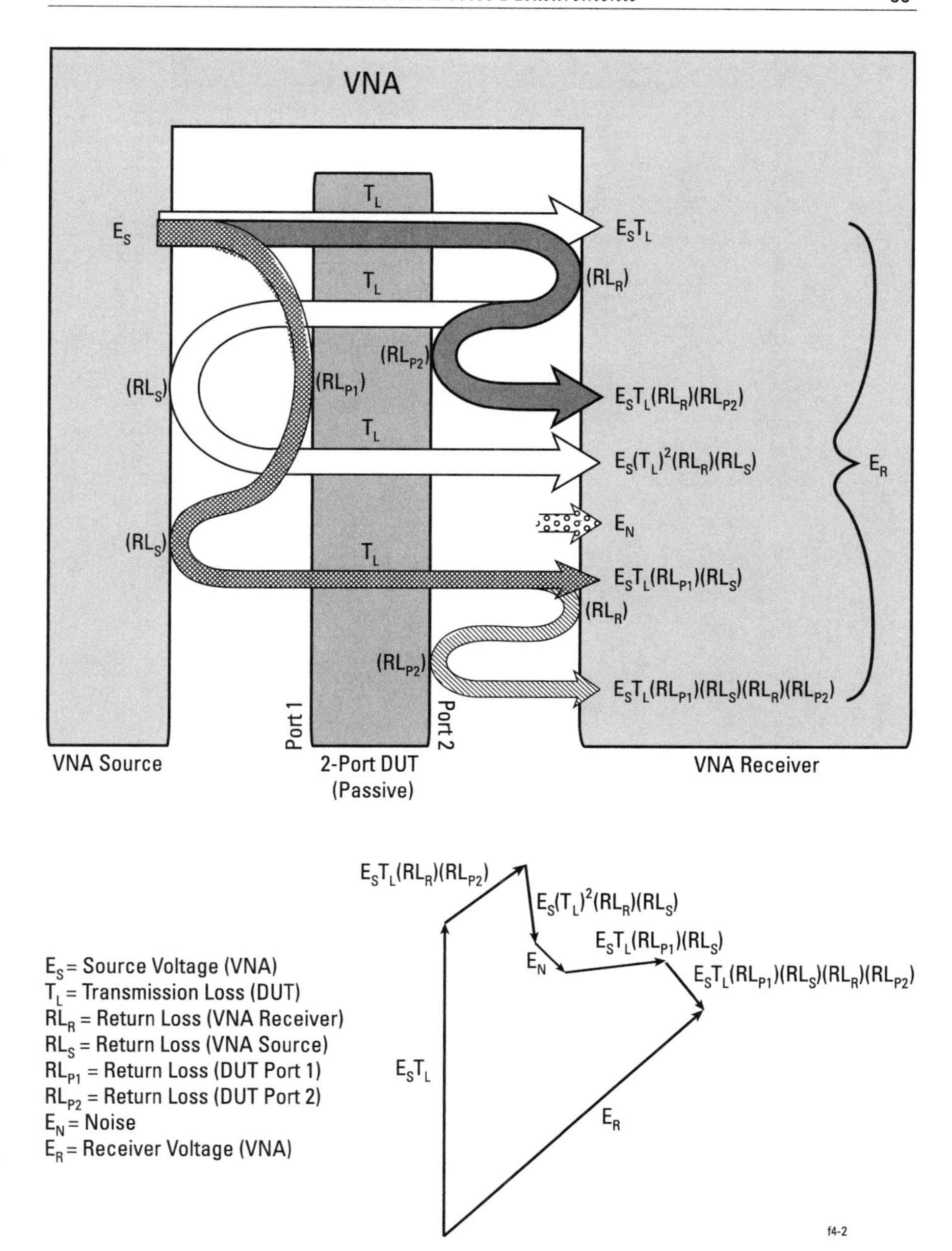

Figure 4.2 Possible reflections in a two-port DUT measurement.

line loss may range from low (<0.1 dB) to moderate (tens of decibels). Delay lines must deal with all the same error terms as previously discussed (Figure 4.2), but since electrical delay is of utmost importance, we must consider how

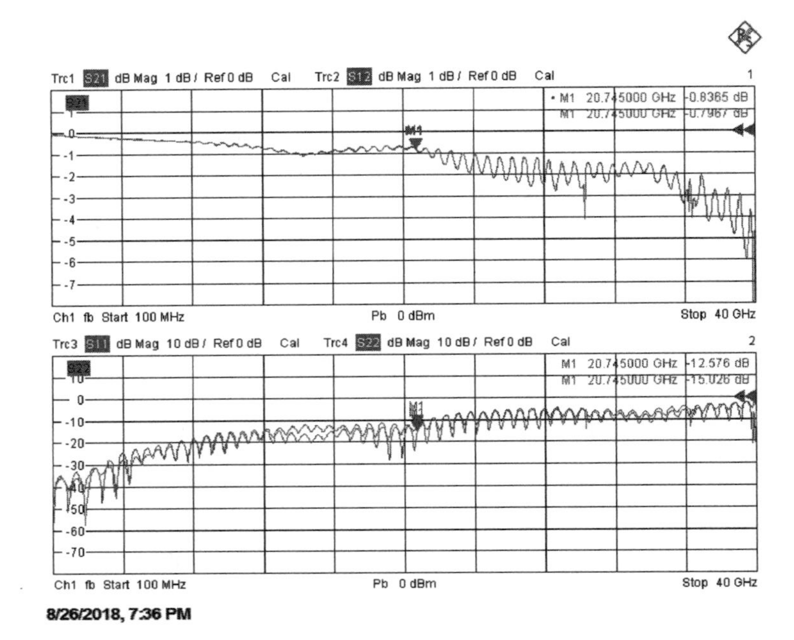

Figure 4.3 Cable measurement (with ripple due to reflections).

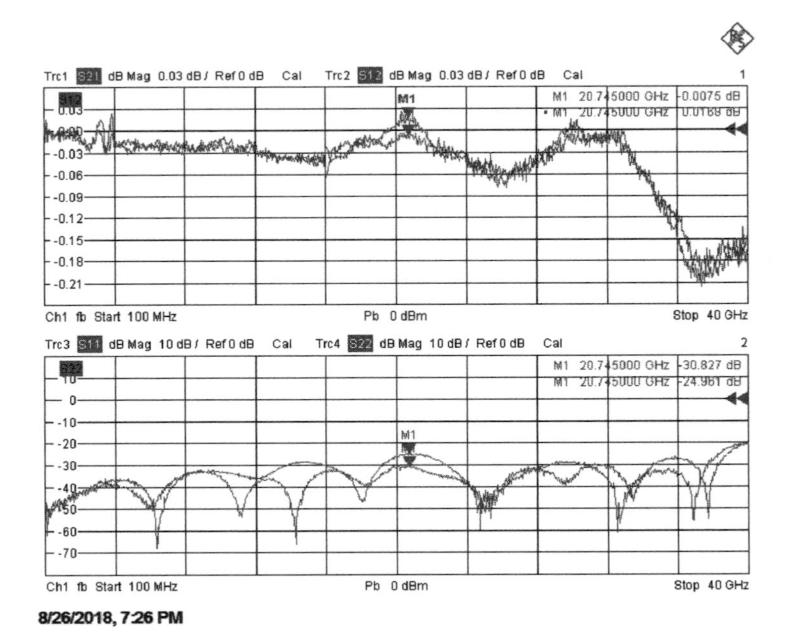

Figure 4.4 Low-loss cable measurement with gain observed at 20.7 GHz.

it is impacted by multiple reflections. Fortunately, amplitude and phase ripple are related, so ripple on one should allow the other to be calculated easily using Table 3.3 (Chapter 3).

Delay measurements are directly related to both distance and physical length of a cable. The fundamental relationship is expressed by:

$$v_p = f\lambda \tag{4.1}$$

where v_p = velocity of propagation in a medium = $c_o/\sqrt{\varepsilon_r}$, c_0 = speed of light in a vacuum (2.998e11 mm/s), ε_r = relative permittivity of a dielectric-loaded medium,[1] f= frequency (Hz), and λ = wavelength (mm).

If you can measure the electrical delay at a particular frequency, then you can determine the length from:

$$L = \left(v_p/f\right)\left(x/360\right) \tag{4.2}$$

where L = length (mm) in free space, x = delay in degrees at frequency f(Hz), and v_p = speed of propagation (mm/s).

For example, in Figure 4.5, the display format "Unwrapped Phase" in the upper right display area shows Marker 1 at 5.16 GHz with a phase value of –5,430.0°. Inserting these values into (4.2) yields a free-space cable length of 876.4 mm.

The phase response of this cable appears to be a straight line (therefore linear) with a constant slope. In reality, there will be some deviation from a perfectly linear response, but you will never see it with this phase unwrap format because the *y*-axis scaling is 2,500° per division. Instead, you can get a sense of the deviations by comparing the length calculation at the three marker frequencies (Table 4.2).

In many instances, it is important to know the deviation from linearity, and the VNA provides a built-in function for this. (In the ZVA, it is under "Trace" > "Trace Function" "Linearity Deviation"). Simply click the Auto button and the analyzer calculates a regression line based on all trace points and shows deviation from this ideal line. The function tabulates the slope of the calculated regression line, offset constant, and equivalent electrical length and shows how the phase varies from this ideal slope across the frequency span. This is shown in the lower left window (Figure 4.5). Finally, another important function associated with delay measurements is group delay, developed in Chapter

1. In the case of coaxial cable, velocity factor is more commonly used than dielectric constant. The relationship between relative permittivity (ε_r) and velocity factor (VF) is given by: $VF = 1/\sqrt{\varepsilon_r}$ so $c = c_o \times VF$.

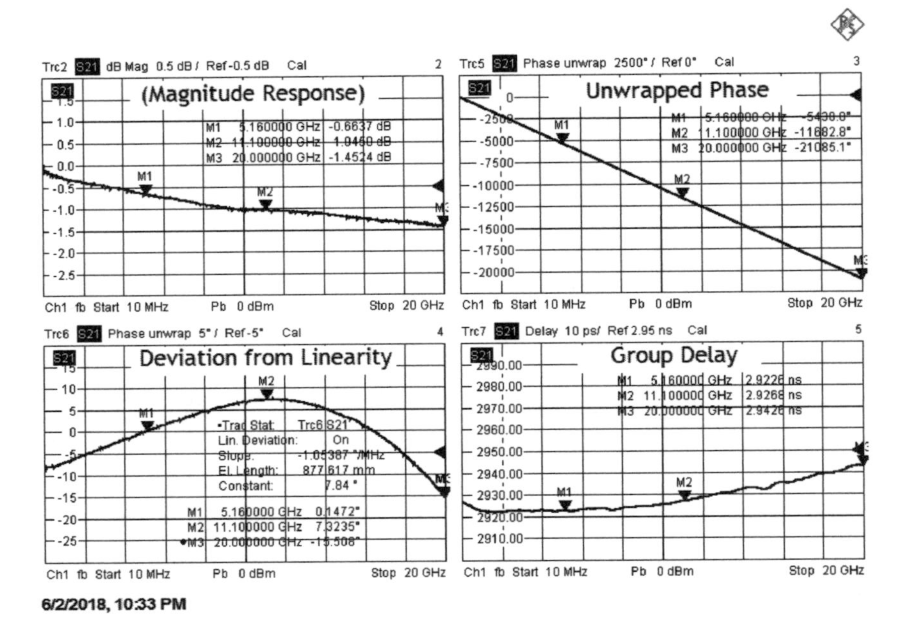

Figure 4.5 Multiple ways of making a delay measurement on a two-port device.

Table 4.2

Length Calculations Based on Marker Values Using the Phase Unwrap Display Function

Marker Frequency (GHz)	Marker Phase Value (Degrees)	Calculated Phase Delay Length (mm)	Group Delay (ns)	Calculated Group Delay Length (mm)
5.16	5,430.0	876.4	2.9226	876.2
11.10	11,682.8	876.5	2.9268	877.5
20.00	21,085.1	878.0	2.9426	882.2

3. Since we are measuring a two-port device, the group delay function directly yields electrical length.

These three measurements (unwrapped phase, deviation from linearity, group delay) provide unique insights into the frequency response of a device. The basic Unwrapped Phase measurement gives an intuitive sense of the phase linearity based on the slope of the unwrapped phase response. Also, it is easy to compare two cables at a glance: the longer cable will have a steeper negative slope.

Linearity deviation uses a regression function to reveal departures from ideal phase linearity over a frequency range. While linearity deviation is not generally an issue for cables in traditional narrowband communications applications, it is growing in importance for wider (multiple gigahertz) modulation bandwidths, where equalization may be required to correct the phase response to achieve demanding error vector magnitude (EVM) requirements. Finally, group delay, defined as the change of phase over a given change in frequency, provides important insight into propagation delay as a function of frequency. This dispersion can have detrimental effects on a communications link if different frequencies arrive at different times, leading to distortion or even self-jamming.

A fourth method for evaluating delay in cables or delay lines, the calibration offset method, was introduced in Section 3.3.1. It is equally useful for determining the electrical length of two-port devices for phase-matching applications.

4.1.3 Filter Measurements

Filters have played an essential role in over-the-air communications since the earliest days of radio. Their importance has only grown during the past 120 years, and today they are truly ubiquitous RF components, found in applications covering audio to terahertz frequencies. The four most common filter topologies are lowpass (LP), bandpass (BP), highpass (HP), and bandstop (or notch) with widely varying architectures and associated specifications. High-performance filters present difficult measurement challenges for modern VNAs because of the need for wide dynamic range. Figure 4.6 illustrates the problem by identifying typical performance requirements for a bandpass filter: The passband region usually requires low loss and low ripple, while the stopband region(s) exhibit very high attenuation. The transition region changes from low to high attenuation over a narrowly prescribed frequency range. Let's examine these requirements from the VNA's perspective.

Passband measurements are easy. Signal levels are high (very little attenuation), so there is plenty of signal getting to the reference and measurement receivers and the VNA S_{21} trace will be noise-free. Filters are usually well matched (15 to 20-dB RL or better) in the passband region, so return loss ripple due to poor source match or limited directivity is unlikely, provided that a good calibration, good test cables, and proper operator technique are used. Default instrument settings are fine for the passband region, with care taken to ensure a sufficiently small IF BW and sufficiently large number of measurement points to discern passband ripple (peaks and valleys).

Stopband measurements typically involve very-low signal levels at the measurement receivers. Low signal level means low SNR and large errors due to noise fluctuations. Three things can be done to reduce the impact of these fluctuations:

Filter Wizard® Part: 13ED10-13000/H2500-O/O

This part is a product of Dover Corporation's Microwave Products Group. Visit www.klmicrowave.com for information on all K&L Microwave® brand product offerings.

	Spec:	Typical:
Center Frequency:	13000 MHz	12975.5 MHz
0.5 dB Bandwidth:	2500 MHz	2650.0 MHz
Insertion Loss:	0.8 dBa	0.65 dBa
Stopband Atten. (10500 MHz):	* 93 dBc	* 104.33 dBc
Stopband Atten. (15500 MHz):	* 91 dBc	* 101.33 dBc

*** Attenuation only guaranteed to 60 dBc.**

Filter Type:	Cavity
Spec Return Loss:	11.7 dB (1.7:1 VSWR)
Typ Ult Rej:	27300 MHz

Inches:	2.28 x 0.50 x 0.50 inches
Millimeters:	57.83 x 12.70 x 12.70 mm

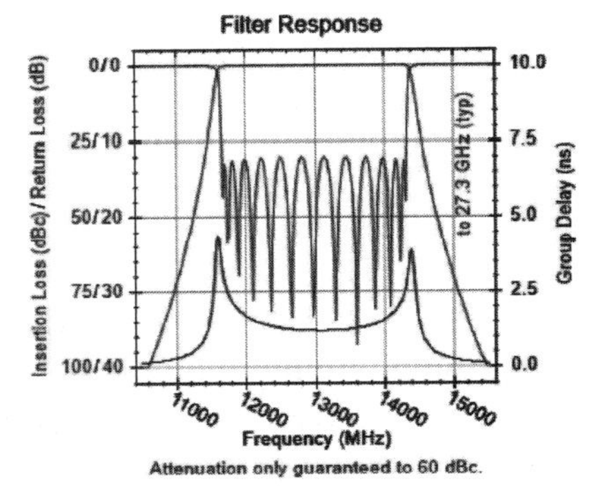

For all standard products, due to manufacturing variations, actual attenuation at any given frequency point may vary by +/- 10%, actual bandwidth may be wider than specified, and passband insertion loss may exceed given value by 10%. For return loss information, see specifications above. Typical ultimate rejection -60dBc out to 27.3 GHz. For custom requirements and details of spurious response, please contact the factory. [Browser: Type = InternetExplorer11 | Name = InternetExplorer | Version = 11.0 | Major Version = 11 | Minor Version = 0 | Platform = WinNT]

Figure 4.6 Microwave bandpass filter specifications. (Courtesy K&L Microwave.)

1. *Reduce the IF BW.* Reducing the IF BW also reduces the thermal noise entering the VNA receivers while preserving the amplitude of the desired signal. This improves the SNR at the expense of sweep speed. Generally, reducing the BW by a factor of ½ (e.g., 10 kHz to 5 kHz) doubles the sampling time. This likely will not double the sweep time, because sweep time depends on many factors, as shown in Figure 4.7, including:
   - T_{PREP}: measurement preparation time to allow RF generator and receiver LO synthesizers to tune and settle at the measurement frequency and also allow the RF generator amplitude to settle to the proper level.
   - T_{POST}: Time required for hardware postprocessing, which includes calculating and applying SEC. It is important to note that a two-port calibration requires measurements to be performed in both the forward (S_{11}, S_{21}) and reverse (S_{12}, S_{22}) directions.
2. *Employ trace averaging or smoothing.* By averaging a trace over several sweeps, random effects (noise) are cancelled out while enhancing deterministic (signal) information. This results in improved SNR. The amount of SNR improvement increases with averaging factor, but at the cost of greater measurement time. Smoothing is an alternative to trace averaging that compensates random effects by averaging adjacent measurement points along a trace. Compared to sweep averag-

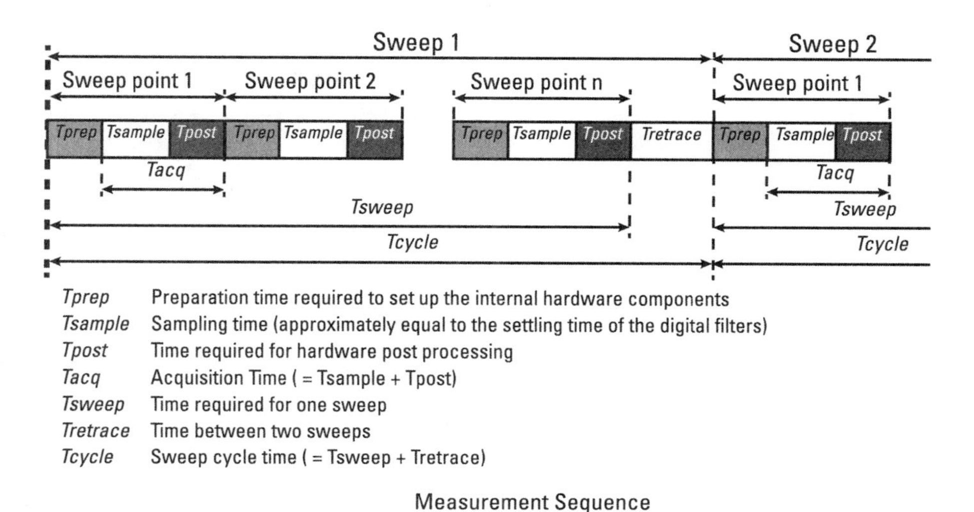

Figure 4.7 Sweep time details.

ing, smoothing does not significantly increase measurement time. Smoothing will diminish or remove narrow peaks or dips, possibly hiding imperfections. (For this reason, smoothing is often avoided, even in situations where, properly applied, it would be beneficial.) Some VNAs define several different types of averaging. For example, the ZNB has a reduce noise mode that operates on the real and imaginary parts of each measurement point, providing effective noise suppression for "Real" and "Imag" formats and polar diagrams. Another mode, called Flatten Noise, applies cumulative trace averages to the linear magnitude and phase values to suppress trace noise for display modes of dB Mag, Phase, Unwrapped Phase, and Lin Mag formats. A third mode, Moving Average, applies a simple moving average to the real and imaginary parts of each measurement. (The effect is similar to the reduce noise mode, but with a finite history.)

3. *Increase the generator power.* Increasing the signal level directly improves SNR without incurring a sweep-time penalty. However, this option requires a VNA to have a wideband high-power source capability. This is generally available only on higher-cost, performance-class instruments. Additionally, high-power operation may cause the measurement receiver to compress in operating regions that do not offer much attenuation (such as in the filter passband).

4. *Lower the noise floor.* In some extreme cases, external preamps may be used to improve the VNA measurement receiver's noise figure and thus lower the noise floor. This has disadvantages as well. By placing

a preamp directly on port 2, full two-port S-parameter measurements can no longer be meaningfully applied, because when the port-2 generator is turned on, the port-2 measurement receiver sees reflections from the output stage of the preamp, not from the DUT. Also, S_{12} of the DUT will be lowered by the reverse isolation of the preamp. Generally, if a very high stopband attenuation must be measured, a setup optimized for this particular measurement is used, which includes a preamp on port 2 and a simple normalization calibration in the forward thru direction (e.g., one-path, two-port calibration).

Another way to reduce the noise floor on instruments equipped with direct generator/receiver access loops is to bypass the port 2 directional coupler and connect the DUT directly to the port-2 measurement receiver. By eliminating the coupler loss, the noise floor can generally be lowered by 10 to 13 dB (i.e., by the coupling factor of the microwave coupler). As mentioned above, care must be taken when using a preamp or direct receiver access (or both) to prevent receiver compression during high-amplitude/low-attenuation measurements.

Depending on the application, one or more of these approaches will be preferred. For example, while tuning a filter, it is desirable to use as wide of an IF bandwidth as possible to ensure fast trace update rates. For this application, high power would be useful, perhaps combined with direct receiver access to facilitate coarse tuning.

If the measurement objective is to archive device performance, high-accuracy S-parameter data is required, so a user should employ a full two-port calibration along with a narrow IF filter BW and perhaps even trace averaging to get the best possible accuracy, even if it means sacrificing measurement time.

Generally, the dynamic range requirements of high-performance filters make it impossible to satisfy all measurement objectives with a single combination of power level, BW, and number of points. This is especially true if measurement speed is also a factor. Under these circumstances, the segmented sweep mode is recommended. Segmented sweep permits frequency segments to be created with unique power levels, IF bandwidths, and number of points (Figure 4.8). This allows users to create a best of all worlds scenario for a particular measurement application. For example, measuring the microwave bandpass filter specified in Figure 4.6 requires all these techniques (Figure 4.9).

4.1.3.1 Filter Group Delay Measurements

Group delay is another important filter parameter. As discussed in Section 3.3.2, group delay is a calculated quantity. It is defined as the change in a DUT's phase response that occurs over a given frequency range. From (3.2),

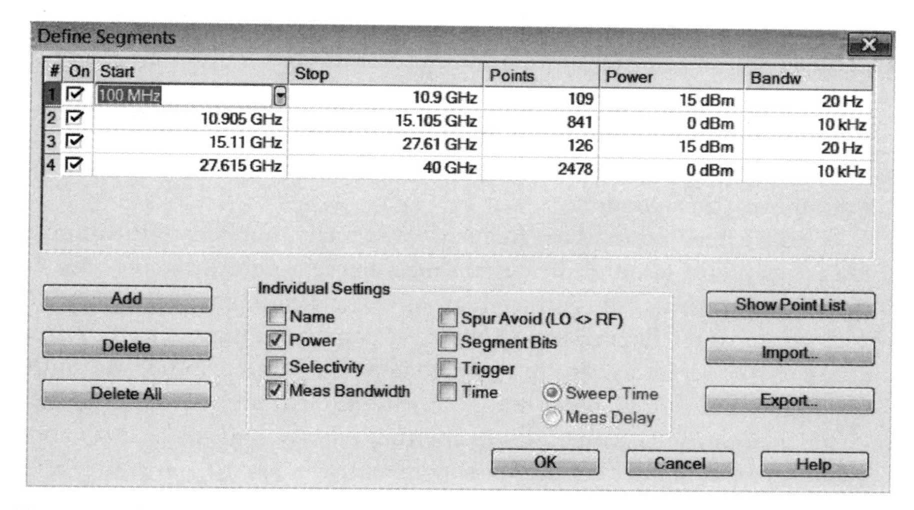

Figure 4.8 Segmented sweep table.

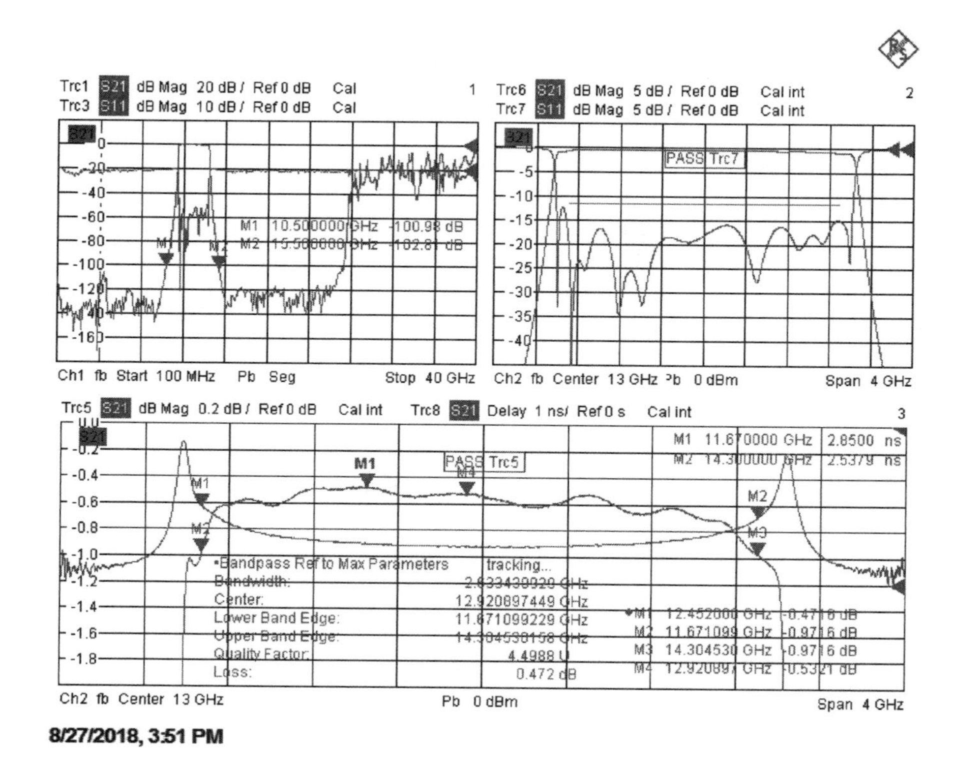

Figure 4.9 Microwave bandpass filter measurement results.

$$\tau_g = \frac{-1}{360°}\left(\frac{\Delta\varphi}{\Delta f}\right)$$

where $\Delta\varphi$ = change of phase in degrees, Δf= change of frequency in hertz, and τ_g= group delay in seconds.

The Δf is referred to as the frequency aperture, and refers to the number of frequency points (around the current measurement point) that are used for the group delay calculation. Although group delay measurements are generally performed across the filter passband (where attenuation is minimal), the resulting trace can be very noisy, in spite of a good SNR. The reason lies within the group delay formula itself: Since the aperture term f is located in the denominator, as it becomes smaller (e.g., fewer aperture points used/smaller evaluation frequency span), the quotient becomes more sensitive, resulting in greater trace noise. The answer is to increase the number of frequency points used to define the aperture, keeping in mind that increasing the aperture will cause greater smoothing of the calculated group delay value (at the expense of fine detail in the phase response). So what is the optimum aperture size? Sometimes the aperture is specified by the customer. Otherwise, the designer must seek a balance between group delay detail and trace noise.

Figure 4.10 provides examples of filter group delay measurements using different aperture values. Here, the frequency sweep uses 1,001 points, with a frequency spacing of 250 kHz. With a 0.1% aperture (only 1 aperture point), the group delay response is noisy. At 1% aperture (10 points), noise is suppressed, and the group delay response reveals fine detail. Setting the aperture to 20% (200 points) causes distortion at the edges of the filter passband where the group delay is changing rapidly.

4.1.4 Passive Multiport Devices Measured as a Two-Port

Passive multiport devices have traditionally been measured using two-port VNAs (largely due to the scarcity and cost of multiport VNAs). Multiport measurements are carried out two-ports at a time, with all remaining ports terminated in 50Ω. By using trace math, different measurements can be made, such as calculating amplitude and phase balance between power splitter arms, directivity of RF couplers, and isolation for circulators/isolators. The basic procedures for these devices are presented below.

4.1.4.1 Power Splitter Imbalance

Start with the "Two-Port Measurement Baseline" (Section 4.2). Connect the power splitter as shown below to measure the isolation between ports (Figure 4.11(a)). Save the results to a memory trace (Trace > Trace Function > Trace

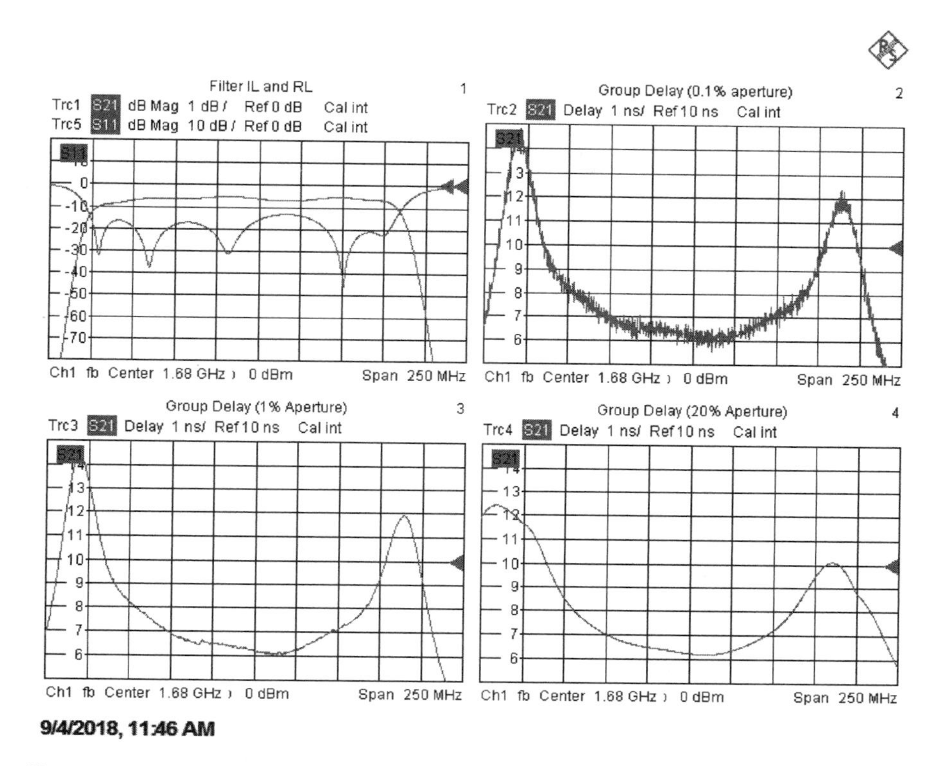

Figure 4.10 Filter group delay measurements using 0.1%, 1%, and 20% apertures.

Mem > Data to new mem). Next, add a new trace in a new display window, and configure the power splitter as shown in Figure 4.11(b) to measure the amplitude imbalance.

Measure S_{21}, and then store the measurement to a trace memory (Trace > Trace Function > Trace Mem > Data to new mem). (If you are also interested in measuring the phase imbalance, create a new trace in a new display window and change the display format to "Phase." Measure S_{21} and store this measurement to a trace memory (Trace > Trace Function > Trace Mem > Data to new mem).)

Next, configure the power splitter as shown in Figure 4.11(c). Measure S_{21} using the same active trace as you used previously. Now add a new trace and a new display window. Enable trace math for the traces of interest (some analyzers turn this function on automatically when you perform a "Data to memory" operation). The trace math function should be "Data(Trc2)/mem(Trc2)). The result is the difference in amplitude (amplitude imbalance) between splitter arms (if the measurement format is "dB(mag)" and phase imbalance (if the measurement format is "Phase"). Figure 4.12 shows an example of these three measurements for a Mini-Circuits ZN2PD-9G-S+ power splitter.

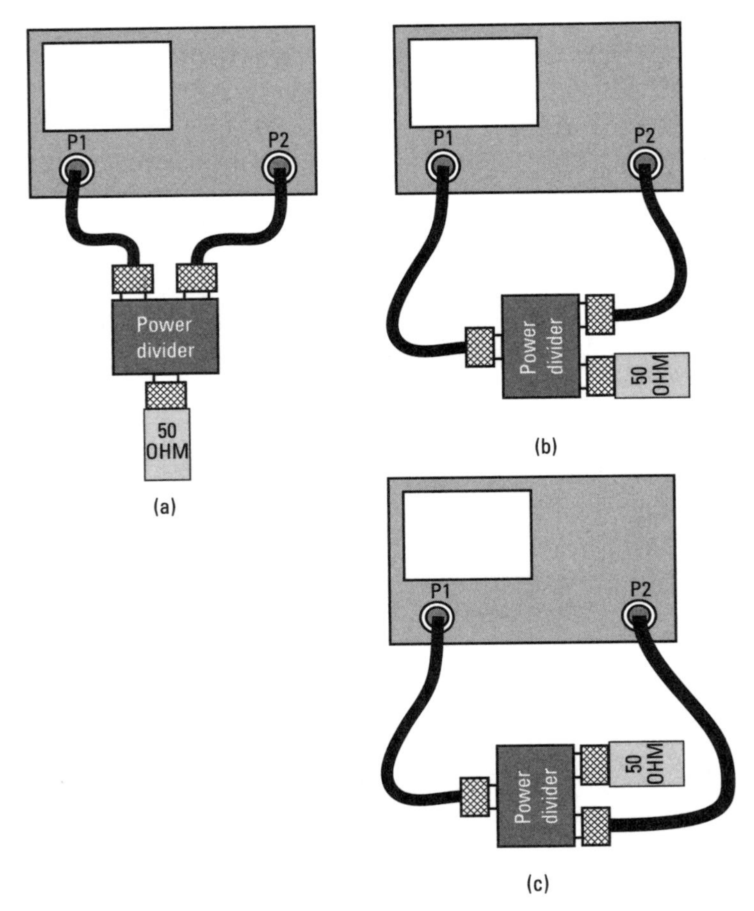

Figure 4.11 Setup for power splitter measurements. (a) Port isolation. (b) Establishing one branch of the power divider as a reference for amplitude and phase measurements. (c) Measuring amplitude/phase imbalance.

4.1.4.2 RF Coupler Directivity

Start with the "Two-Port Measurement Baseline" (Section 4.2). Connect the RF coupler as shown in Figure 4.13(a). Note this is opposite to conventional operation, because port 1 of the VNA is connected to the output port of the coupler, while port 2 of the VNA is connected to the coupled port. In this way, the true directivity (= |Isolation| – |Forward Coupling Factor| – |Insertion Loss|) can be calculated with just two different terminations (an open (or short) and a match standard).

Terminate the coupler input port with an open (or short) termination as shown in Figure 4.13(a) . Measure the forward coupling factor as S_{21} (dBMag),

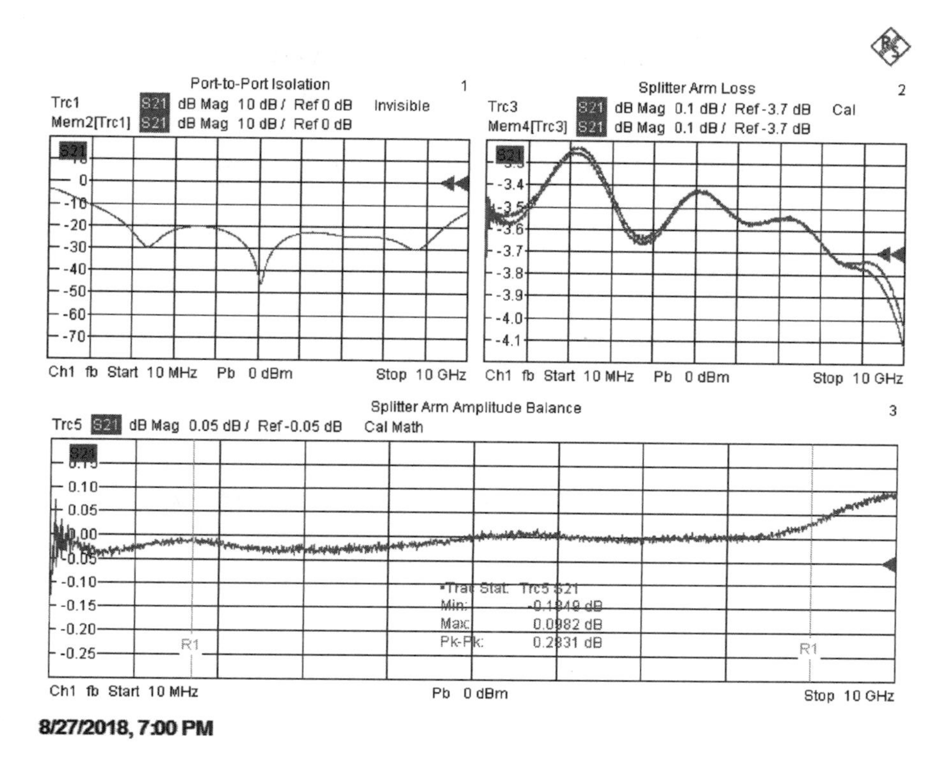

Figure 4.12 Power splitter measurement results.

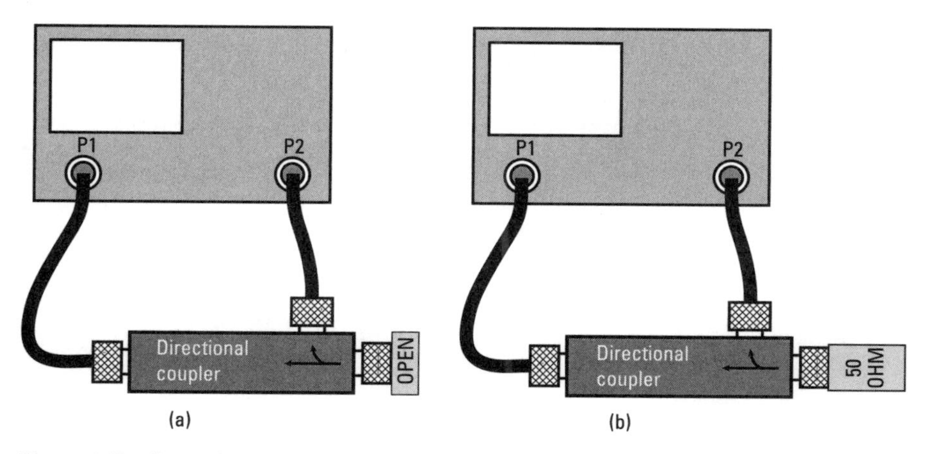

Figure 4.13 Setup for directional coupler directivity measurements. (a) Forward coupling factor and insertion loss (b) isolation.

and then store the measurement to a trace memory (Trace > Trace Function > Trace Mem > Data to new mem).

Next, replace the open (or short) termination with a 50Ω match as shown in Figure 4.13(b). Measure the reverse isolation as S_{21} (dBMag). To calculate the directivity, create a new trace in a new display window and enable trace math for the traces of interest (some analyzers turn this function on automatically when you perform a "Data to memory" operation). The trace math function should be "Mem2(Trc1)/Trc1." The displayed result is the coupler directivity (dBMag format). See Figure 4.14 for a measurement of a Mini-Circuits ZFDC-10-5S coupler. At 50 MHz, the directivity is better than 45 dB, but rolls off significantly with frequency. At the upper end of the specified operating band (1 GHz), the directivity decreases to approximately 23 dB.

One could also make this measurement in the conventional way, but it would take three measurements:

- *Step 1:* Connect VNA port 1 to Directional Coupler (DirCpl) Input port, VNA port 2 to DirCpl Output port, terminate DirCpl coupled port in 50Ω: Meas S21 (DirCpl insertion loss).

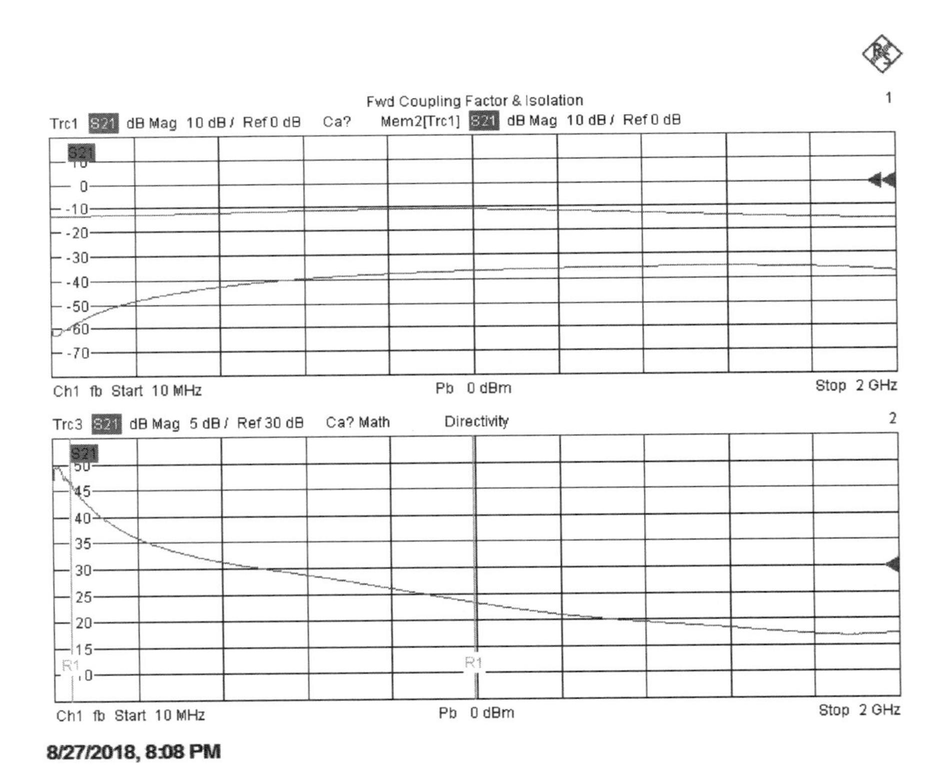

Figure 4.14 Directional coupler measurement results.

- *Step 2:* Connect VNA port 1 to DirCpl Input port, VNA port 2 to DirCpl coupled port, DirCpl Output port terminated in 50Ω: Meas S21 (forward coupling factor).
- *Step 3:* Connect VNA port 1 to DirCpl Output port, VNA port 2 to DirCpl coupled port, DirCpl Input port terminated in 50Ω: Meas S21 (reverse isolation).

Then the directivity would once again be calculated from: |reverse isolation| – |forward coupling factor| – |insertion loss|.

Regardless of the method used, it is important to make sure your calculations include the influence of the coupler's insertion loss, which always acts to reduce directivity. While this factor is usually very small at HF through UHF, it can be considerable at microwave frequencies and may significantly reduce the true coupler directivity.

4.1.4.3 RF Isolator/Circulator Performance

Start with the "Two-Port Measurement Baseline" (Section 4.2). Configure the RF isolator or circulator as shown in Figure 4.15.

Measure S_{21} in trace 1 (Trc1), and measure S_{12} in trace 2 (Trc2). Define a third trace in a different display window (Trc3). Configure the format as S_{21} (dBMag). Then apply a math trace function to Trc3 (=Trc1/Trc2). An example is shown in Figure 4.16.

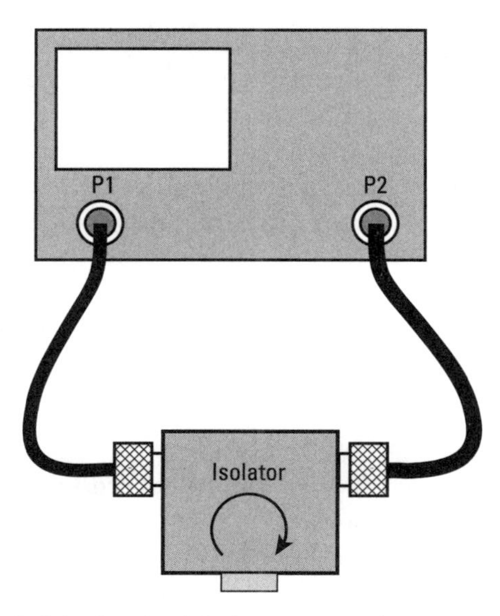

Figure 4.15 Setup for isolator (or circulator) measurement.

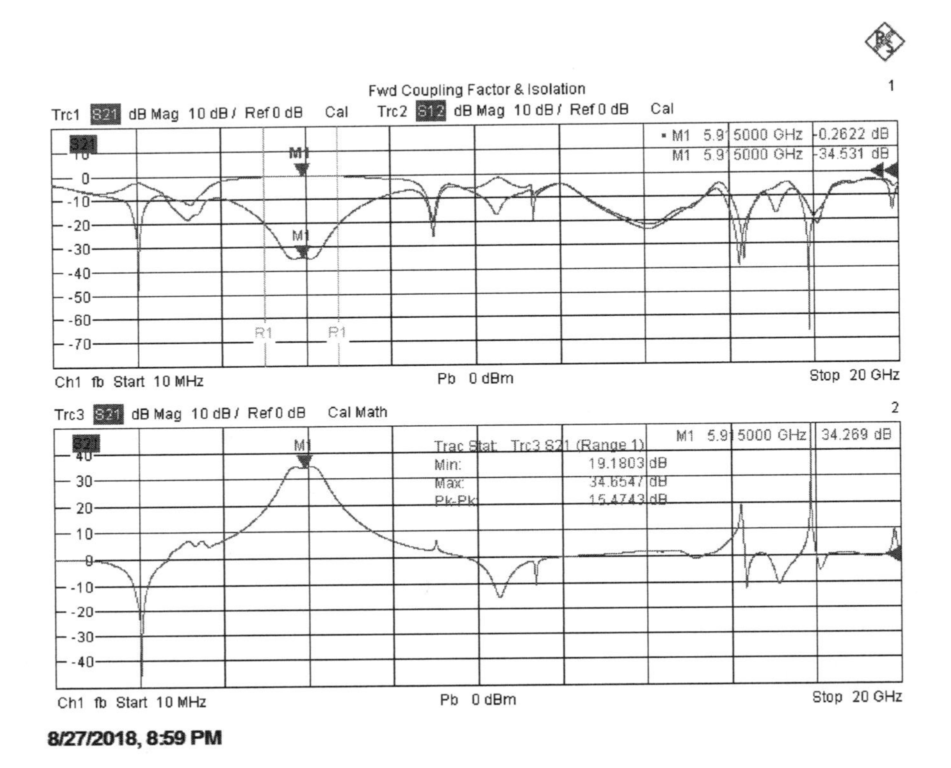

Figure 4.16 Isolator measurement results.

4.1.5 Switches: Time-Domain Transient Measurements

Switches present unique measurement challenges because full characterization often requires both frequency-domain and time-domain (transient) measurements. In particular, transient measurements characterize amplitude and phase settling-time performance. Settling time measurements have traditionally been performed on a separate spectrum analyzer, but there are two drawbacks to this approach:

1. It necessitates a second measurement setup with additional test equipment and components, as well as additional test time to move the DUT from the VNA bench to the spectrum analyzer (SA) bench setup.
2. Spectrum analyzers have traditionally been limited to amplitude-settling measurements only.

Conventional VNAs are generally incapable of settling-time measurements, because their traditional mode of operation is too slow. They take a measurement at a particular frequency point, calculate and apply the associated S-parameter correction terms, display the results on the screen, and then repeat for the next measurement point (refer to Figure 4.7). The cycle time is in the range of 3 to 10 μs per point for most VNAs, far too coarse for measuring switching transients. Additionally, the measurement bandwidth is often limited to 1 MHz or less. However, the ZVA network analyzer has a unique pulse profile mode that eliminates these traditional VNA drawbacks and make transient measurements easy. The operation is shown in Figure 7.8. The analog-to-digital (A/D) converter samples at its maximum rate (80 Ms/s) and stores uncorrected samples in a high-speed buffer (RAM). After acquiring data for a complete sweep, the samples are digitally processed by the software (see arrows in the block diagram). By separating the sampling process from the SEC calculations, switching transients can be measured with 12.5-ns resolution and with up to a 30-MHz IF bandwidth.

Figure 4.17 shows a setup for making transient measurements on an electronic switch. Here, a pulse generator provides a TTL-level switching voltage to the electronic switch. This TTL-level signal is also fed to the external trigger input of the VNA. Figure 4.18 shows the instrument dialog for configuring the Pulse Profile operating mode. Figure 4.19(a) shows the switch measurement results. Channel 1 measures the transient insertion loss characteristics of the switch under off and on conditions. It is interesting to note that the switch takes 4.48 μs after receiving the turn-on control voltage before it switches states. The

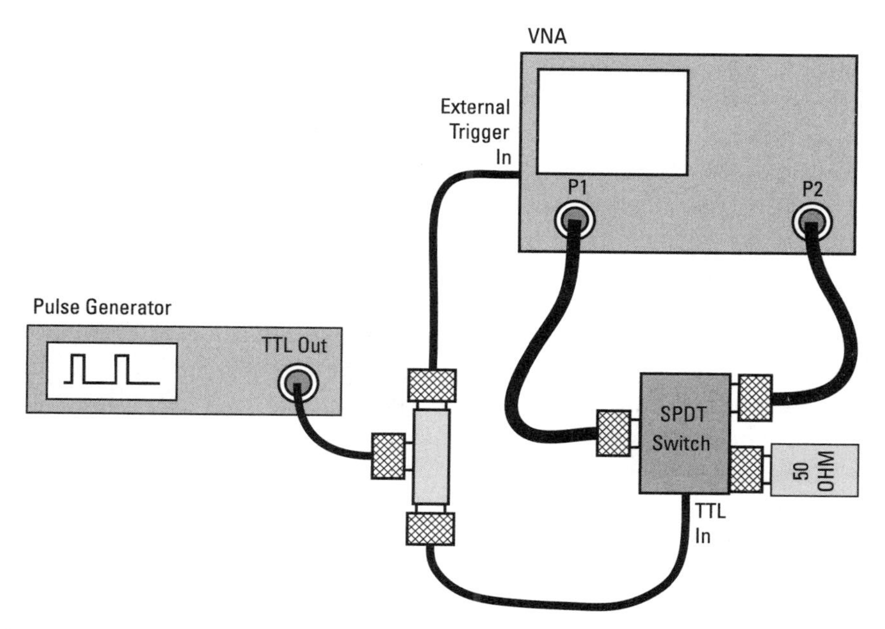

Figure 4.17 Setup for making transient measurements on an electronic switch.

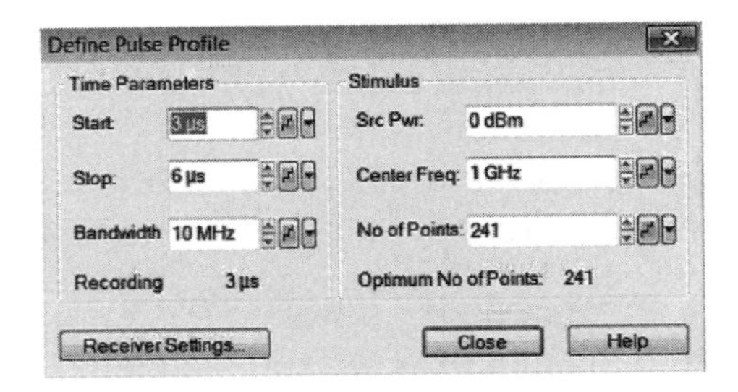

Figure 4.18 Pulse Profile mode dialog.

bottom portion of the screen shows a frequency-domain sweep in the off and on modes. Figure 4.19(b) zooms in on the turn-on transient that occurs around 4 μs after receiving the turn-on control voltage. It is interesting to note that the small spike in amplitude is offset in time from the phase spike.

4.1.6 Phase and Delay-Matching of Complex-Modulated Signals

Emerging technologies are employing multiple antenna elements via phased arrays and/or beamforming components. These systems provide multiple signals with specific amplitude, phase and timing relationships between each element or signal stream (Figure 4.20). The measurement challenges include measuring the relative phase and timing between individual channels, assessing phase stability, and knowing when to repeat calibration. The challenge is more difficult when the systems use frequency conversion or require wideband signal characterization. Fortunately, modern multiport VNAs provide a compelling solution for this challenging measurement problem.

VNA receivers need to be coherent for the measurement of S-parameters. Since receiver coherence implies a fixed, repeatable phase relationship between receivers over time, it is reasonable to appropriate this capability for determining the relative phase of external signals. By calibrating out the phase differences between receiver paths, accurate measurements can be made even when the source frequency is not perfectly aligned with the VNA receiver frequency. However, a VNA does have some limitations. It cannot discern phase differences beyond +/−180° for a CW signal, because each RF cycle is indistinguishable from the next (or previous) RF cycle. So a different technique using an FM chirp waveform allows us to discern timing differences.

Consider an FM chirp signal. The frequency of a linear FM chirp changes with time according to the equation:

$$f(t) = \frac{(f_2 - f_1)}{T} t + f_1 \tag{4.3}$$

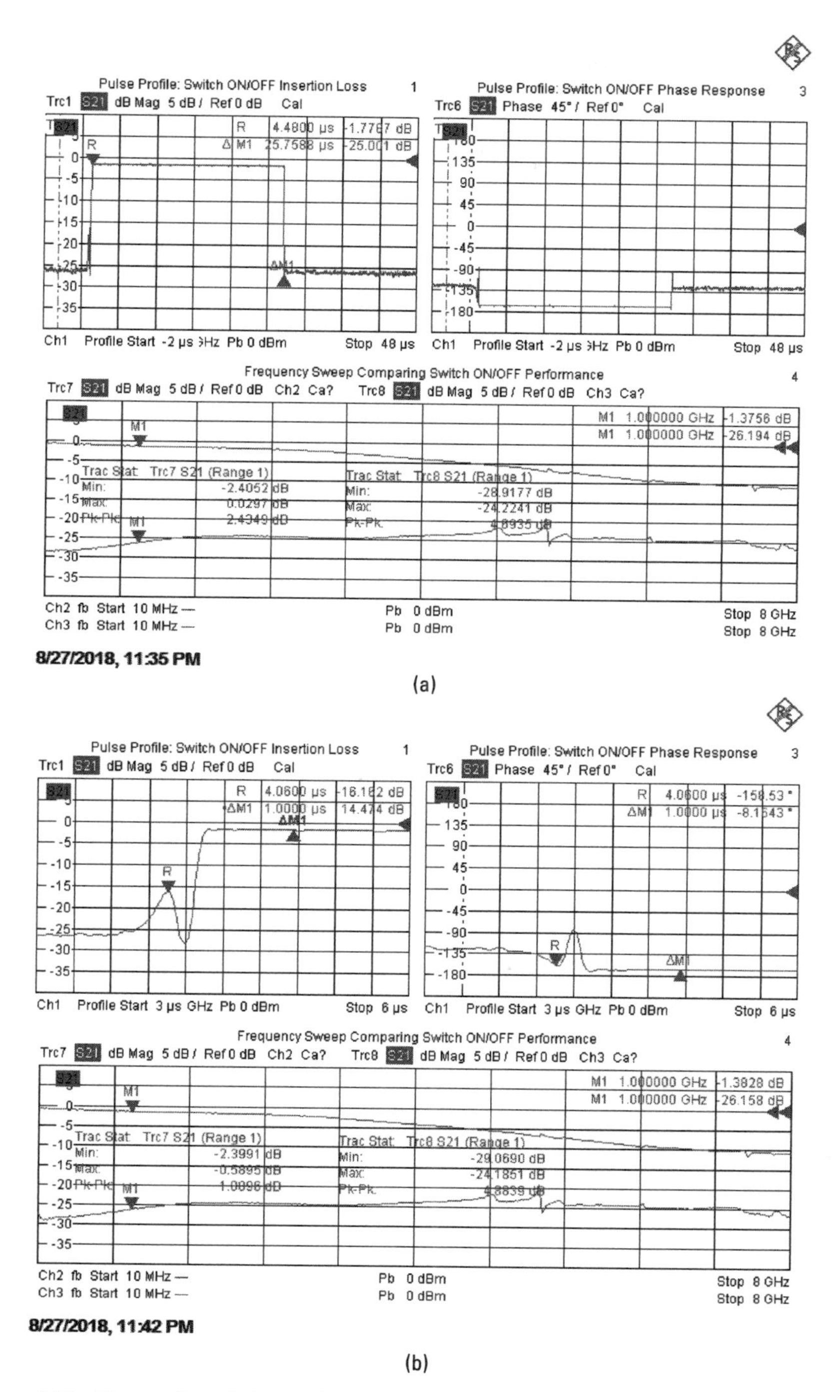

Figure 4.19 Electronic switch transient measurement results using pulse profiler mode. (a) Complete pulse. (b) Examining transients at beginning of switch turn-on.

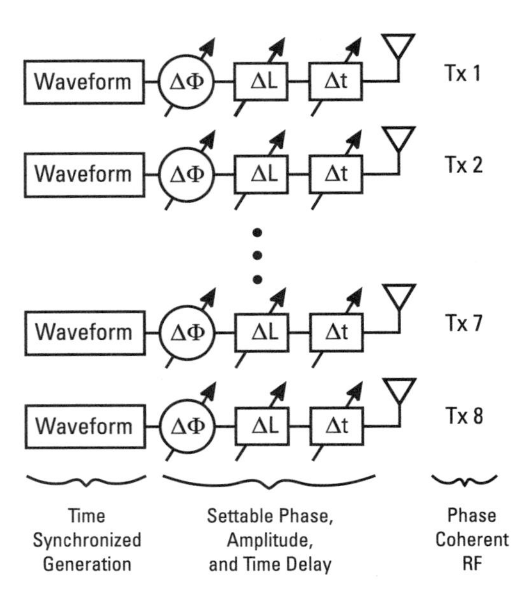

Figure 4.20 Control elements of a multi-element coherent transmission (beamforming) system.

where t is time (independent variable), f_1 is the starting frequency, f_2 is the ending frequency, and T is the time required to chirp from f_1 to f_2.

Frequency is the time derivative of phase. Therefore, if we integrate this chirp equation with respect to time, we obtain the instantaneous phase of the chirp signal:

$$\theta(t) = \frac{\pi(f_2 - f_1)}{T} t^2 + 2\pi f_1 t + \Theta_o \tag{4.4}$$

Where

$\theta(t)$ is the instantaneous phase of the chirp signal,

Θ_o is a constant representing the starting phase (offset from 0°),

f_1, f_2, T are same as in (4.3),

t is the independent variable time.

We now have two equations. We can use these to address our two unknowns (phase shift and timing offset).

We obtained (4.4) by integrating (4.3). Conversely, if we started with a quadratic expression for phase (4.4) and differentiated, we would obtain (4.3). We know from calculus that any time you differentiate a function with respect to an independent variable, you end up with an expression for slope. Here, the function is phase, the independent variable is time, and so (4.3) provides the phase slope at any instant in time. Equation (4.4) reveals that the instantaneous phase of a linear FM chirp signal is quadratic in time, while (4.3) tells us that the slope of the phase is a linear function of time, and in particular suggests the phase slope of a linear FM chirp will have different values at different points in time. Thus, if two identical chirp waveforms are observed at different points in time (i.e., time-shifted), they will have different linear phase slopes. If we measure these phase slopes and compare them, the difference will also be linear (e.g., a line whose slope represents the amount of time shift between the chirps). If the phase slopes of the two chirps are identical, the difference between them will be zero, resulting in a horizontal flat line.

4.1.6.1 Measurement Approach

In this VNA technique, one DUT path is designated the reference for amplitude and phase. All other DUT paths are measured relative to the reference path. To solve this particular problem, we use a FM chirp signal to compare the phase responses of two DUT paths at a time. Specifically, if we subtract one phase response from another (reference) phase response, we can determine the relative phase delay (from the mean value of the phase), and the time delay (from the phase slope).

4.1.6.2 VNA Setup

Two signals applied to VNA port 1 and port 2 can be detected by measurement receivers b_1 and b_2, respectively. Assuming the reference DUT path is connected to the VNA's b_1 receiver, measuring the complex ratio b_2/b_1 allows the relationship between both carriers to be displayed (magnitude and phase.) By taking advantage of direct receiver access (Option R&S ZVA-B16), both the measurement (b_1 through b_4) and reference (a_1 through a_4) receivers behind each test port become available for these measurements, so a four-port instrument has eight receivers available for coherent signal measurements. The internal receivers share a single LO for coherence, but it is a good idea to use a common 10-MHz reference signal between the VNA and the external signal generator. Otherwise, the IF bandwidth has to be selected wide enough to accommodate the frequency uncertainty between the source and receivers.[2]

4.1.6.3 VNA Calibration

As with any VNA measurement, calibration is required to obtain meaningful results. In this instance, a simplified calibration needs to be performed only

2. It does not matter if the source frequency varies slightly during the measurement. It only has to remain within the receiver's IF bandwidth.

once at the beginning of the measurement. A well-matched symmetrical power splitter is highly recommended as a calibration standard.[3] For higher accuracy requirements, the power splitter imbalance can be measured with the network analyzer and used for further correction. For this measurement example, the phase imbalance of the power splitter is small and can be neglected (Figure 4.21).

An additional source of measurement error is caused by reflections between the VNA, DUT, and power splitter. Reflections can be reduced by

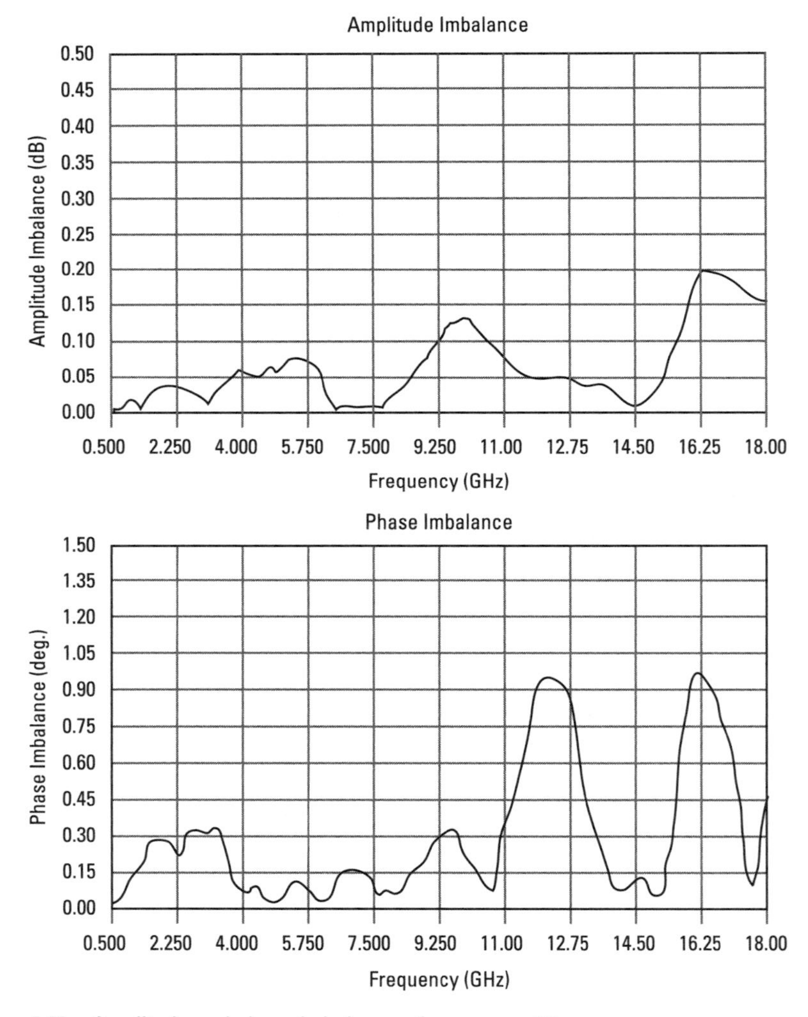

Figure 4.21 Amplitude and phase imbalance of a power splitter.

3. An example is the ZFRSC-183 from Minicircuits. This power splitter has negligible imbalance for magnitude and phase.

improving the port match of the VNA. By installing a well-matched 10-dB attenuator[4] at the end of a test cable (Figure 4.22), a 15-dB port match can be reduced to better than (15 dB + 2 × 10 dB) = 35 dB, so any reflections are due entirely to the attenuator match (in this case, attenuator VSWR = 1.15:1 which corresponds to a 23-dB return loss.) By placing this attenuator at the end of a test cable approach, the phase error associated with a DUT having 15-dB port match is reduced from 1.8 to 0.7°.

After connecting the test cables (and attenuators), we are ready to begin the calibration process. First, the receivers must be calibrated over the desired measurement range.

- *Calibration Step 1:* Connect the common port of the power splitter to the VNA source that will be used for the calibration sweep. Connect one arm of the power splitter to the VNA receiver that will serve as the reference receiver for all measurements (e.g., b_1). Connect the other arm to the first receiver that needs to be calibrated (here, assumed b_2).
- *Calibration Step 2:* Set up a trace to measure the ratio b_2/b_1. Select display format "Phase." Use trace mathematics (Data/Mem) to normalize the phase difference between these two traces to zero. (Using this method, we remove the influence of the cables and the attenuators as well.)

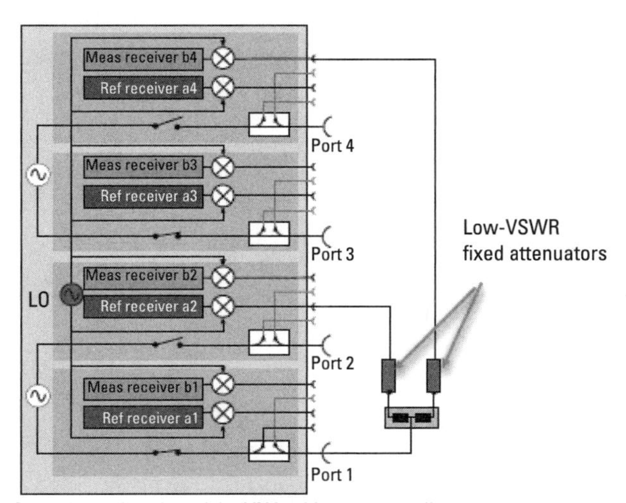

Figure 4.22 Installing attenuators at the ends of the test cables to improve the VNA raw port match.

4. An example is the BW-S10W2 from Minicircuits. This precision fixed attenuator has VSWR < 1.15:1 up to 12 GHz.

After this calibration step, move the cables around a bit and note the phase stability. If the phase changes by more than a few degrees, consider using better cables (high-quality phase stable cables) or troubleshoot to find which cable (or connection) is poor.

- *Calibration Step 3:* Keeping the reference cable attached to the power splitter, remove the other cable from the power splitter arm and attach the next cable/path that needs to be calibrated. Assuming that it is the cable attached to measurement receiver b_3, set up a new trace to measure the ratio b_3/b_1. Select the display format "Phase." Use trace mathematics (Data/Mem) to normalize the phase difference between these two traces to zero. As in step 2, check for phase stability. Repeat step 3 for the remaining cables or paths.

Note that the memory traces used for calibration must have the same number of points (as well as start and stop frequency) as the current measurement. If you change any of these, you will invalidate the calibration for all receiver paths.

4.1.6.4 Chirp Considerations

After calibrating the VNA using a CW source, the actual measurement is performed with an externally supplied FM chirp signal. To analyze the phase between two chirped signal paths, the receivers of the network analyzer have to sweep across the desired frequency span. To perform the measurements, the sweep repetition frequency f_{rep} and the frequency span f_{span} of the chirp have to be known. f_{span} is the frequency range from the start frequency f_1 to the stop frequency fn of the chirp. f_{rep} equals 1/repetition time (Figure 4.23).

The sampling time of the network analyzer's receivers has to be at least as long as the chirp period to ensure that the VNA receivers can detect the signal when it sweeps past the currently tuned frequency. As a rule of thumb, the sampling time is about 1/IFBw, where IFBw is the VNA's measurement bandwidth. Therefore, we should choose an IF filter with a sampling time that is equal to or longer than the chirp period. Using this approach, only one frequency point will be measured by the VNA for each chirp pulse, so you will need at least as many chirps as frequency points in the VNA sweep as a bare minimum (Figure 4.23(b)). However, it is advantageous to use a narrower measurement bandwidth (in the order of $f_{rep}/10$) to provide averaging over multiple chirps.

A second concern is chirp desensitization. This is similar to the well-known pulse desensitization problem, where the pulse energy is dispersed across the frequency spectrum, reducing the amplitude of the individual spectral lines. In a pulse, desensitization of the central line is calculated from 20 $\log_{10}$ (pulse duty cycle), and the amplitude of adjacent lines roll off according to the familiar

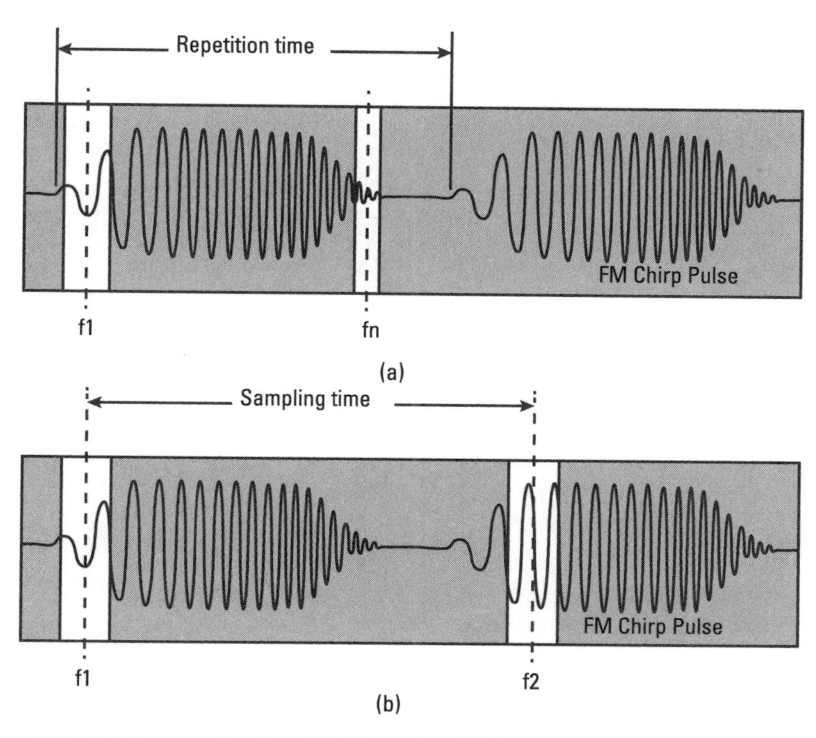

Figure 4.23 (a) An example of an FM chirp pulse. (b) Sampling-time requirements for chirp detection.

Sin(x)/x function. A chirp signal also has spectral lines (spaced according to the reciprocal of the duty cycle), but their amplitude is nearly constant across the chirp bandwidth (no Sin(x)/x roll-off).

For this chirp example, we use a 25-MHz chirp (total bandwidth 50 MHz) with a 50% duty cycle (5 μs for the chirp, 5 μs off time for a 10-μs period). The spectral line spacing is the reciprocal of the chirp period, or 1/10 μs = 100 kHz. So the bulk of the signal energy is spread across (50-MHz BW/100-kHz spacing) = 500 spectral lines, resulting in desensitization of 10log10(500) = 27 dB. In addition, because of the 50% duty cycle, there is an additional attenuation of 10log10(0.5) = 3 dB, for a total of 30-dB desensitization. In other words, applying FM chirp modulation to a 0-dBm unmodulated CW signal reduces the level of individual spectral lines to –30 dBm. Since the VNA measures these individual spectral lines, there is a loss of 30-dB dynamic range. This is shown in Figure 4.24. In Figure 4.24(a), the entire 50-MHz chirp is displayed, with an amplitude of –30 dBm. In Figure 4.24(b), the trace shows the amplitude and 100-kHz spacing of individual spectral lines. In Figure 4.24(c), the time-domain representation of the chirp is shown, where the aggregate of

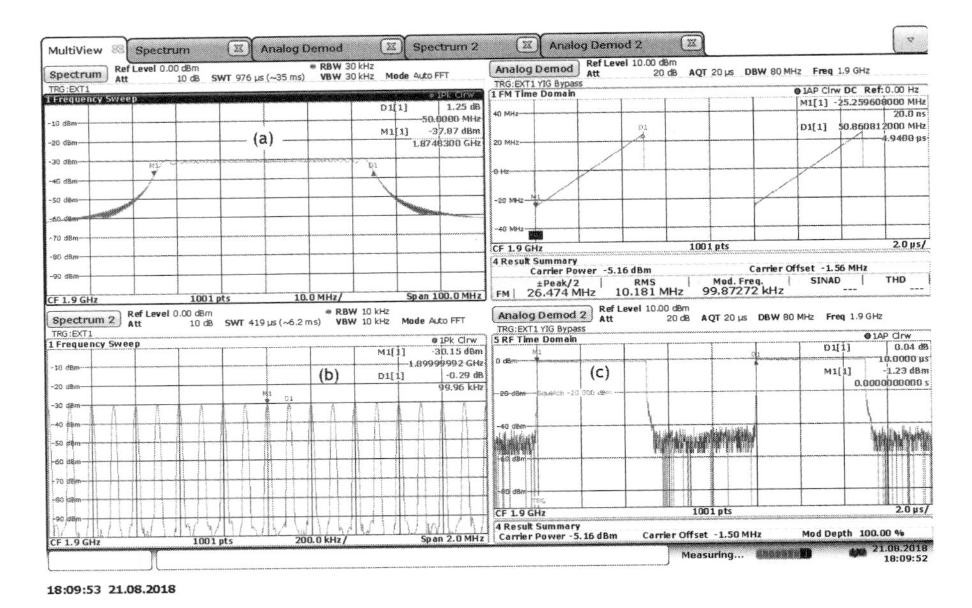

Figure 4.24 Chirp analysis with a spectrum analyzer (50-MHz chirp with 5-μs duration and 10-μs duty cycle). (a) Frequency spectrum of entire chirp, showing amplitude desensitization. (b) Amplitude and spacing of individual spectral lines. (c) Time-domain response of chirp signal.

all the spectral lines in the frequency domain provides a chirp amplitude of approximately 0 dBm in the time domain.

Therefore, it is important to consider desensitization when deciding on an appropriate chirp for this application. As a general rule of thumb, use the maximum duty cycle (to minimize the number of spectral lines and power desensitization) and minimum chirp bandwidth to meet the application requirements.

4.1.6.4 Phase Measurement with a Chirp Signal

In this final section, we replace the internal CW source with the external chirp source to compare the phase responses between DUT paths.

- *Measurement Step 1:* Turn off the VNA's internal generators (in a ZNB, select "Channel" > "Power Bw Avg" > "Power" tab and click the "RF Off All Channels" checkbox).
- *Measurement Step 2:* Apply the external FM chirp source to the input of the DUT paths. (At least two paths need to be stimulated with the chirp source simultaneously so that meaningful phase comparisons can be made at their outputs.) Connect the outputs to the calibrated VNA measurement receivers via the test cables plus attenuator pads.

- *Measurement Step 3:* Measure the relative phase and/or amplitude responses. For the phase response, turn on Trace Statistics. Specifically, for the ZNB, select "Trace" > "Trace Config" > "Trace Statistics" > "Mean/Std Dev/RMS On" for the phase measurement, and "Trace" > "Trace Config" > "Trace Statistics" > "Phase/Electrical Length On" for the timing delay measurement. Figure 4.25 shows the results of initial measurements for two paths. In Figure 4.25(a), the amplitudes of the two paths are compared. Path b_2 is about 2 dB higher than path b_1. Figure 4.25(b) shows the phase and timing delay. Here, the mean phase delay is 567.64°, and the timing delay is –50.686 ns. Figure 4.25(c) shows the amplitude of the individual paths across the 50-MHz chirp BW. Here, you can also clearly see the desensitization. (Note that VNA power calibration has not been performed, so the absolute amplitudes are approximate.)
- *Measurement Step 4 (optional):* The final step is to adjust the phase and/or timing delay of the individual paths, if this is possible (and required). In this example, the two DUT paths are actually two baseband-to-RF paths of an R&S SMW vector signal generator. Identical chirp waveforms are running in the dual baseband generators. By using the Digital Impairments capability of the I/Q modulator, IQ delay and phase offset values can be entered directly (Figure 4.26). Here, I entered –567° to

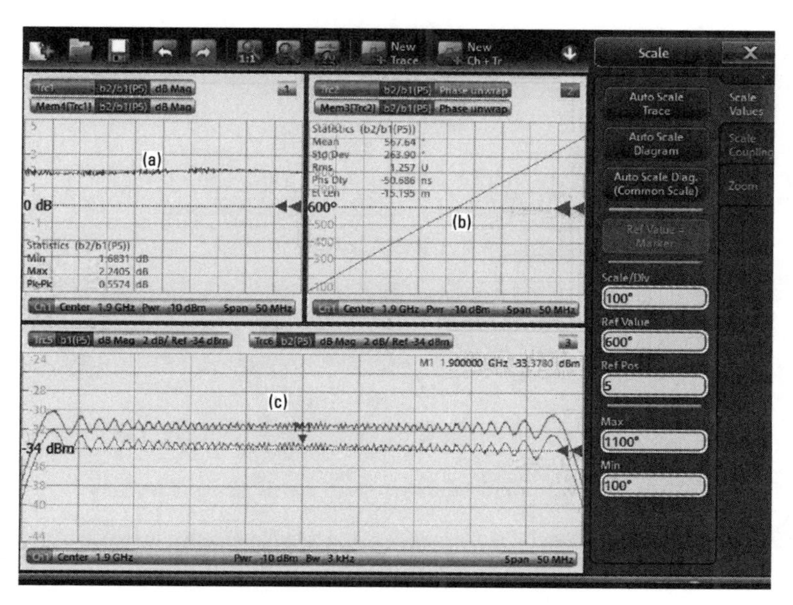

Figure 4.25 Chirp analysis with VNA: initial phase and timing. (a) Amplitude imbalance between chirps. (b) Initial phase and timing offset. (c) Power (dBm) of each chirp signal.

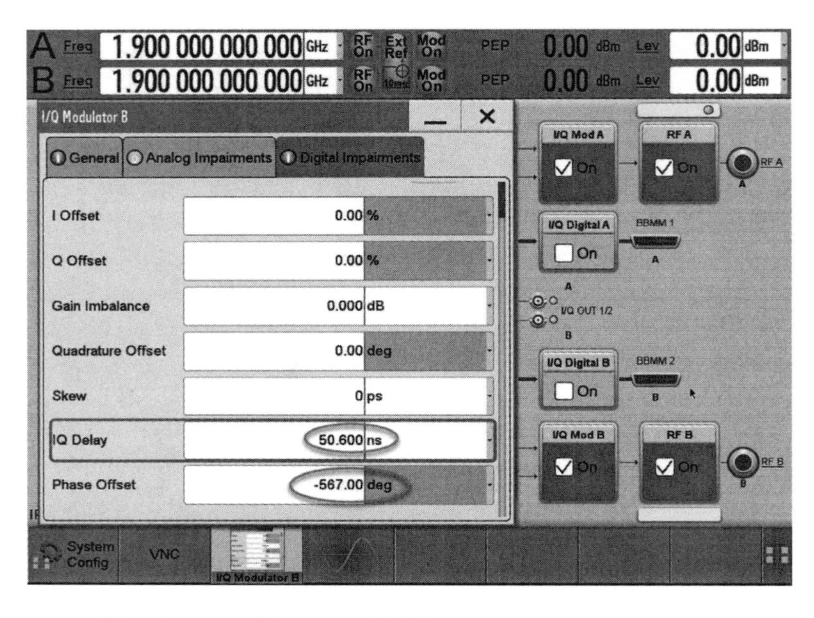

Figure 4.26 IQ impairments dialog of the R&S SMW vector signal generator.

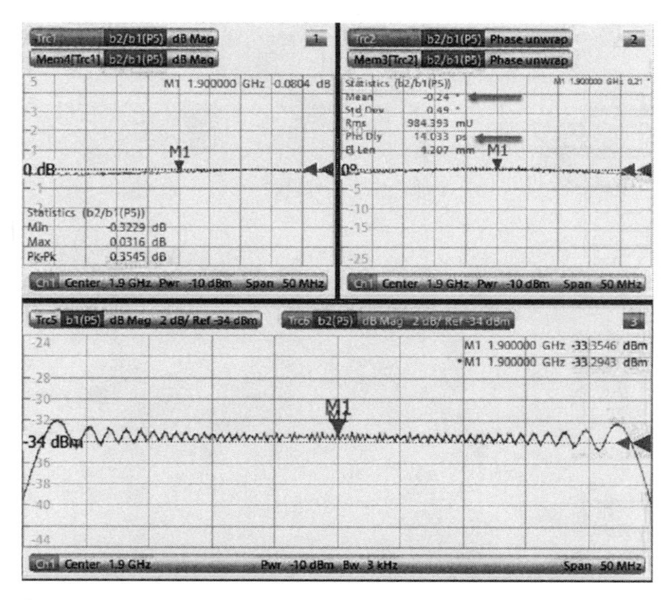

Figure 4.27 Chirp analysis with VNA: results after amplitude phase and timing alignment.

compensate for the phase offset and 50.6 ns to compensate for the timing delay of the b_2 path.

By entering smaller increments for the IQ delay, you can immediately see the effect on the slope of the phase. If adjusting the delay in the wrong direction, the slope will increase. If adjusting in the right direction, the slope will decrease, until it becomes a horizontal (flat) line. Figure 4.27 shows results after adjusting the phase offset and IQ timing delay between the paths. The phase offset was reduced to –0.24° and the timing offset to less than 15 ps (see arrows). The amplitudes of the two signals were also adjusted.[5]

Selected Bibliography

Application Note, "Reflection Measurements – Revisited," Anritsu Microwave Measurements Division, Morgan Hill, CA, April 2000, Rev. C.

Jorgesen, D. and C. Marki, Application Note, "Directivity and VSWR Measurements: Understanding Return Loss Measurements," Marki Microwave, Morgan Hill, CA, 2012.

5. More details are available in the R&S App Note 1GP108, "Generating Multiple Phase Coherent Signals – Aligned in Phase and Time."

5

Active Two-Port Device Measurements

An active device provides the ability to electrically regulate the current flowing through it. Typically, an active device requires an external power source to perform its design function(s). From an RF perspective, the broadest and most important class of active devices is the amplifier. These fundamental RF components make appearances in many different stages of communications equipment, from baseband through millimeter-wave frequencies, with power-handling capability ranging from picowatts to megawatts. Fortunately, in spite of their many different operating characteristics, amplifier measurements fall into two broad categories: linear and nonlinear measurements.

In this chapter, we consider amplifier applications from the VNA standpoint, focusing on the different testing challenges posed by particular amplifier types. Table 5.1 provides an overview of the active two-port device topics covered in this chapter. First, we examine the general-purpose amplifier, which places minimal demands on the VNA and allows measurement setups and techniques similar to passive devices. Then we look at high-power amplifiers, which often require external test sets to avoid damaging VNA circuitry. High-gain amplifiers represent another interesting corner case that introduces unique power calibration challenges due to the wide amplitude range requirements between DUT input and output ports. Finally, we consider low noise amplifiers and the unique performance requirements they place on VNAs, particularly for noise figure measurements. The end goal is to help the user to make accurate amplifier measurements while avoiding common pitfalls and expensive mistakes.

Table 5.1
Amplifier Categories and Typical Measurements

Category	Measurements	Amplifier Type General Purpose	 High Power	 High Gain	 Low Noise
Linear	Gain	5.1.1	5.1.1	5.1.1	5.1.1
	Return loss	5.1.1	5.1.1	5.1.1	5.1.1
	Hot S_{22} measurements	—	5.2.4	—	—
	Group delay	5.1.1	5.1.1	5.1.1	5.1.1
	Noise figure	—	—	—	5.4
Nonlinear	Harmonics	5.1.2.4	5.1.2.4	5.1.2.4	5.1.2.4
	1-dB compression point	5.1.2.2	5.1.2.2	5.1.2.2	5.1.2.2
	AM/PM conversion	5.1.2.3	5.1.2.3	5.1.2.3	5.1.2.3
	Intermodulation	5.1.2.5	5.1.2.5	5.1.2.5	5.1.2.5

5.1 General-Purpose Amplifier Measurements

For our discussion, amplifiers of the general-purpose category have moderate gain (10 to 20 dB), relatively low output power (P1dB approximately +10 dBm), and moderate noise figure (3 to 6 dB). A good example is Mini Circuits' ZJL-3G+, which operates over a frequency range of 20 to 3,000 MHz (Table 5.2).

5.1.1 Linear Measurements

Ideal linear two-port networks provide distortion-free transfer of signals from input to output. The associated voltage transfer function is

$$v_{out}(t) = g_v v_{in}(t) \tag{5.1}$$

Where $v_{out}(t)$ = network output voltage, $v_{in}(t)$ = network input voltage, and g_v = network voltage gain.

Table 5.2
Specifications for Mini-Circuits' ZJL-3G+

Frequency Range (MHz)		Gain (dB)			1-dB Compression Point (dBm)	IP3 (dBm)	Noise Figure (dB)	VSWR (typ.)	
Low	High	Typ	Min	Flatness (typ)				In	Out
20	3,000	19	14	+/–2.2	8	22	3.8	1.4	1.6

Passive devices are primary examples of such ideal networks. However, under controlled bias and excitation conditions, active devices may be constrained to exhibit linear behavior.

Typical linear measurements for general-purpose amplifiers include gain, return loss, and group delay. When an amplifier is operated in its linear region (typically at least 10 dB below its 1dB compression point), these measurements are conducted in a similar fashion to passive devices. Simply follow the steps in Chapter 4 for establishing a two-port measurement baseline (Section 4.1.1). This baseline uses the default VNA power level (usually between −10 and 0 dBm) to ensure the best accuracy for the system error correction (SEC). After completing calibration, we set the VNA power level to a value that ensures that: (1) the DUT will be operating in its linear range, and (2) the VNA receivers are also operating in their linear range. Both of these criteria must be met for accurate measurement results. In the case of the Mini-Circuits' ZJL-3G+, the output 1-dB compression point is given as 8 dBm. To compute a conservative input power that meets criterion (1) for linear DUT operation, we start with the compression point value, subtract the highest expected device gain, and then apply an additional 10-dB back-off. From Table 5.3, this would be 8 dBm (compression point) – 19 dB (typical gain) – 10 dB (rule-of-thumb backoff) = −21 dBm input power.[1] Next, we need to check criterion (2) for linear VNA receiver operation. Here, the VNA port-2 measurement receiver will be seeing the full brunt of the amplifier's output power, which will be the input power level plus the typical gain, or −21 dBm + 19 dB = −2 dBm. We need to check this value against our VNA compression point specifications. According to Table 5.3, we more than meet our 10-dB rule-of-thumb margin for a ZVA8, but we will be operating at only 5 dB below the ZVA50's compression

Table 5.3
VNA Receiver Compression Point
(by Frequency Model)

Typical ZVA/B Properties		
1-dB Compression Point		
Device	Port Input $L_{P,1dB}$	Direct Receiver Access Input $L_{P,1dB}$
ZVA8	10 dBm	–5 dBm
ZVA24	6 dBm	–10 dBm
ZVA40	3 dBm	–10 dBm
ZVA40	3 dBm	–10 dBm

1. Without power calibration, VNA power level accuracy can vary +/–3 dB. This is another reason for operating at a conservatively low excitation level (but not so low that trace noise becomes a problem).

point. For the ZVA50, we may either: (1) reduce the input power by another 5 dB to –26 dBm, or (2) add 5 dB of attenuation to the port-2 measurement receiver. (All ZVA models have optional internal mechanical step attenuators for their measurement receivers with 5 dB step size.) The advantage of the first method is that it costs nothing, but has the drawback that since a lower power level is applied to the input, the SNR will be lower at the port-1 (and port-2) reference and measurement receivers, possibly leading to greater trace noise. The advantage of the second method is that it preserves the SNR of the port 1 receivers. Also, by experimenting with the mechanical attenuator settings, you may find that there is actually no receiver compression in your measurement frequency range of interest. The downside is the additional cost of the mechanical attenuator.

There is a third alternative: adding a fixed attenuator somewhere between the amplifier output and the VNA port-2 input. The downside of this approach is that all port-2 power levels will be reduced. A new calibration will also be required (with the attenuator in place) to compensate for this attenuation, and if full two-port calibration is employed, this attenuation will tend to increase the trace noise of all four measured S-parameters. This extra attenuation will also impair calibration stability (over time and temperature) due to port 2's reduced raw coupler directivity. Alternatively, the attenuator can be measured separately and exported as an .s2p file. Then the attenuator can be installed on the DUT output for the gain measurement, and the attenuator file imported as an S_{21} memory trace to correct the gain with trace math. Unfortunately, with the attenuator in place, S_{22} measurements of the amplifier will be impossible.

Before connecting the DUT to the VNA, it is a good idea to terminate the DUT's RF input and output in appropriate impedances (typically 50Ω loads) and verify that there is no DC present. Alternatively, a VNA equipped with bias tees provides a way of introducing DC via the VNA's coaxial center pin(s), and simultaneously prevents DC from harming sensitive RF subassemblies (receivers and generators) within the VNA.[2]

After setting the power level of the VNA and checking for the presence of DC bias on the DUT RF lines, power down the DUT and connect it to the VNA. Once connected, apply operating voltages to the DUT. Small-signal gain, return loss, and group delay measurements can be performed in a manner analogous to passive components, with S_{21} (dB Mag) for measuring gain, S_{11} (dB Mag) and S_{22} (dB Mag) providing input and output return loss, respectively, and S_{21} (delay) presenting group delay in the forward direction (refer to Section 4.1.2.1 for details).

2. For example, the ZVA's built-in bias tees and the ZNB's optional bias tees can handle a maximum nominal DC voltage of 30V.

5.1.2 Nonlinear Measurements

Measuring nonlinear devices with a VNA requires careful attention to signal levels. To understand why, it is necessary to review the mathematics of nonlinear functions before diving into measurements.

Nonlinear behavior is represented by a power series:

$$v_{out}(t) = \pm\sum_{n=1}^{\infty} a_n V_{in}^n(t) \tag{5.2}$$

where $v_{out}(t)$ = network output voltage as a function of time, t, a_n = voltage gain coefficient of the nth power series term, and $V_{in}^n(t)$ = network input voltage.

Expanding (5.2), we obtain:

$$v_{out}(t) = a_1 V_{in}(t) + a_2 V_{in}^2(t) + a_3 V_{in}^3 + a_4 V_{in}^4 + \ldots \tag{5.3}$$

Term a_1 is equivalent to the voltage gain g_v of (5.1), and is responsible for the linear component of an amplifier's response to RF excitation. Conversely, we attribute the nonlinear behavior to higher-order coefficients. Limiting the power series to the third (cubic) term keeps the mathematics manageable (for illustrative purposes). In practice, the linear, squared, and cubic terms have the strongest impact on device operation.

If we introduce a sinusoidal signal $V_{in}(t)$ to a device governed by (5.3):

$$V_{in}(t) = U_{in,1}\cos(2\pi f_{in,1} \cdot t) \tag{5.4}$$

where $U_{in,1}$ = peak voltage value of $V_{in}(t)$ and $f_{in,1}$ = frequency of $V_{in}(t)$.

Then the output response becomes (truncated to the third term):

$$\begin{aligned} v_{out}(t) &= a_1 U_{in,1}\cos(2\pi f_{in,1} \cdot t) + \\ &a_2 U_{in,1}\cos^2(2\pi f_{in,1} \cdot t) + a_3 U_{in,1}\cos^3(2\pi f_{in,1} \cdot t) \end{aligned} \tag{5.5}$$

After expanding this equation, we obtain harmonics of the input frequency $h_n = n \cdot f_{in,1}$ as illustrated in Figure 5.1.

The amplitude of the resulting harmonics depends on the coefficients in (5.5) as well as the respective harmonic order (n) and the input level. When the input level increases by ΔdB, the output level increases by $n \cdot \Delta$dB. This is illustrated in Figure 5.2, which shows the output level of both the fundamental tone and the second harmonic versus input power level for the ideal case. Due to the small coefficient a_2, the amplitude of the second harmonic starts out much lower than the fundamental at low drive levels, but as power increases, the second harmonic increases twice as fast as the fundamental.

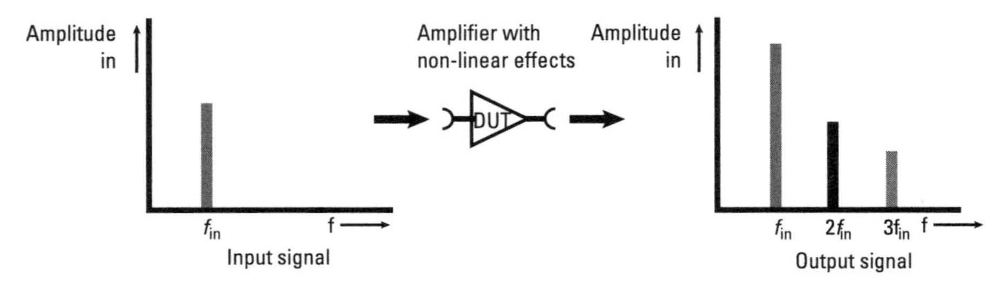

Figure 5.1 Input and output spectrum of amplifier showing nonlinear effects on single-tone excitation.

One immediately obvious revelation is that the level of the second harmonic directly depends on the level of the excitation. So how should we specify an acceptable harmonic level? Data sheet specifications for harmonic distortion usually refer to the output power of the second harmonic relative to the fundamental, measured at a particular input drive level. Alternatively, level-independent specifications are possible using the concept of intercept point. As seen in Figure 5.2, if we extrapolate the fundamental and second harmonic traces to the point where they intersect, we obtain a metric that is independent of excitation level. This intercept point cannot be reached in practice due to compression effects, but it can be calculated using measurements at lower power levels (where the fundamental tone is operating in its purely linear range). This resulting second harmonic intercept (SHI) must be specified relative to either the input level (SHI_{in}) or output level (SHI_{out}). It is calculated as follows:

$$SHI_{out} = \Delta_{k2} + L_{out} \tag{5.6}$$

where SHI_{out}= output second harmonic intercept point (dBm), L_{out} = level of output tone (dBm), and Δ_{k2} = difference between second harmonic and fundamental signal levels at L_{out} (in decibels).

Similarly,

$$SHI_{in} = SHI_{out} - G \tag{5.7}$$

where SHI_{in} = input second harmonic intercept point (dBm) and G = device power gain (decibels).

This second harmonic discussion helps illustrate a critical point about nonlinear device measurements: knowing the absolute power levels of input and output signals is crucial for obtaining meaningful results. This brings us to the VNA. While a VNA does a great job of measuring relative signal levels (in decibels) thanks to system error correction (SEC), it alone is insufficient for nonlinear measurements. An additional power calibration step is required

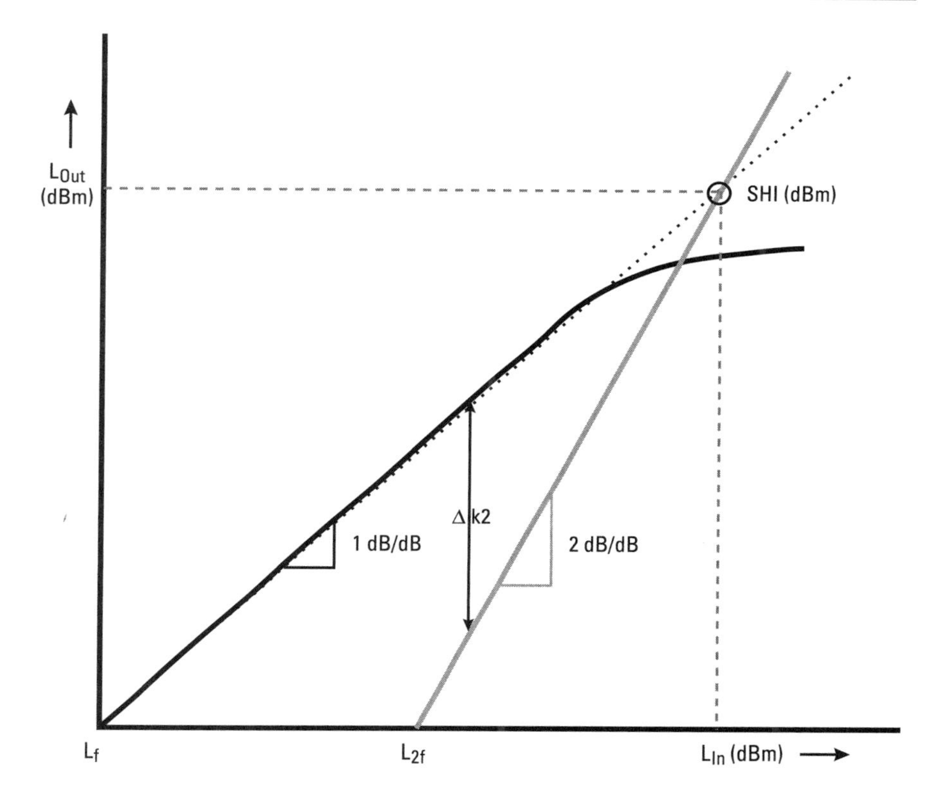

Figure 5.2 Second harmonic and fundamental tone output power versus fundamental tone input power.

to enable the reference and measurement receivers to make accurate absolute power level measurements (in dBm) at the calibration plane. Additionally, it is often desirable to calibrate the VNA's generator as well so that it can be conveniently set to provide a specified power level across at all points of a frequency sweep or over a range of power levels during a power sweep.

5.1.2.1 VNA Power Calibration

Most mid-range and high-end VNAs allow power calibration to be performed at the same calibration plane as the SEC. Sometimes these calibration procedures are combined, as in the case of Smarter Cal with the R&S ZNB or ZNBT. However, in all cases the power calibration is distinct from SEC and may be used with or without an active SEC. For a typical amplifier measurement, VNA port 1 provides stimulus to the amplifier's input port and VNA port 2 measures the signals emerging from the amplifier's output port. In this scenario, the VNA port-1 generator and reference receivers both require power calibration, whereas

VNA port 2 requires only power calibration of its measurement receiver. The following are typical steps for this process.

1. Set up the VNA for the desired sweep conditions (start/stop frequency, number of points, IF BW, power level). It is a good idea to perform power calibration at a reasonably high-power level, where the VNA SNR is favorable (0 dBm or −10 dBm are typical values). Additionally, you may choose a lower point density for power calibration, since the VNA generator typically retains power accuracy over a wider swath of bandwidth than the system error correction. For example, if you are going to measure a linear frequency sweep extending from 10 MHz to 3.01 GHz with 5-MHz steps, you might perform the SEC using this same 5-MHz grid (601 points), but then perform the power calibration using 50-MHz steps (61 points). This will reduce overall calibration time (due to the slower measurement speed of typical wideband power sensors) without sacrificing accuracy. After completing the power calibration, you can set the step size back to 5 MHz. The instrument will interpolate power calibration values between calibrated frequency points for the higher point-density measurement sweep.
2. Connect the power sensor to the end of the VNA port-1 test cable. (It is assumed that the VNA has already established communications with the power sensor or power meter.)
3. To configure the source power calibration on the ZVA, press "Channel" > "CAL" > "Start Power Cal" > "Source Power Cal…" In the resulting dialog box, select "Modify Settings…" In this next dialog, you can fine-tune the power calibration process. For example, we want to power-calibrate the reference receiver and perform a flatness calibration for the VNA source/generator, so we make sure the checkboxes are checked for both "Reference Receiver Cal" and "Flatness Cal" in the "Includes" section (default condition). Since the flatness calibration is an iterative process, we change the "Maximum Number of Readings" to 10. We also want +/0.1 dB source power accuracy, so we change the "Tolerance" to 0.1 dB. Now click "OK" to close the "Modify Source Power Cal Settings" dialog.
4. Start the power calibration by pressing "Take Cal Sweep." You will soon see a trace marching across the dialog screen. The first sweep is slowest as the analyzer sets the power of the source to the requested value at each frequency point and waits for the power sensor to complete a measurement and return the measured value. Based on this result, it creates two internal calibration tables: one for the reference receiver, and one for the VNA generator. After this initial sweep, the

analyzer uses the now-calibrated VNA reference receiver for all subsequent power measurements (based on settings in the "Use 'reference receiver after' '1' Power Meter Readings" in the source power calibration dialog). Thus, subsequent sweeps occur much faster as the VNA adjusts the port-1 source power to meet the user-defined tolerance limits. Once the source power is within the tolerance window for all sweep points, the ZVA performs a final verification sweep and displays the results on the screen. Press "Close" to close the Source Power Cal dialog. (If calibrated power levels are required in the reverse direction, change the Source to "Port 2," connect the power sensor to the end of the port-2 test cable, and repeat steps 3 and 4 to power calibrate the port-2 reference receiver and generator.)

5. To calibrate the port-2 measurement receiver, connect a thru standard between port 1 and port 2. Then, click "Channel" > "CAL" > "Start Power Cal" > "Receiver Power Cal..." In the resulting dialog, set the "Wave Quantity to calibrate" to "b2" and ensure that the calibration source is set to "Port 1." The "Reference Power Value" should be set to "Reference Receiver." Then pressing "Take Cal Sweep" causes the ZVA to perform a power calibration sweep using the calibrated power accuracy of the a_1 reference receiver to calibrate the b_2 measurement receiver, a very quick process. Click "Close" to exit the dialog. This completes the power calibration process for VNA generators and receivers.
6. To verify the power calibration, overlay a trace of wave quantity a_1 (source port 1) and b_2 (source port 1). Use coupled markers to compare the power values along the traces. The traces should have the same values (within a few hundredths of a decibel).
7. The ZVA will automatically append this power calibration to any preexisting S-parameter calibration that may be active. To see this, go to "Channel" > "Calibration" > "Cal Manager" and examine the calibration properties for the active calibration. (It would be a good idea to save this combined calibration in the calibration pool for future use or to guard against accidental loss (e.g., accidentally pressing "preset").

5.1.2.2 AM/AM Conversion (Compression-Point) Measurements

AM/AM conversion measurements represent the first major departure from VNA operational orthodoxy. Up until now, the VNA has operated in a mode where its receivers and generators provide ratio measurements, such as b_2/a_1 or b_1/a_1 (for S_{21} and S_{11}, respectively). For AM/AM and AM/PM measurements (and for all nonlinear measurements), we need absolute power measurements as well as ratio measurements to obtain meaningful results.

AM/AM conversion is a metric for quantifying gain compression in an amplifier. As drive levels increase, the amplifier will eventually reach a point where it can no longer accommodate input power increments in a linear fashion. The amplifier's power gain begins to decrease as it approaches its maximum output level, a phenomenon referred to as gain compression. This kind of AM/AM conversion is an important amplifier metric because it distinguishes the linear from the nonlinear operating range. The compression point may be measured by configuring a power sweep at a single frequency and noticing how the gain changes with increasing power. By using a low starting power level (–30 dBm for a general purpose amplifier) and initially a fairly low ending power level (–20 dBm), we can avoid overdriving or damaging the DUT. After taking a sweep, any gain compression will be immediately apparent as a reduction in gain. The power level at the end of the sweep can be increased in small increments (1 to 3 dB) until a desired amount of gain compression is observed. Some analyzers provide a way to limit the maximum allowed drive level to prevent overdriving a DUT. Most analyzers also have built-in trace functions that allow the compression point to be automatically calculated and displayed.

However, it is a good idea to add additional traces (besides S21) for the DUT input and output power. There are several reasons for this:

1. By plotting input power, you can see where the VNA's generator reaches maximum output. This is important information, because in the midst of making measurements, we may lose sight of the available VNA drive power, particularly at higher frequencies, and wonder why we cannot see gain compression in our DUT.
2. The built-in compression point trace function bases its results on the input power (*x*-axis value at the 1-dB compression point), and calculates the equivalent output power value by simply adding the measured gain to this value. This would be fine if the *x*-axis value represented the actual input power. Regrettably, this number represents only the power level that has been set, not the delivered power. Even after power calibration, the value at, say, –15 dBm, may not be exactly –15 dBm.[3] It could be off by several tenths of decibels, resulting in unacceptable measurement error for certain amplifier types (like traveling wave tube amplifiers). Trace markers allow the actual input and

3. Another way to deal with this issue would be to turn ALC on. ALC uses a feedback loop to ensure the power requested at each point (i.e., the *x*-axis value) is approximately equal (within several hundredths of a decibel) to the power delivered.

output levels to be used in concert with the 1-dB compression point marker information to obtain greatest measurement accuracy.

Below is a step-by-step process for setting up a compression point measurement on a ZVA. It is assumed that a valid S-parameter calibration and power calibration have been performed for a two-port measurement.

1. Remove the DUT from the test setup. Configure a small-signal linear frequency sweep for the DUT, and establish an S_{21} (dB Mag) gain measurement. Insert the DUT and make sure it is operating correctly under small-signal excitation. (This gives instant feedback that the DUT is turned on and functioning properly, always a good starting point.) Then power down the DUT.
2. It is a good idea to turn off the VNA sources for the remainder of the VNA configuration, ("Channel" > "Power Bandwidth Avg" > "RF Off (All Chans)") This will prevent accidentally overdriving the DUT during the power sweep setup.
3. Add a new Channel, Trace, and Diagram area. (This will be used for displaying power sweep results for a gain compression measurement.) "Channel" > "Channel Select" > "Add Channel + Trace + Diag Area"
4. Configure a power sweep. "Channel" > "Sweep Type" > "Power..." Change the "Channel Base CW Frequency" to a mid-band frequency for your DUT (here, 2 GHz).
5. Change the number of sweep points to 101 for small power steps and good gain compression resolution: "Channel" > "Sweep" > "Number of Points" 101.
6. Set the power sweep range to −30 to −15 dBm: "Channel" > "Stimulus" > "Start" −30 dBm and "Channel" > "Stimulus" > "Stop" −15 dBm.
7. Turn on the 1-dB compression point trace function: "Trace" > "Trace Funct" > "Compression Point" (by default the value is "1 dB").
8. Change the scale for Trc2 to 0.5 dB/division: "Trace" > "Scale" > "Scale/Div..." 0.5 dB.
9. Add a new trace to observe the input power: "Trace" > "Trace Select" > "Add Trace" and "Trace" > "Measure" > "Wave Quantities" > "a1 Src Port 1."
10. Add a new trace to observe the output power: "Trace" > "Trace Select" > "Add Trace" and "Trace" > "Measure" > "Wave Quantities" > "b2 Src Port 1."

11. Link the scaling of the a_1 and b_2 traces (such that the difference between the traces will be the DUT gain): “Trace” > “Trace Select” > “Trace Manager…” (Under “Scale”, select “Trc3” for both “Trc3” and “Trc4”), and then “Close”.
12. Add a coupled marker. Assuming either Trc3 or Trc4 is active, select “Trace” > “Marker” > “Marker 1” and then “More” and click “Coupled Markers.”
13. Connect the DUT, apply DC power, and then re-enable the VNA RF: “Channel” > “Power Bandwidth Average” > de-select “RF OFF (All Chans).”

You can adjust the power sweep “Stop” value in small increments until you observe just over 1-dB compression. See Figure 5.3 for an example of expected measurement results.

Often, we only need to characterize the 1-dB compression point at low, middle, and high frequencies within the DUT operating range. Separate channels can be configured to measure these three points, or the frequency of the existing channel can simply be changed. For more comprehensive data, some VNAs have a built-in function for swept compression point that performs a

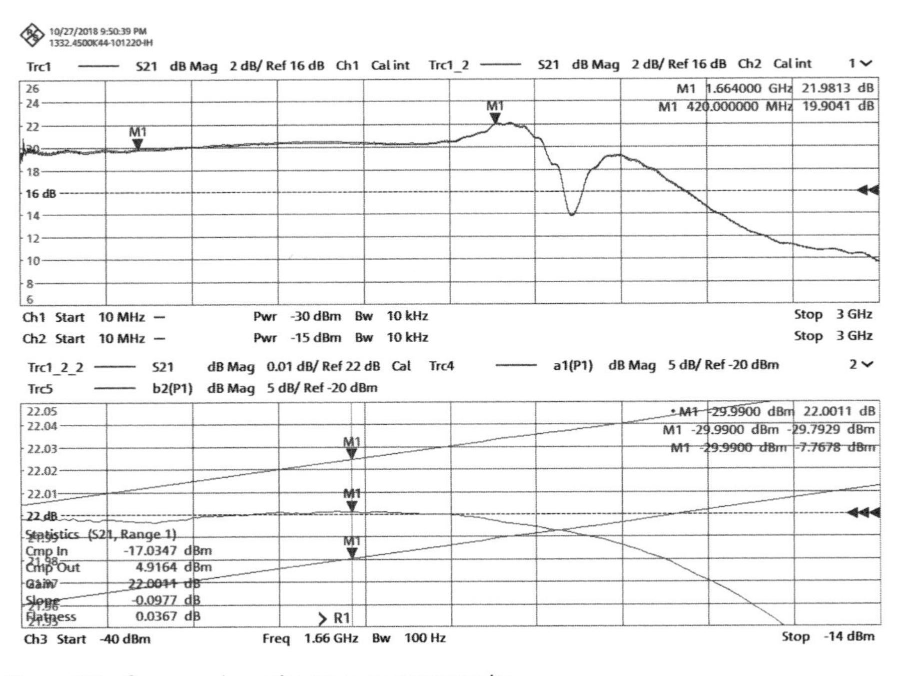

Figure 5.3 Compression point measurement results.

power sweep at each frequency point within a frequency sweep and displays the resulting compression point as a function of frequency.

One important consideration for compression point measurements is making sure that any observed compression is due entirely to DUT compression and not to the VNA. VNA receivers are designed to remain linear over the power range of the instrument's own generators. Amplifiers can result in power levels that exceed this range and may cause the VNA's measurement receivers to compress. This is one instance where it is very useful to have built-in VNA receiver attenuators. By incrementing the b_2 receiver attenuator in 5-dB steps, we can instantly diagnose the compression. If the gain changes, we know the VNA receiver was in compression. If there is no gain change after increasing the attenuator, we know that the VNA receiver is operating in its linear range.

AM/AM with Gain Expansion

Some amplifiers exhibit gain expansion during a power sweep, where the gain increases until the onset of output compression. This effect may be very subtle, as shown in Figure 5.3 (marker M1), or it may be pronounced (several decibels). In many cases, designers want to characterize amplifiers for compression with reference to the maximum gain rather than the small-signal gain. In this situation, the left edge of the n-dB compression point evaluation range is set to the maximum gain value. In Figure 5.3, this evaluation range is designated by the vertical line that is denoted as R1. The Evaluation Range dialog shows the associated starting and stopping power levels.

5.1.2.3 AM/PM Conversion Measurements

From a VNA perspective, setting up a measurement of AM/PM conversion is very similar to the AM/AM conversion measurement of Section 5.1.2.2. The only difference is that the trace is set to display S_{21} (Phase) rather than S_{21} (dB Mag). In practice, AM/PM conversion is often required at the 1-dB compression point, and is measured relative to the phase value at the start of the power sweep (where the device is operating in its linear range). This information can be easily obtained by displaying both traces are using markers to read the required AM/PM conversion value.

5.1.2.4 Harmonic Distortion Measurements

Harmonic distortion measurements introduce the second major departure from VNA operating orthodoxy. Whereas AM/AM compression measurements added absolute power measurements to the well-established portfolio of ratio measurements (for S-, Y-, and Z-parameters), harmonic distortion forces an entirely new VNA operating concept, where the receivers and generators are tuned to different frequencies. For this reason, harmonics (and other nonlinear measurements) are often treated as an extended capability by equipment manufacturers. Typically, this capability is available without investment in additional hardware. Dialogs are often provided to simplify setups that otherwise can be tricky if not downright unwieldy to configure. Harmonic measurements are a good case in point. Since a VNA (unlike a spectrum analyzer) measures only at discrete frequency points, we immediately see that the required grid for the fundamental and harmonic frequency measurements is different. For example, if your fundamental frequency range of interest is 10 MHz to 1.5 GHz, and you establish a frequency spacing of 5 MHz, bear in mind that the second harmonics will extend from 20 MHz to 3 GHz, in 10-MHz steps. This means the overall frequency range for fundamental plus second harmonic measurements should be 10 MHz to 3 GHz. If you decide to power-calibrate at each measurement point, your frequency grid must have 5-MHz steps up to 1.5 GHz, and 10-MHz steps from 1.51 GHz up to 3 GHz. However, it is often easier to just establish a 5-MHz step spacing across the entire range and not worry about the extra calibration points that will not be used between 1.5 and 3 GHz. Alternatively, there are some measurement situations where the extra power calibration time may become burdensome, and a nice solution would be to use a segmented sweep to cover the two ranges.

Assuming that the required SEC and power calibrations have been performed, you can then configure the analyzer to make the desired measurements. It is generally a good idea to start with well-known measurements to make sure everything is working correctly and then add the more advanced measurements. In this case, we start with a gain measurement over the fundamental frequency range of 10 MHz to 1.5 GHz, and set a power level that we determine is well within the DUT's linear operating range. After confirming that everything is working correctly, we add the harmonic measurements. There are two ways that we can display harmonics. The first is as a frequency sweep, where we set the input power and measure the fundamental and harmonic levels, and calculate the second-order intercept. The second method is by performing a power sweep, similar to the 1-dB gain compression point measurement (Section 5.1.2.2), and observing how the fundamental and second harmonic levels change as a function of the drive level at a single frequency. We use the former method (frequency sweep) for this example.

1. Add a new trace to the existing channel to measure the fundamental output power: "Trace" > "Trace Select" > "Add Trace and Diag Area" (Trc2) and "Trace" > "Meas" > "Wave Quantities" > "b2 Src Port 1."
2. Add a new channel and trace for the second harmonic measurement. (A new channel is required because the VNA port 2 measurement receiver must be tuned to the second harmonic frequency, not the generator frequency.) "Channel" > "Channel Select" > "Add Channel

+ Trace" (Trc3) and "Trace" > "Meas" > "Wave Quantities" > "b2 Src Port 1."

3. Configure the VNA measurement receivers to sweep at twice the fundamental frequency: "Channel" > "Mode" > "Port Config", then under "Receiver Frequency," click on the "..." next to "fb" for port 2. In the frequency control dialog that appears, change the highlighted "1" to "2" and then click OK. You will see a message "Set arbitrary mode active?" Click "yes." Then, back in the Port Configuration dialog, make sure that "Measure Source Port Waves at" is set to "Source Frequency." Click OK to close the dialog. Now we have traces that show the absolute levels of the second harmonic and the fundamental.
4. Add a new trace that provides the second harmonic level relative to the fundamental ("dBc") using trace math: "Trace" > "Trace Select" > "Add Trace + Diag Area" (Trc4) and "Trace" > "Trace Funct" > "User Def Math..." then, select "Trc3" then "/" then "Trc2" and "OK" to close the dialog.
5. Activate Trace Math: "Trace" > "Trace Funct" > "Math = User Def."
6. Calculate and display the second-order intercept point (if desired) by adding a new Trace and Diagram Area: "Trace" > "Trace Select" > "Add Trace + Diag Area" (Trc5). "Trace" > "Measure" > "b2 Src Port 1" and ensure that the properties are displayed as "Power" (default setting) because the desired second-order quantity will be displayed as an absolute power value (in dBm), not a relative value (in decibels): "Trace" > "Trace Funct" > "User Def Math..."
7. Since second-order output intercept was presented in (5.5) as $SHI_{out} = \Delta_{k2} + L_{out}$, the value Δ_{k2} is represented by Trc4, and L_{out} is given by Trc2, the required trace math equation is Trc4 × Trc2 (because trace math operates on the underlying linear values, not the logarithmically displayed values). Additionally, it is necessary to click "Result is wave quantity" to provide the proper offset conversion between dBm (linear absolute power) and dB (relative power) for the trace math calculations.[4]

---

4. If "Result is Wave Quantity" is disabled, the analyzer assumes that the result of the mathematical expression is dimensionless (e.g., a ratio of wave quantities). However, if it is enabled, the analyzer assumes that the result represents an absolute power relative to 0 dBm Converting 0-dBm absolute power to voltage, we obtain:

$$0.001W = \frac{V^2}{50\ ohms} \rightarrow V = 0.2236\ volts$$

To display this absolute voltage value on a logarithmic scale, we apply logarithmic conversion:

$$20\log_{10}(0.2236) = -13.01\ dB$$

VNA Generator Harmonics

One potentially serious problem with VNA-based harmonic measurements is that a VNA generator can have considerable harmonic content (on the order of –20 to –30 dBc). Because of this, VNA manufacturers often provide additional filtering options that can reduce harmonic levels to –60 dBc or lower. These options are potentially important for broadband amplifiers, which offer little harmonic rejection at the device input. Alternatively, when making measurements on narrow-bandwidth amplifiers, the DUT's input matching network provides a degree of harmonic rejection. It is best to check this by temporarily placing a harmonic filter in front of the DUT, and comparing harmonic results against measurements made without the filter.[5]

5.1.2.5 Intermodulation Distortion

Intermodulation distortion is similar to harmonic distortion in that they are both nonlinear phenomenon, but intermodulation distortion arises from nonlinear interactions between two or more signals. Assuming that the same nonlinear behavior represented by (5.3), if we now substitute two CW signals with different frequencies and amplitudes (5.8), we get the spectral components shown in Table 5.4 and displayed in Figure 5.4:

$$V_{in}(t) = U_{in,1}\cos(\omega_1 \cdot t) + U_{in,2}\cos(\omega_2 \cdot t) \tag{5.8}$$

The amplitudes of the fundamental components are strongly determined by the linear coefficient a_1. Similarly, a_2 determines the amplitude of the second harmonics and second-order intermodulation (as well as the DC component), while a_3 does the same for the third harmonics and third-order intermodulation. Figure 5.4 shows the output spectrum for the linear and nonlinear components, along with relative amplitudes where defined mathematical relationships exist. For example, the diagram shows that the third harmonics are 9.54 dB below the third-order intermodulation products, assuming the two input signals have equal amplitude.[6]

Fortunately, the equation required to define the third-order intercept point is very straightforward. It is given by:

This value is appended to the trace math result for calculations involving display of absolute power values.

5. Lowpass or bandpass filters are often used as harmonic filters. Alternatively, a notch filter tuned to the second-harmonic frequency rejects the second harmonic before it enters the DUT.
6. From Table 5.3, the amplitude of the third harmonic is: $a_3 \cdot 0.25 \cdot U_1^3$. The amplitude of the third-order intermodulation is $a_3 \cdot 0.75 \cdot U_1^2 \cdot U_2$. If $U_2 = U_1$, then the difference in amplitudes is obtained from the ratio of these coefficients: $20 \cdot \log_{10}\left(\frac{0.25}{0.75}\right) = -9.54\ dB$.

Table 5.4
Mixing Products at Amplifier Output Due to Two-Tone Excitation (Nonlinear Device Characteristics)

Two-tone excitation given by: $V_{in}(t) = U_1 \cos(w_1 \cdot t) + U_2 \cos(\omega_2 \cdot t)$	**Power Series Representation of Nonlinear Characteristics: $V_{out}(t) = a_1 V_{in}(t) + a_2 V_{in}^2(t) + a_3 V_{in}^3(t)$**		
Spectral components:	First-order products	Second-order products	Third-order products
DC		$a_2 \cdot 0.5 \cdot (U_1^2 + U_2^2)$	
Fundamental frequencies	$a_1 \cdot U_1 \cos(\omega_1 t)$		$a_3 \cdot 0.75 \cdot U_1^3 \cos(\omega_1 t)$
	$a_1 \cdot U_2 \cos(\omega_2 t)$	$a_3 \cdot 1.50 \cdot U_2^2 \cdot U_1 \cos(\omega_1 t)$	$a_3 \cdot 1.50 \cdot U_2^2 \cdot U_1 \cos(\omega_1 t)$
			$a_3 \cdot 0.75 \cdot U_2^3 \cdot U_1 \cos(\omega_2 t)$
			$a_3 \cdot 1.50 \cdot U_1^2 \cdot U_2 \cos(\omega_2 t)$
Second harmonics		$a_2 \cdot 0.5 \cdot U_1^2 \cos(2 \cdot \omega_1 t)$	
		$a_2 \cdot 0.5 \cdot U_2^2 \cos(2 \cdot \omega_2 t)$	
Second-order intermodulation		$a_2 \cdot U_1 \cdot U_2 \cos(\cdot \omega_2 - \omega_1) t$	
		$a_2 \cdot U_1 \cdot U_2 \cos(\omega_2 + \omega_1) t$	
Third harmonics			$a_3 \cdot 0.25 \cdot U_1^3 \cos(3 \cdot \omega_1 t)$
			$a_3 \cdot 0.25 \cdot U_2^3 \cos(3 \cdot \omega_2 t)$
Third-order intermodulation			$a_3 \cdot 0.75 \cdot U_1^2 \cdot U_2 \cdot \cos(2 \cdot \omega_1 + \omega_2) t$
			$a_3 \cdot 0.75 \cdot U_1^2 \cdot U_2 \cdot \cos(2 \cdot \omega_1 - \omega_2) t$
			$a_3 \cdot 0.75 \cdot U_2^2 \cdot U_1 \cdot \cos(2 \cdot \omega_2 + \omega_1) t$
			$a_3 \cdot 0.75 \cdot U_2^2 \cdot U_1 \cdot \cos(2 \cdot \omega_2 - \omega_1) t$

$$IPn_{out} = \frac{a_{IMn}}{n-1} + L_{out} \tag{5.9}$$

where IPn_{out} = output intercept point of nth order, in dBm, a_{IMn} = level difference between the intermodulation product of nth-order and the fundamental of the output signal, in decibels, and L_{out} = level of one of the two output signals, in dBm.

For the third-order intercept, this becomes:

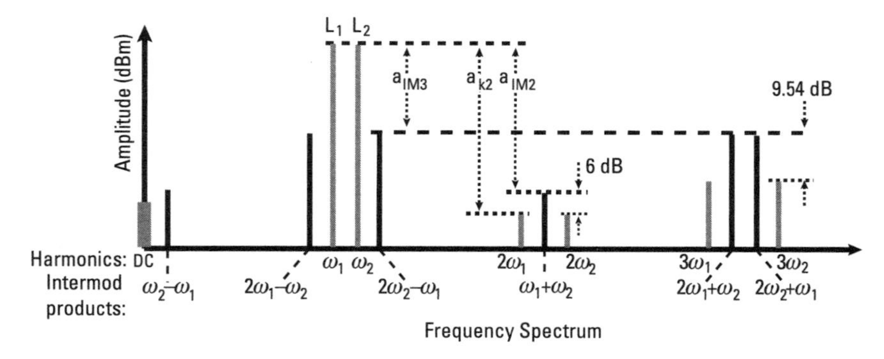

Figure 5.4 Amplitude versus output spectrum for harmonics and intermodulation distortion products.

$$IP3_{out} = \frac{a_{IM_3}}{2} + L_{out} \tag{5.10}$$

IMD Measurements with a VNA

Fortunately, many VNAs offer special dialogs for simplifying the setup, calibration, and measurement of intermodulation products. For example, the ZNB provides a dialog that guides the user through each step in the process as presented here.

1. Combine ports 1 and 3 together using an external power combiner (Figure 5.5).
2. Launch the Intermodulation Wizard. Channel > Channel Config > Intermodulation > Intermodulation Wizard…
3. Configure the IMD measurement. (Note that the dialog is preconfigured for a third-order measurement consisting of two CW tones positioned 1 MHz apart.[7]) Click "Next."
4. Configure the frequency sweep range for the IMD measurement. Click "Next."
5. Determine the IMD spectral components and calculated values to be displayed. Initially, we will view only the lower tone input (LTI) and

7. Other possibilities exist. For example, a user may wish to evaluate IMD based on a fixed center frequency and sweep the tone spacing instead, or maintain one tone at a fixed frequency and sweep the other tone.

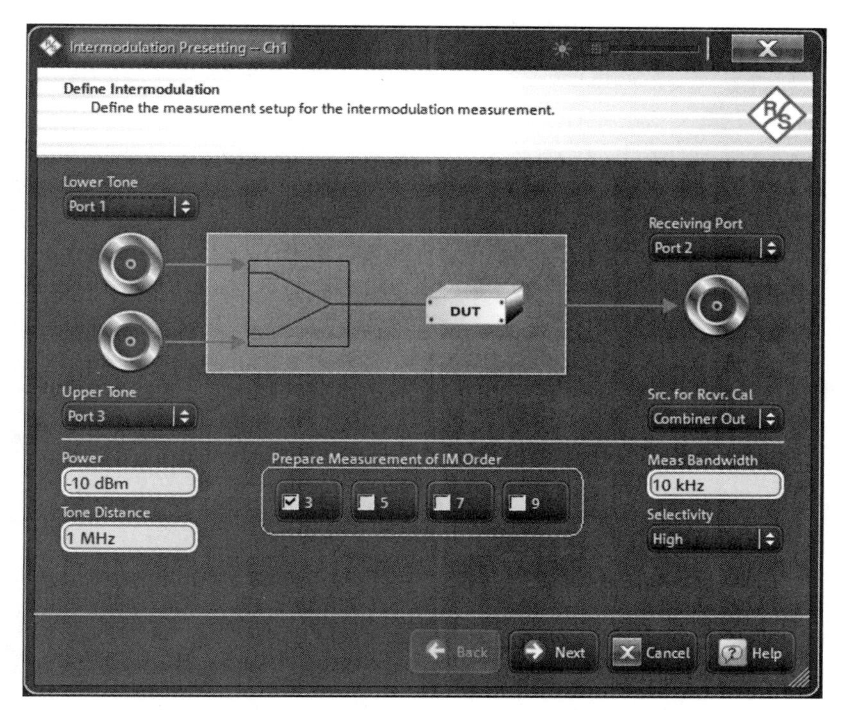

Figure 5.5 ZNB Intermodulation Wizard dialog for IMD measurements.

lower tone output (LTO) power, and add other measurements incrementally to ensure that everything is working correctly. Click "Next."

6. Select "Continue with a Power Calibration." Click "Next."
7. Change the "Max Iterations" to 10 and the "Tolerance" to 0.1 dB. Click "Next."
8. Click "Cal Power" at the top right of power calibration dialog and, if you are using a resistive power combiner with 6-dB loss, apply a "Cal Power Offset" (representing the power splitter) of –6 dB and apply a "Port Power Offset" of +6 dB to compensate. This will boost the port power at the start of the power calibration by 6 dB, allowing it to converge more quickly to the desired value of –10 dBm at the DUT input. Click "Close."
9. With a power sensor connected at the VNA port 1 test plane, click "Start Cal Sweep." This will calibrate the lower tone. Click "Next."
10. After completing the lower tone sweep, click "Start Cal Sweep" to power calibrate the upper tone. Click "Next."

11. Connect a thru at the DUT test plane and click "Start Cal Sweep" to power calibrate the VNA port 2 receivers.
12. At the completion of the power calibration process, connect the DUT and apply DC bias. Check the sweep of the input and output power. They should both be flat. Add markers and verify that LTI (lower tone at input) has a value of –10 dBm and LTO (lower tone at output) is separated from LTI by the gain of the amplifier.
13. Now add two new traces (Trace > Trace Config > Add Trace) and change the measurement quantities to UTI and UTO (Trace > Meas > Intermodulation Distortion > "Upper Tone at DUT In" and "Upper Tone at DUT Out." Note that UTI should overly LTI, and UTO should overlay LTO.
14. Add three new traces to the screen and drag them to a new diagram area. Change the measurements to "Lower IM3 Product at DUT Out," "Upper IM3 Product at DUT Out," and "Noise at DUT Out." The first two measurements should overlay each other, and the noise trace should be 10 dB (or so) lower to ensure that you are actually measuring IMD components and not simply bottoming out on the noise floor of the DUT. If you are not seeing IMD values higher than the noise floor, raise the input power level ("Channel" > "Power Bw Avg") by a few decibels until you do.
15. Finally add two new traces in a new display area and, from the "More Intercept points" menu, select the "Upper" third-order intercept point, which calculates the IP3 based on the levels of the upper tone and upper third-order products. Similarly, for the other trace, select the "Lower" third-order intercept point. You can also add a third trace for the "Major" tone, which denotes the lower or upper intercept point, whichever is smaller. (The "Major" intercept point reveals the worst-case performance of the DUT.)
16. For completeness, the ZNB also has a CW mode, which allows the user to view the IM3 as it would appear on a spectrum analyzer. Here, it applies two fixed tones and sweeps the receiver across a frequency band that includes the associated IMD products. Channel > Channel Config > Intermodulation > Add CW Mode..." Drag the resulting trace to a new diagram area.

5.2 High-Power Amplifier Measurements

From Table 5.1, it is clear that high-power amplifiers share many measurements in common with general-purpose amplifiers. The biggest difference lies in the power level, which requires an external test set comprising high-power components. Figure 5.6 provides a block diagram of a typical test set for high-power applications. Here, the DUT is assumed to be a 100-W amplifier with 13-dB power gain.

In this block diagram, power levels are presented at different points in the test setup. Couplers and attenuators are sized to provide slightly less than −10 dBm to the VNA receivers to avoid compression of the VNA receivers using direct access (Table 5.3). All external test set components have conservative power ratings. For example, since the impedance of a termination may shift as a function of temperature, a 500-W 50Ω termination is used to provide plenty of thermal margin.

This setup includes components for two-tone testing (shown inside the dotted line in the lower left corner). Isolators are included to prevent the signals from one driver amp from mixing in the output stage of the other driver amp, creating their own intermodulation distortion products. If two-tone testing is not required, the components in this box can be eliminated.

The driver amp on port 2 ensures that measurements made in the reverse direction have sufficient SNR to provide a noise-free calibration and stable measurement results. Keep in mind that gain measurements in the forward direction (S_{21}) use error correction derived from both forward (S_{11}, S_{21}) and reverse (S_{12}, S_{22}) measurements. The power level at the port-2 reference receiver may seem low for reverse drive (–46 dBm). This attenuator value was selected to prevent receiver compression (or damage) in the forward direction for the situation where the load might become disconnected or damaged (high reflection condition) during high-power testing.

One of the unspoken keys to full two-port, high-power measurements is direct access to the VNA measurement and reference receivers at each port. This capability is provided as a hardware option on some VNAs (particularly higher-performance models). However, there are instances where full two-port measurement capability is not required. This is particularly true when measuring amplifiers with high reverse isolation (e.g., S_{12} is at least 30 dB lower than S_{21}). In this case, only measurements in the forward direction (S_{11}, S_{21}) are required and a one-path, two-port calibration is sufficient. A mid-range analyzer without direct receiver access may be used for this type of high-power measurement, if its ports can be reconfigured (functionally remapped) by the firmware. For example, Figure 5.7 shows the dialog for redefining the physical ports of a four-port ZNB for use with an external high-power test set.

5.2.1 Power Considerations for High-Power Calibration

When dealing with high-power measurements, it is best to proceed methodically to avoid damaging components and expensive test equipment. Besides

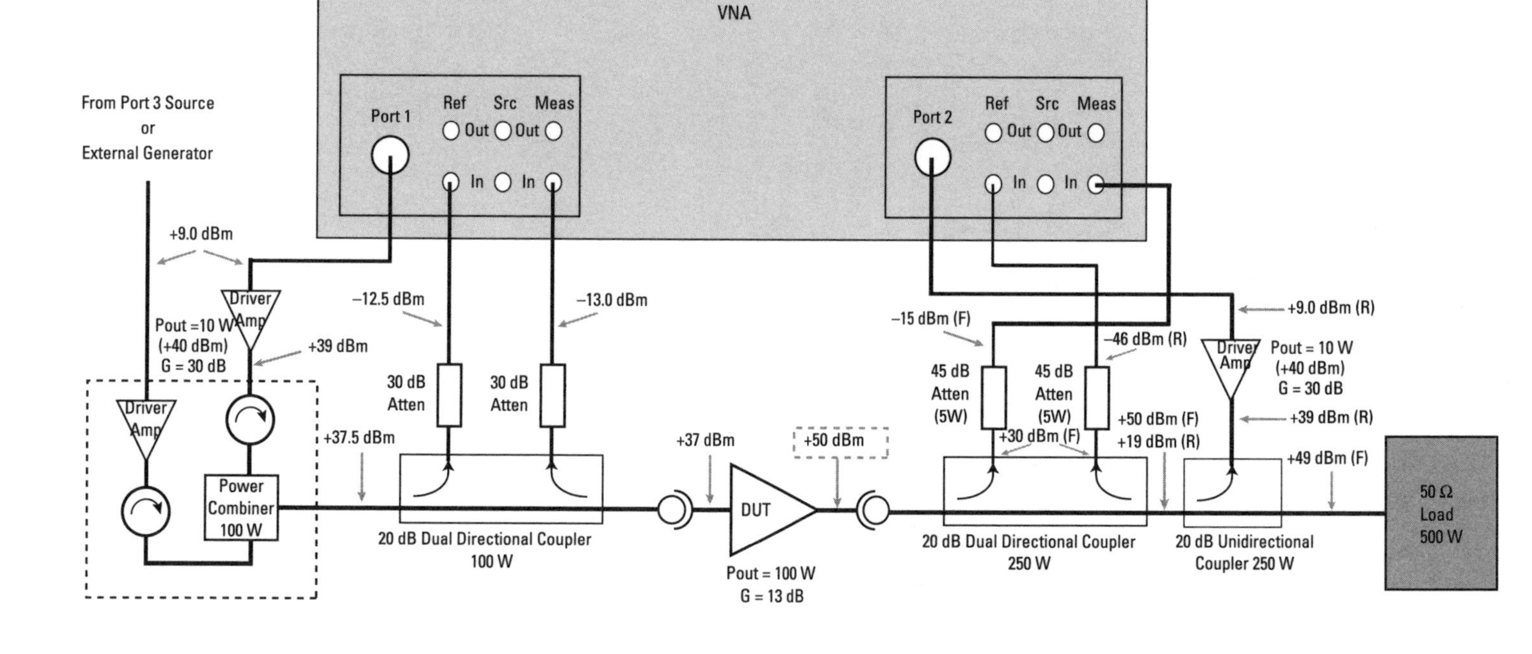

Figure 5.6 High-power amplifier measurement setup.

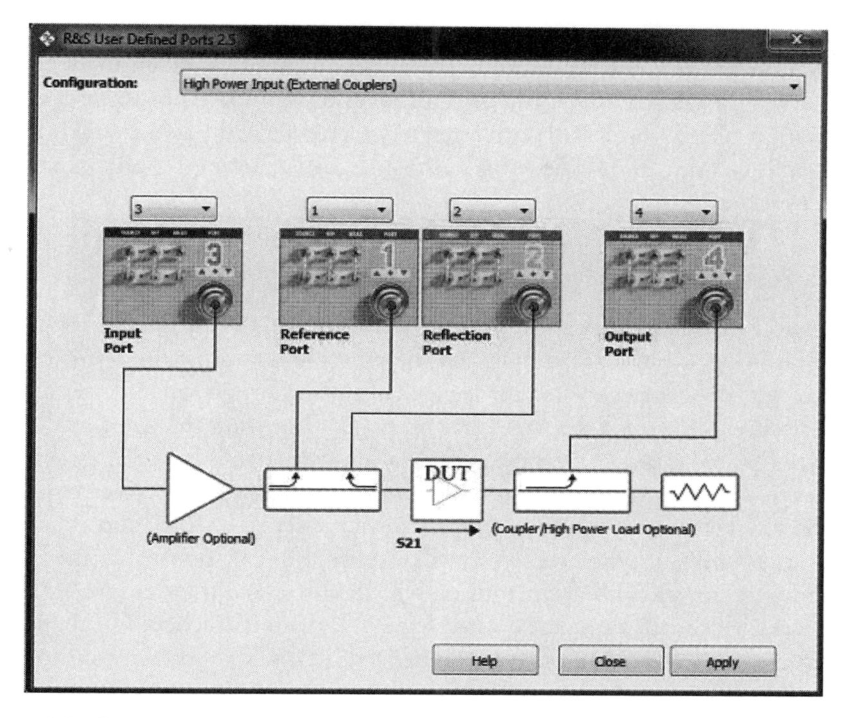

Figure 5.7 Port redefinition using the ZNB VNA.

carefully mapping out power levels throughout the system for normally expected operation (Figure 5.6), care must also be taken to map out a successful calibration strategy. Power ratings of manual calibration kits cannot be ignored. While the full reflection aspect of open and short standards provides some inherent protection against accidental high-power stimulus, the match standard can easily be destroyed. To prevent this, make sure that the power level at the test plane is limited to no more than +10 dBm. (Some VNAs allow the user to limit the source power at specific ports.) In high-power situations where such little power at the test plane causes unacceptably low SNR at the receivers, one could instead substitute a well-matched high-power load for the calibration match standard (assuming the calibration kit was using modeled data, not measured data), or change to a calibration technique that avoids a match standard altogether, like TRL (see Section 2.4.4.3.)

Similarly, power sensors with appropriate high-power attenuators should be used for power calibration. For best SNR, power calibration should be performed with port 1 power levels set to the value required to drive the DUT to maximum output. (In the example presented in Figure 5.7, this would be +37 dBm.) This level ensures the VNA port 1 reference receiver will see good SNR. Avoid the temptation of performing the power calibration at a low level

(0 dBm) and using a handy low-power sensor. Inevitably, at some point in the calibration process, the power will accidentally be set to a much higher level, destroying the sensor. (Even if a user manages to maintain a 0-dB level at the calibration plane, the VNA port-1 reference receiver will have a much lower SNR and require more averaging/narrow IF BW filtering to obtain a good power calibration.)

5.2.2 Power Calibration for High-Power Applications

Whenever inserting an external network between the VNA port and DUT test plane, it is a good idea to compensate the power calibration with the expected gain or loss of that network. This is particularly important when driver amplifiers are involved, since the VNA has no way of knowing the associated gains or losses prior to an initial power calibration sweep. (This initial sweep is a reference sweep only and is used to calibrate the reference receiver based on the power sensor readings.) If the VNA power is set to 0 dBm, and the driver amp has 30-dB gain, the first sweep will result in 30 dBm applied to the power sensor. The power calibration routine will iteratively adjust the source power on successive sweeps until the desired level of 0 dBm is reached, but the initial sweep can apply far more power than desired. (This is another reason to map power levels with a block diagram first to ensure that, even under maximum excitation, power levels experienced by every component in the high-power setup, including the VNA receivers, will remain within acceptable limits.) By employing a power calibration offset, we can tell the VNA what power level to expect, and set its port level accordingly. The required steps for the ZNB network analyzer are presented below. (These steps are very similar for other R&S analyzers as well.)

1. Set up sweep parameters (start and stop frequencies and number of points) and power level.
2. From "Channel" > "Cal" > "Power Cal," set up the tolerance and maximum number of iterations as desired. (Here, we choose 10 iterations and +/–0.1 dB tolerance.)
3. Click "Cal Power…" and on the dialog under the picture of the amp, enter the expected gain of the system (e.g., between the VNA port and the DUT test plane). In this example, assuming a driver amplifier with 30 dB (max) gain, enter 30 dB. You can see that the power result is given as "30 dBm" when the Channel Base Power (Pb) is 0 dBm (Figure 5.8(a)).
4. Set the port power offset to eliminate the difference between the source power level and the power level at the DUT. In practice, this simply

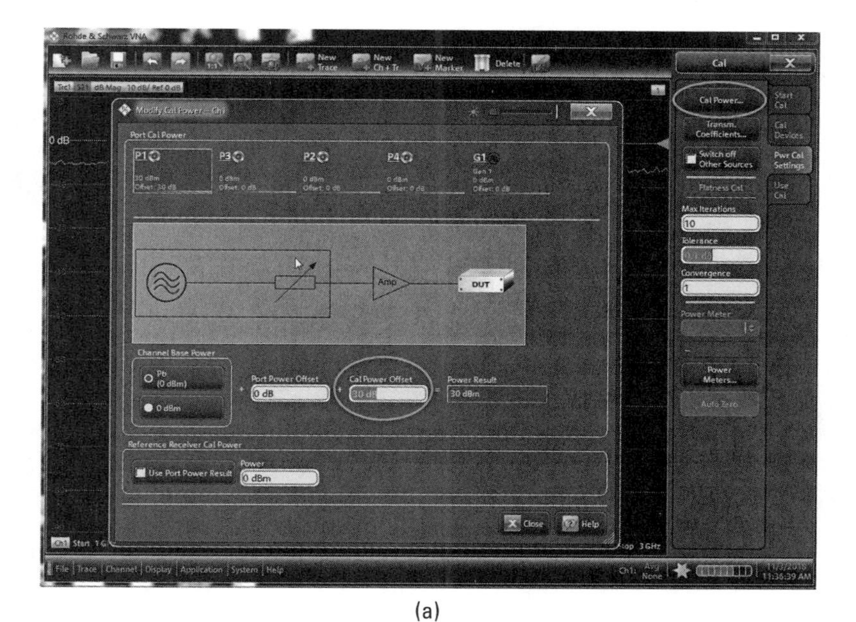

(a)

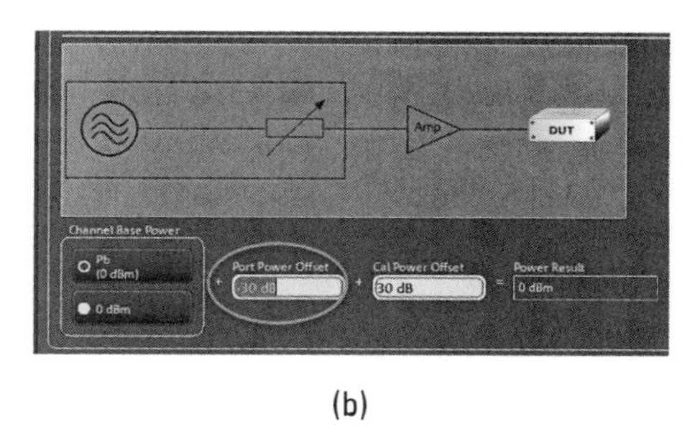

(b)

Figure 5.8 (a) Calibration power offset applied for a 30-dB gain driver amplifier. (b) Port power offset applied for a 30-dB gain driver amplifier.

means applying the opposite of the amplifier gain value (–30 dB) into the port power offset box as shown in Figure 5.8(b). By taking this step, you no longer have to mentally keep track of the driver amp gain. Instead, the base power (Pb) will approximate the power delivered to the DUT at the beginning of the power calibration process. (After completing the power calibration, Pb will be the power applied to the DUT (within the power calibration tolerance).)

5. Repeat this process for the other VNA generator (if two-tone measurements are required).

5.2.3 Adding an Attenuator to a Low-Power Sensor for High-Power Measurements

Occasionally, one is faced with making high-power measurements with a low-power sensor. Rather than risk destroying the sensor by accidentally applying excessive power, R&S VNAs make it easy to add the frequency response of a high-power attenuator via the power calibration dialog. Thus, a high-power attenuator may be installed to protect a low-power sensor, along with automatically compensating the attenuation variation across the measurement range of interest. For example, consider a frequency sweep from 10 GHz to 26.5 GHz. An available 10-dB high-power attenuator is only rated to 18 GHz and has the frequency characteristics shown in Figure 5.9.

By capturing and saving the attenuator's two-port S-parameters, they can be used for an accurate power calibration up to 26.5 GHz. Below is the process for the ZNA.

1. Set up the basic power calibration per the steps in Section 5.1.2.1. Here, a sweep from 10 GHz to 26.5 GHz will be used, with 100-MHz steps for the power calibration. (The S-parameters for the 10-dB attenuator were obtained every 5 MHz, which is 20× more points than is necessary for this example.)
2. Include any required calibration offsets per Section 5.2.2. Specifically, if the 10-dB attenuator of Figure 5.9 is attached to the power sensor,

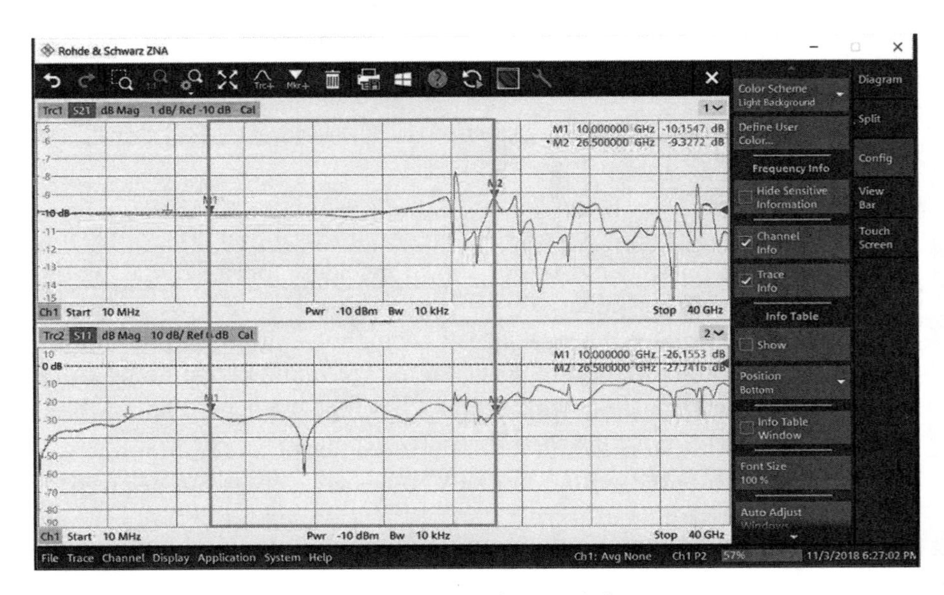

Figure 5.9 The 18-GHz high-power attenuator characteristics.

and a 30-dB driver amp is attached to port 1 of the VNA, then the calibration power offset value should be +30 dB gain – 10 dB loss = +20 dB (net power gain) with a corresponding port power offset of –20 dB. In the following example, I have no driver amplifier, so only the 10-dB attenuator is used, leading to a calibration power offset of –10 dB, and a corresponding port power offset of +10 dB.

3. From "Channel" > "Cal" > "Power Cal," select "Power Cal Settings" tab and "Transm. Coefficients..." In the dialog, select "Two Port at Power Meter" and then click the "Two Port Config..." button (Figure 5.10(a)).
4. Click on the "Import File..." button, select the .s2p file containing the previously-measured attenuator coefficients, click "Open" and then select "Trace S21" and then "OK" (Figure 5.10(b)). Then click "Close" twice.
5. Click on "Power" under the "P1" icon to begin the power calibration process.
6. Repeat the process for other VNA ports requiring power calibration. Then click "Apply" to exit the power calibration dialog.
7. If the measurement receiver calibration (b_2) is desired, remove the 10-dB attenuator/power sensor combination and install a thru adapter at the end of the test cables.[8] Then from "Channel" > "Start Cal" > "Power Cal...", click on the measurement receiver for port 2. Click on "Start Cal Sweep" and then click "Apply" at the end of the calibration sweep to exit the calibration dialog.

5.2.4 Hot S22 Measurements

Depending on an amplifier's design, there can be a considerable difference in its output impedance as a function of drive level. This is particularly important to keep in mind when making S-parameter measurements, because in a classic VNA only one port is driven at a time. Thus, the S_{22} value will always return the amplifier's output match under the condition of no input drive. Hot S_{22} measurements get around this limitation by ensuring there is a signal applied to the input while measuring the output match (S_{22}). The "hot" in Hot S_{22} emphasizes that a drive signal is applied to the DUT input, ensuring that there is considerable power at the DUT output during the measurement. This opens up a number of issues, starting with how to handle this output power. We certainly do not want to expose VNA port 2 to more than about +10 dBm of power;

8. Calibration of the b_2 measurement receiver will be required to accurately measure the power at the output of the DUT.

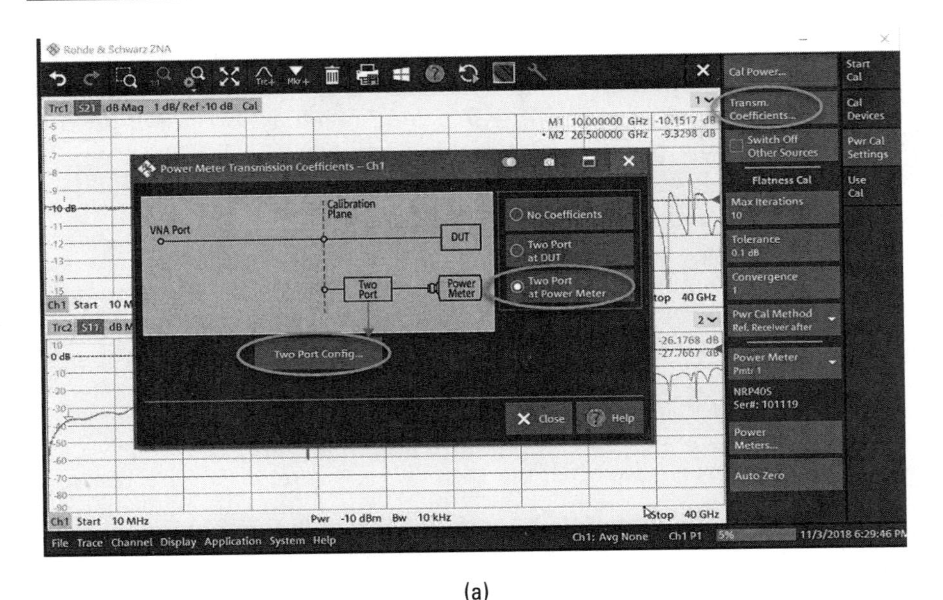

(a)

Power Meter Transmission Coefficients – Ch1

Two Port Configuration – Ch1

	Frequency	Transm. Coefficients
1	10 MHz	-9.48 dB
2	15 MHz	-9.83 dB
3	20 MHz	-9.92 dB
4	25 MHz	-9.99 dB
5	30 MHz	-9.93 dB
6	35 MHz	-9.88 dB

Insert Delete Delete All

Get Trace... Import File... Recall... Save...

Close Help

(b)

Figure 5.10 (a) Power meter transmission coefficients dialog. (b) Dialog for importing 10-dB attenuator S-parameters as transmission coefficients.

+10 to +20 dBm will cause receiver compression, and levels above +27 dBm risk damaging the VNA. So an external test set is recommended. Another issue is how we should handle the hot excitation tone at the amplifier input. We do not want it to be at the same frequency as the S_{22} measurement, because the signal reflected from the amplifier output will be obscured by the much larger amplified input tone, leading to nonsensical results. One choice is to park the port-1 excitation tone somewhere within the operating frequency range of the amplifier. If this approach is selected, the user should choose a frequency that is not on the measurement grid. For example, if the S_{22} measurement sweeps from 100 MHz to 1 GHz in 100-MHz steps, a good candidate would be 550 MHz, midway between two S_{22} measurement points. A different approach would be to allow the excitation tone to track the S_{22} measurement frequencies with a fixed offset (e.g., 10 MHz). The offset should be several times wider than the IF BW to avoid degrading the measurement. However, ensure that the frequency offset does not extend beyond the normal operating range of the amplifier, particularly at the start or end of the sweep.

The amplified excitation signal may be orders of magnitude higher than the stimulus signal applied to the DUT output for the S_{22} measurement. For example, in Figure 5.6, the S_{22} stimulus signal from VNA port 2 applied to the amplifier output connector is +19 dBm. Assuming even a moderate return loss (10 dB), the reflected signal will be +9 dBm, while the hot stimulus signal from port 1, amplified by the gain of the DUT, is +50 dBm: a dynamic range difference of 41 dB. We want to ensure that the larger signal is properly attenuated before entering the port-2 VNA measurement receiver, but not excessively attenuated, because it will also impact the signal reflected from the output of the amp for the S_{22} measurement. In this example, the signal attenuation is 65 dB, so the reflected signal will have a level of only –56 dBm when it enters the receiver. For an amplifier with a better match (20 dB), this signal level drops to –66 dBm, which will require a narrow IF BW or trace averaging to reduce the trace noise due to low SNR.[9]

5.2.4.1 Hot S22 Measurement Procedure

Here are the steps required for setting up a Hot S_{22} measurement on a ZVA. A similar approach is used for the ZNB. We assume a measurement sweep range from 100 MHz to 2.5 GHz with 5-MHz steps.

1. Perform a power calibration at port 1 using the procedure from Section 5.1.2.1. (While the frequency grid for S-parameter measurements will use 5-MHz steps, 50 or 100-MHz steps should be sufficient for the power calibration.) Since we are going to use a +10-MHz offset frequency for the port-1 stimulus signal, extend the power calibration stop frequency to at least 2.51 GHz. For Hot S_{22} measurements, you only need to calibrate the port-1 reference receiver and the generator levels. (Port 2 will strictly be measuring S-parameters.)

9. An alternate approach would place a high-power isolator (or circulator) at the output of the DUT so that high power from the excitation tone would be reduced by the isolator's attenuation. This would also allow the VNA port-2 drive to be applied to the DUT output port without significant attenuation (but the S_{22} reflected signal would still be reduced by the isolator attenuation).

2. Change the stop frequency back to 2.5 GHz, and change the frequency step size to 5 MHz. Perform a one-port OSM S-parameter calibration at port 2 using enough power to obtain good SNR at the port-2 reference and measurement receivers. (If necessary, decrease the IF BW to improve the SNR.)
3. From "Channel" > "Port Config" > "Port Settings," > "Arb Frequency" tab, turn the excitation tone at port 1 "on" by clicking the port-1 checkbox in the "Gen" column and apply a frequency offset of +10 MHz (Figure 5.11). Click "Apply" and "OK."
4. From the "Channel" > "Port Config" > "Port Settings," > "Arbitrary Power" tab, click the "Channel Base Power" and apply a port power offset value that will drive the amplifier to an appropriate output level. (If you apply a port power offset of +5 dB to the Pb value of −10 dBm, the port-1 power stimulus value will become −5 dBm.) Click "Apply" and "OK."
5. Connect the DUT and measure S_{22} (this will be a Hot S_{22} measurement).

5.2.4.2 Hot S22 Measurement Results

Figure 5.12 shows composite measurement results for an amplifier, including power gain, input match, and output match (Hot S_{22}). From the gain plot, it is clear that a stimulus level of −5 dBm is driving the amplifier into compression

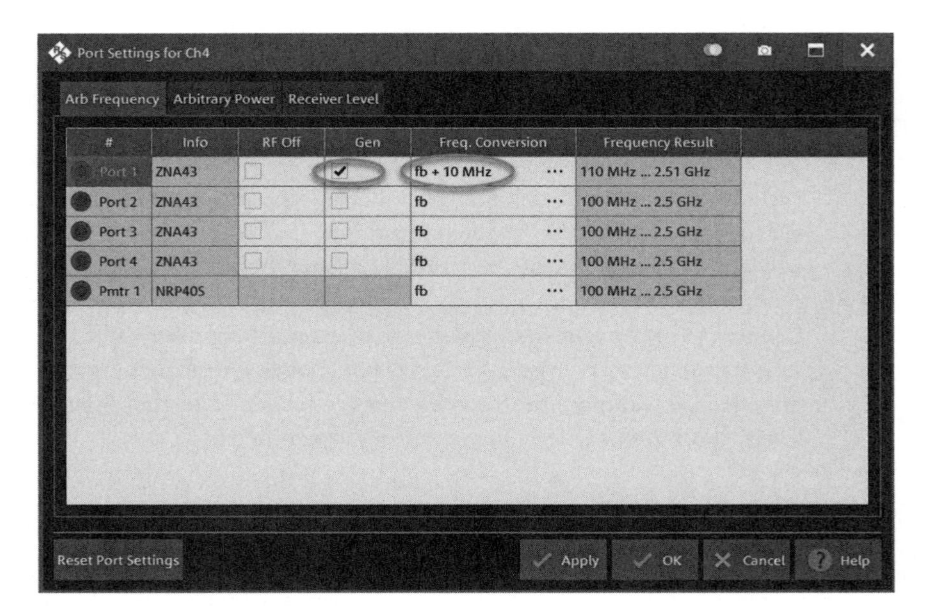

Figure 5.11 Applying a stimulus tone at port 1 for a Hot S22 measurement.

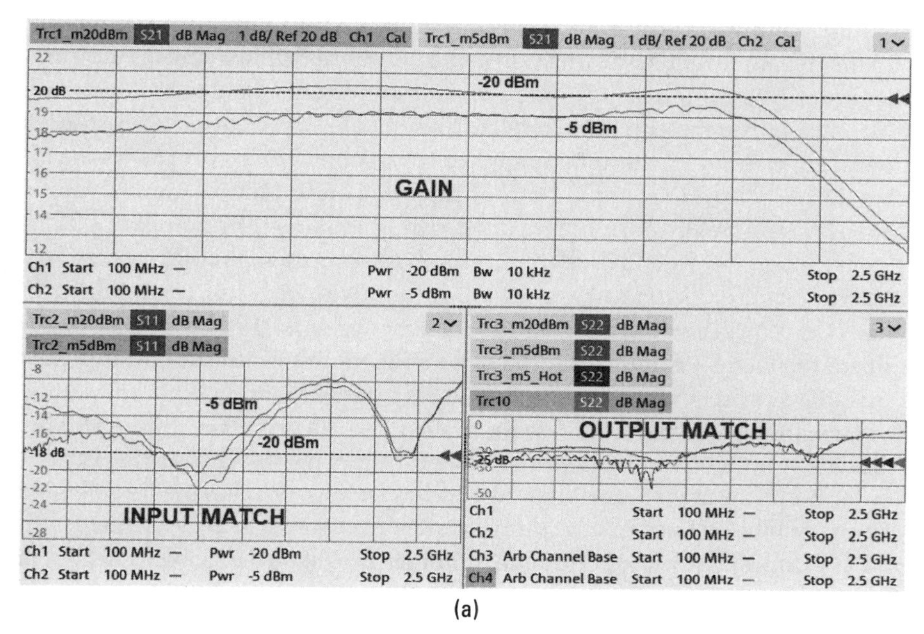

(a)

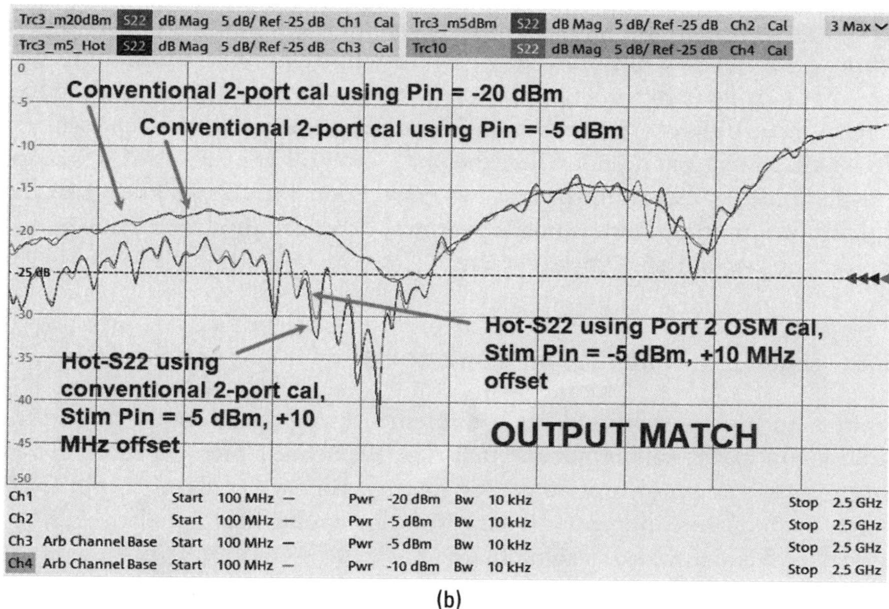

(b)

Figure 5.12 (a) Amplifier measurement results. (b) S22 results using different techniques.

(approximately 1-dB gain compression) over the entire frequency sweep. The input match also changes with the drive level.

In the output match window of Figure 5.12(b), several different measurement techniques are shown. When two conventional S-parameter measurements are made using a full two-port error correction, there is absolutely no difference in the measurement results. This should not come as a surprise; as previously mentioned, stimulus is applied to only one DUT port at a time with conventional S-parameters, so both of these S_{22} measurements were obtained with zero stimulus at port 1. The other two traces show the amplifier output match under Hot S_{22} stimulus conditions, but there are slight differences. This is because one measurement was based on full two-port error correction with port 1 set as a generator with an offset frequency of +10 MHz. While this is fine for the Hot S_{22} stimulus, it produces erroneous results because the two-port S-parameter system error correction requires measurements at both ports using the same frequency. Here, because we applied a +10-MHz offset for the Hot S_{22} stimulus, it is going to use this same offset for the S-parameter error correction. Under these conditions, the forward and reverse thru measurements will yield essentially infinite loss (because source and receiver are at different frequencies), and any directly associated error correction terms will be negligible. The fourth trace was produced using the proper procedure (presented above), where only an OSM calibration was performed at port 2, and the frequency offset stimulus tone was applied at port 1. This is a critical point and reveals that full two-port error correction does not necessarily give the best results in all cases.

The astute reader will note that the power levels for the two Hot S_{22} traces appear to be different. Keep in mind that these displayed values represent Pb (the base power) and do not reflect the port power offset of +5 dB that was applied at port 1. Also, from the first two conventional traces, we know that the DUT's output impedance is unaffected by the comparatively miniscule port-2 power applied for the s_{22} measurement.

5.3 High-Gain Amplifier Measurements

Similar to high-power amplifier measurements, high-gain amplifiers present challenging power calibration requirements. We define a high-gain amplifier as one having a power gain of 60 dB (or greater). For these devices, output levels are easy to measure, but input levels prove challenging because relatively small signal levels are involved, leading to low SNRs.

Figure 5.13 shows an example. Here, we measure a thru adapter using full two-port calibration. Results are shown in the left column using −10 dBm at VNA port 1, which is the same power level used for calibration. In th middle column, port-1 power is reduced to −70 dBm. Notice that S_{11} and S_{21} are both heavily corrupted by noise. S_{22} and S_{12} show more noise as well, since the two-port error correction routine uses a complete set of four measurement

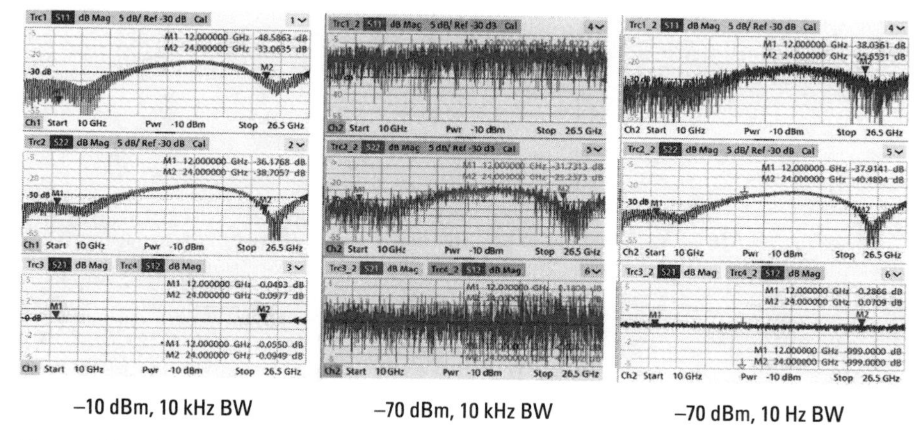

Figure 5.13 S-parameter results for high-level stimulus (−10 dBm), low-level stimulus (−70 dBm), and low-level stimulus with 10-Hz BW.

quantities (raw values of S_{11}, S_{12}, S_{21}, S_{22}) to deliver error-corrected S-parameter results. This low SNR problem can be mitigated by reducing the IF BW from 10 kHz to 10 Hz, as shown in the right column. However, the user pays a large penalty in sweep time, which increases from 720 ms to 1 minute and 6 seconds for 3,301 points.

5.3.1 Attenuator Position Versus Coupler Design

There are other ways of improving S-parameter measurements for cases where low generator levels are required. One important factor is the location of the generator attenuator within the VNA test set. Figure 5.14(a) shows the block diagram of an R&S high-end VNA test set. The source (or generator) attenuator is located before the coupler. This implies that both the reference and measurement receivers will be equally impacted by changes in attenuation level, and hence the S-parameter correction will be maintained. However, it has the disadvantage that both the reference receiver and measurement receiver will be starved for signal at low generator levels. Figure 5.14(b) shows the block diagram essentials of an alternative approach where the source attenuator is placed between the forward and reverse coupler arms so that the reference receiver always sees a healthy SNR, regardless of source attenuation. Its chief disadvantage is that changing the generator attenuation affects the signal level received by the measurement receiver, necessitating a new S-parameter calibration each time the attenuation is changed (or even momentarily changed). In the present example of a high-gain amplifier, this latter approach (Figure 5.14(b)) would nevertheless be favored for its ability to deliver a good SNR at the reference receiver. Fortunately, a VNA of the former variety (Figure 5.14(a)) equipped

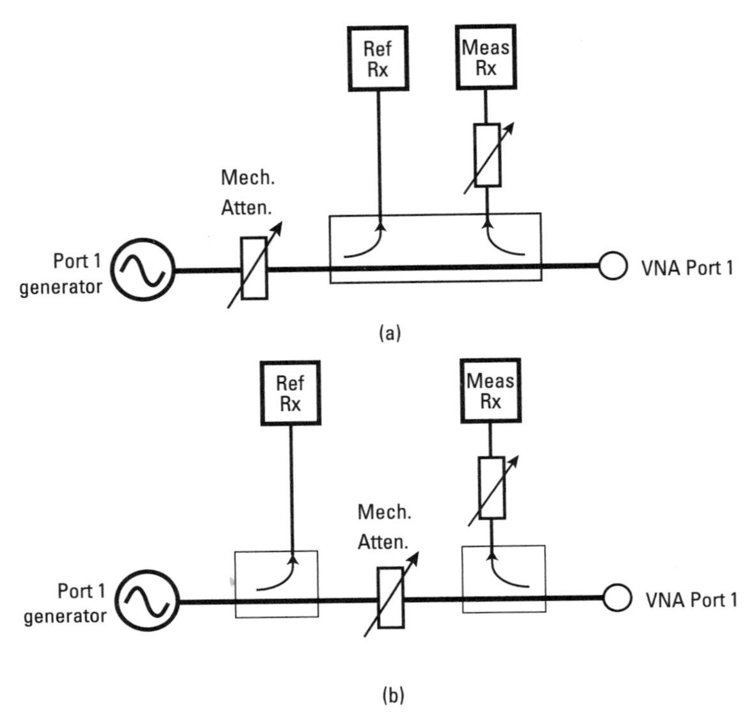

Figure 5.14 Different VNA approaches for location of generator mechanical attenuator. (a) Reference receiver located after source attenuator and (b) reference receiver located before source attenuator.

with direct generator/receiver access can be configured to provide high SNR at the reference receiver by using an external 3-dB splitter and 60-dB attenuator (Figure 5.14(c)).

When making S-parameter measurements with low generator levels, some compromise is necessary in the quest for low trace noise and acceptable measurement time. This involves trade-offs between the number of measurement points required, power levels, bandwidth, VNA test set configuration, calibration method, and trace averaging.

5.3.2 Power Calibration Strategy

Introducing power measurements for high-gain amplifiers is tricky. How do you calibrate the input level (VNA reference receiver and VNA generator) when signal levels are below the range of a typical thermal power sensor? There are several things that can be done to reduce this measurement challenge. The easiest fix would be to exchange a thermal sensor (such as the NRP18T) with a three-path diode sensor (NRP18S). This would instantly improve your minimum power level sensitivity from –35 dBm to –70 dBm.

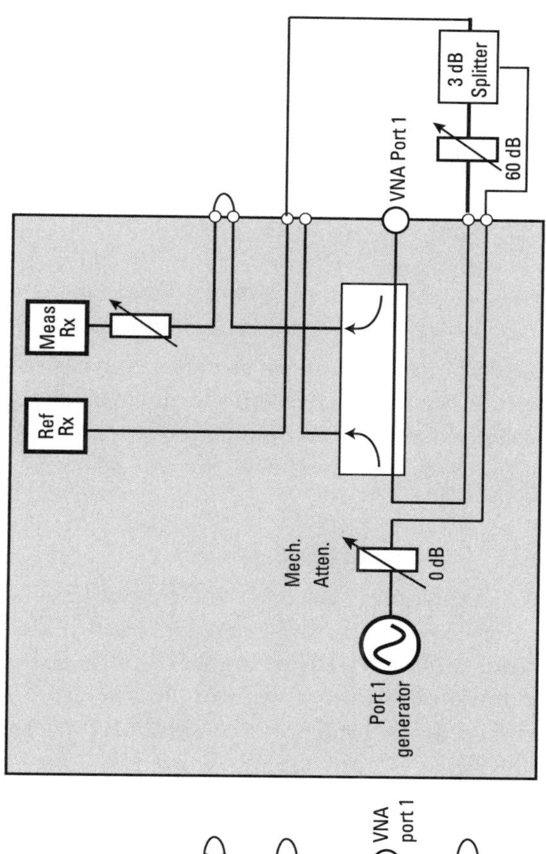

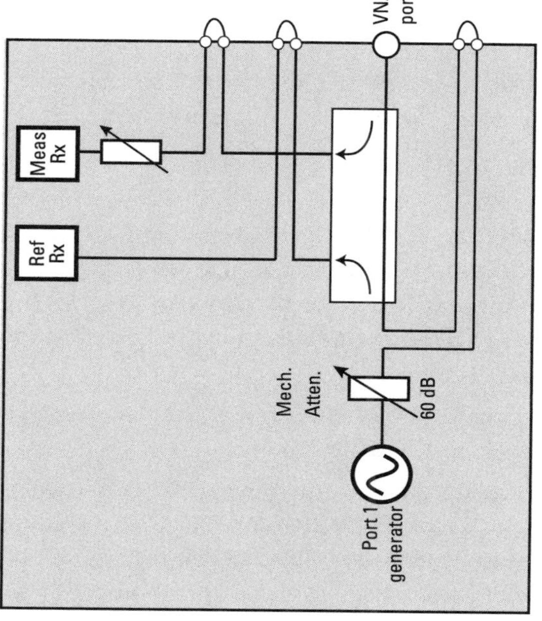

(c)

Figure 5.14 (continued)

Using a power sensor with greater sensitivity is only a partial solution at best. This is because a power sensor is a wideband measurement device, and must operate in the presence of thermal noise that occupies its entire measurement range. For an 18-GHz sensor, the thermal noise can be calculated from $kTB = (1.38 \times 10^{-23} W/K\text{–}Hz) \times (290K) \times (18 \times 10^{9}\ Hz) = 7.2 \times 10^{-HW} =$ –71 dBm. However, a VNA receiver with a 10-kHz BW has a noise floor of –134 dBm, coupled with a measurement dynamic range of greater than 130 dB. Further, VNA receivers maintain 0.1-dB magnitude and 1-degree phase accuracy over a measurement range of at least 60 dB. Given this performance, it becomes immediately clear that a successful power calibration strategy should take advantage of the power sensor's accuracy at higher signal levels and then transfer that accuracy to the VNA reference and measurement receivers. Once power-calibrated, the reference receivers can be used to accurately set the VNA source at the extremely low-power levels required for high-gain amplifier measurements.

5.3.2.1 Power Calibration Procedure for Low Signal Levels

For conventional power calibration at relatively high signal levels (–10 to +5 dBm), both the reference receiver and generator are calibrated at the same level. For the case of a high-gain amplifier, we need to break this procedure into two parts. Part 1 transfers the accuracy of the power sensor to the VNA receiver at a relatively high signal level (to assure good power sensor SNR for accurate power calibration). Part 2 enables the VNA generator to deliver the desired low output power using the calibrated VNA receiver. This procedure is described below for the ZVA (but is similar for the ZNB).

1. Start from a preset. Configure the desired measurement (start/stop frequency, number of points, IF BW). Leave the power at the default value (usually –10 dBm or 0 dBm).
2. Connect the power sensor to the end of the port-1 test cable.
3. From "Channel" > "Cal" > "Power Cal," set "Max Iterations" and "Tolerance" (in this example, 10 and 0.1 dB, respectively), and then click "Power" for port 1 and "Start Cal Sweep" to commence the power calibration. This calibrates both the port-1 reference receiver and the source to –10 dBm.
4. Next disconnect the power sensor from port 1 and use a thru adapter to connect VNA port 1 to VNA port 2.
5. Click on "Meas Receiver" in the power calibration dialog and then click "Start Cal Sweep." This transfers the power sensor accuracy to the port-2 measurement receiver. Click "Apply" to (temporarily) exit the power calibration dialog.

6. Now change the VNA port 1 power to the desired drive level for the high-gain amplifier. (Here, we are using –60 dBm.) Change the IF BW to 100 Hz.
7. From "Channel" > "Cal" > "Power Cal," set the "Pwr Cal Method" to "Ref. Receiver Only." This will ensure that only the previously calibrated reference receiver (step 3) will be used to power calibrate the VNA generator at the low level of –60 dBm (Figure 5.15(a)). You should still have port 1 connected to port 2. Click "Source Flatness" and "Start Cal Sweep." (Figure 5.15(b)). Click "Apply" to exit the power calibration dialog.
8. Create two traces A1 (Src Port 1) and B2 (Src Port 1) and tie the scales together using the Trace Manager. This shows the accuracy of the power calibration at –60 dBm.

5.4 Low Noise Amplifier (LNA) Measurements

The conventional VNA architecture is poorly suited for noise figure measurements. This is why it has long remained an orphan among common amplifier measurements (gain, compression point, Pin, Pout, IMD, harmonics) and prevented the entire suite of required LNA measurements from being performed in a one-touch fashion (e.g., within a single DUT attach/disconnect cycle). To better understand the noise figure challenge, it is worth reviewing basic noise figure theory and relating it to the most common measurement techniques before addressing VNA-based noise figure measurements.

5.4.1 Noise Figure Theory

In a now classic paper, H. T. Friis defined the noise figure of a two-port device as "the ratio of the available signal-to-noise ratio at the signal generator terminals to the available signal-to-noise ratio at its output terminals" [1]. We begin our discussion of noise figure with this expression, but anchor our mathematics in the modern convention of using capital letters to denote noise figure, *F*, and gain *G*, in decibel units and lowercase letters to represent noise factor, *f*, gain, *g*, signal power, *s*, and noise power, *n*.

Assuming that the signal generator has a conjugate impedance match to the input of the DUT, we can write:

$$f = \frac{\left(s_{in}/n_{in}\right)}{\left(s_{out}/n_{out}\right)} \tag{5.11}$$

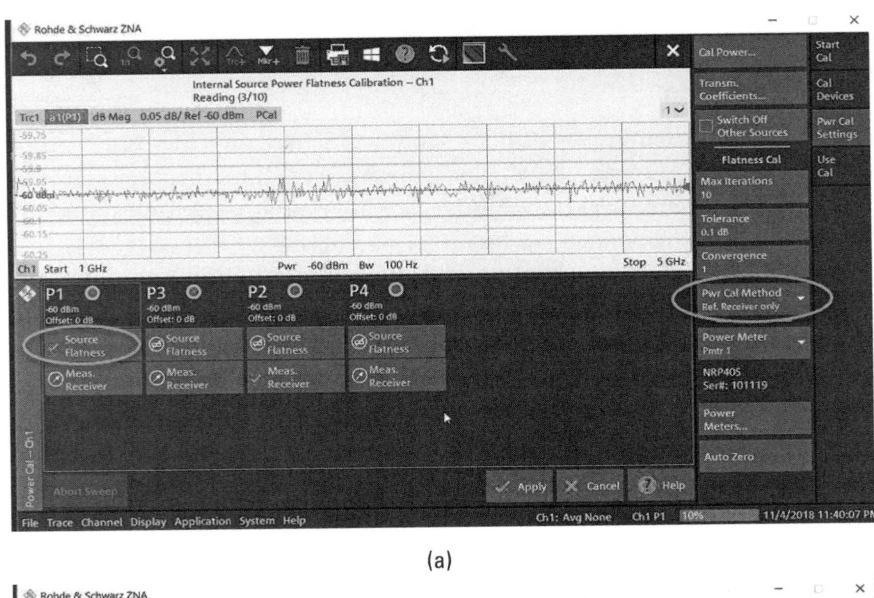

(a)

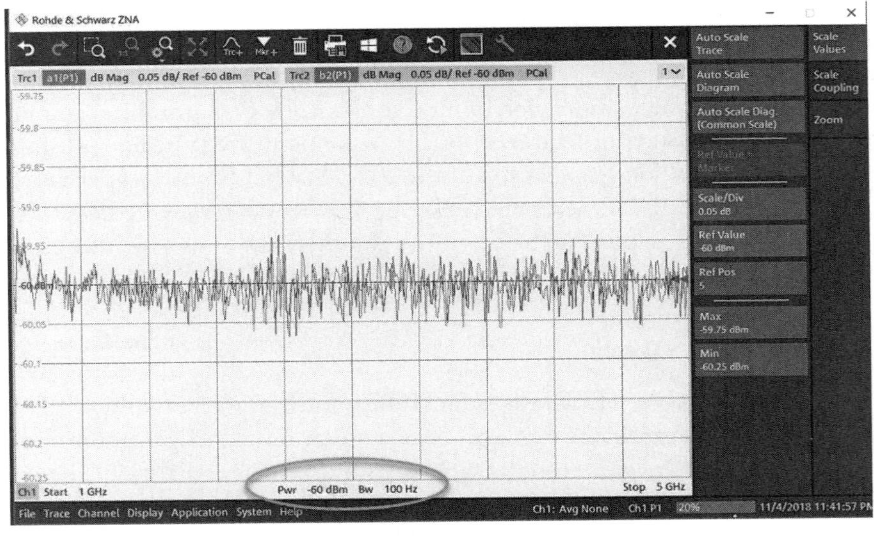

(b)

Figure 5.15 (a) Power calibrating the source at –60 dBm using only the calibrated VNA reference receiver. (b) Power calibration verification (–60-dBm VNA source level).

Figure 5.16 shows the signal and noise levels at the input and output of an amplifier. The In graph depicts the signal level entering the device, shown with the background (thermal) noise floor. The Out graph depicts the signal and noise levels at the device output. The signal level has increased by the gain, g, of the device, but the overall SNR at the output is actually smaller than at the

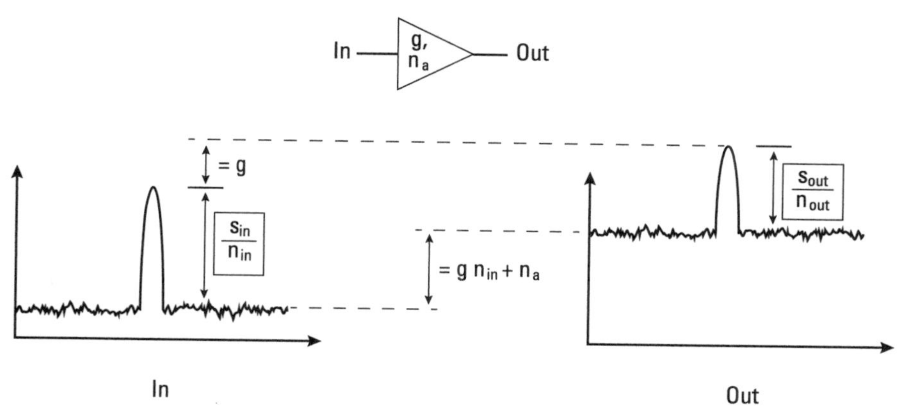

Figure 5.16 Amplifier example for definition of noise factor.

input. This is because the noise level has not only increased due to the amplifier's gain ($g \cdot n_{in}$), but the amplifier adds some of its own noise (n_a). Equation (5.11) can be rewritten, yielding:

$$f = \frac{\left(\dfrac{s_{in}}{n_{in}}\right)}{\left(\dfrac{g \cdot s_{in}}{g \cdot n_{in} + n_a}\right)} = 1 + \frac{n_a}{g \cdot n_{in}} \tag{5.12}$$

It is clear that the noise factor f can be reduced by reducing the additive noise (n_a), increasing the device gain (g), or increasing the noise at the input (n_{in}). Noise figure (F) is simply the noise factor (f) depicted in logarithmic units:

$$F = 10\log_{10}(f) \tag{5.13}$$

A companion term to noise power is noise temperature, since a physically hot device will generate more thermal noise than a cold device. The equation governing this relationship is given by:

$$n = kTB \tag{5.14}$$

where n = noise power (watts), k = Boltzmann's constant (2.38×10^{-23} W/K-Hz), B = bandwidth (Hz), and T = temperature (Kelvin).

If we normalize noise power to a 1-Hz bandwidth, then we can comfortably discuss thermal noise in terms of either power or temperature. A simple proportionality constant (Boltzmann's constant) allows us to switch back and forth between units of power (watts) and temperature (Kelvin). In particular, a noise temperature will always map to a unique noise power. Additionally, noise figure is related to a device's equivalent noise temperature via the following equation:

$$F(dB) = 10\log_{10}\left(\frac{t_{dut}}{t_o} + 1\right) \tag{5.15}$$

where F(dB) noise figure (in decibels), t_{dut} = equivalent noise temperature of the DUT (in Kelvins), and t_0 = 290K.

Finally, by linearizing (5.15), we obtain noise factor:

$$f = \frac{t_{dut}}{t_0} + 1 \tag{5.16}$$

Noise factor and noise temperature are important concepts for the remainder of the noise figure discussion.

While Figure 5.16 provides a good conceptual picture, it is of little practical value. If you have ever tried measuring noise figure this way with a spectrum analyzer, you are likely to find that SNR_{out} is greater than SNR_{in}. The reason is that this picture fails to take into account the noise contribution of the (nonideal) spectrum analyzer itself. A more accurate representation is shown in Figure 5.17. Every noise figure measurement technique must grapple with this problem and apply methods to remove the influence of the measurement equipment from the desired DUT noise figure result.

5.4.2 Y-Factor Method

In the Y-factor method, a noise source is used that consists of a special diode that produces significant broadband noise when it is biased on and a much smaller amount of noise when it is biased off. The amount of noise power produced in the on state relates to an equivalent amount of noise generated by a conductor heated to a high temperature t_h. The noise generated in the off state relates to the equivalent amount of noise produced by the conductor at ambient (frequently, room) temperature t_c. The difference in noise powers between the on and off diode states (or t_d^{on} and t_d^{off}) is the excess noise ratio (ENR) and is generally expressed (in decibels) as:

$$ENR_{dB} = 10\log_{10}\left(\frac{T_d^{on} - T_d^{off}}{T_0}\right) \tag{5.17}$$

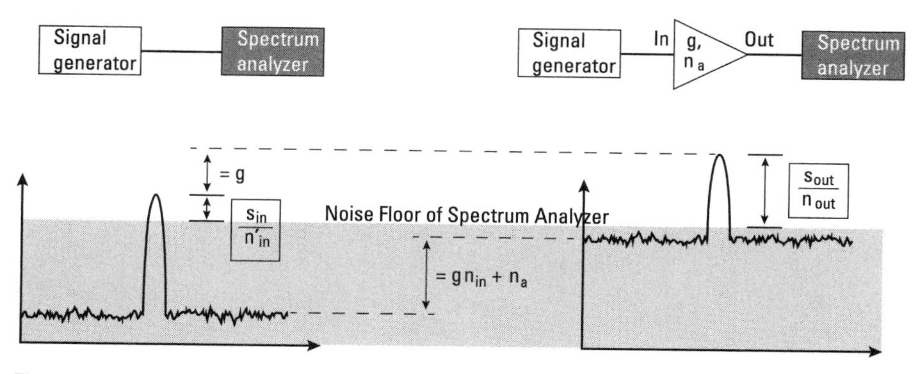

Figure 5.17 Impact of spectrum analyzer's internal noise on SNR measurement.

By knowing the ENR for the on and off states of a diode connected to the input of an amplifier, and measuring the corresponding ratio of output powers, the noise figure can be calculated. The measurement concept is illustrated in Figure 5.18. Notice that the general form of the equations for t_n^{on} and t_n^{off} are simple linear equations where the slope is the amplifier gain, g, and the y-intercept represents the noise added by the amplifier (expressed as a noise temperature, t_a). By using two different noise diode temperatures t_d^{on} and t_d^{off}, we obtain two output values (t_n^{on} and t_n^{off}) and hence two equations that allow us to solve for the two unknowns g and t_a.

The process is as follows, and revolves around the mysterious Y-factor term, which is simply a measured power ratio expressed as:

$$Y(dB) = 10\log_{10}\left(\frac{t_n^{on}}{t_n^{off}}\right) \tag{5.18}$$

In linear terms, this becomes:

$$y(linear) = \left(\frac{t_n^{on}}{t_n^{off}}\right) = \frac{\left(g \cdot t_d^{on} + t_a\right)}{\left(g \cdot t_d^{off} + t_a\right)} \tag{5.19}$$

Solving for t_a, we get:

$$t_a = \frac{g\left(t_d^{on} - yt_d^{off}\right)}{(y-1)} \tag{5.20}$$

To determine the gain, we simply need to determine the slope of the line:

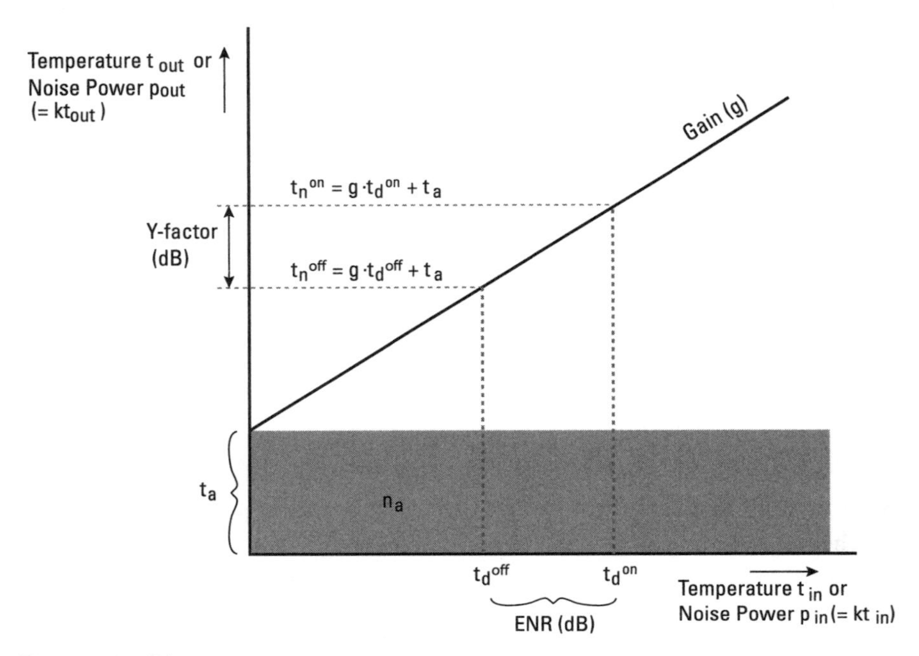

Figure 5.18 Y-factor concept.

$$slope = \frac{\Delta y\ axis}{\Delta x\ axis} = \frac{t_n^{on} - t_n^{off}}{t_d^{on} - t_d^{off}} = \frac{\left(g \cdot t_d^{on} + t_a\right) - \left(g \cdot t_d^{off} + t_a\right)}{t_d^{on} - t_d^{off}} = g \quad (5.21)$$

Armed with the measured Y-factor and the calculated value of g from (5.21), we can easily calculate t_a from (5.20) and determine the noise figure of the amplifier from (5.15).

The actual Y-factor noise figure calculation is more complicated because we cannot directly observe the true diode on and off noise power, because the instrument we are using contributes its own noise to the measurement. The situation is depicted in Figure 5.19.

The measuring device (depicted here as a spectrum analyzer) has a considerable noise contribution t_{sa}. Direct application of the Y-factor method to the measurement step yields the noise of the amplifier plus the spectrum analyzer. An additional calibration step is required to calculate and remove the spectrum analyzer contribution.

5.4.2.1 Calculating Noise Figure Without Knowing Absolute Power

Before jumping into the two step Y-factor measurement process, it is worth noting that the graphical analysis presented in Section 5.4.2 assumed the ability to measure absolute noise power, rather than just the ratio of powers to determine

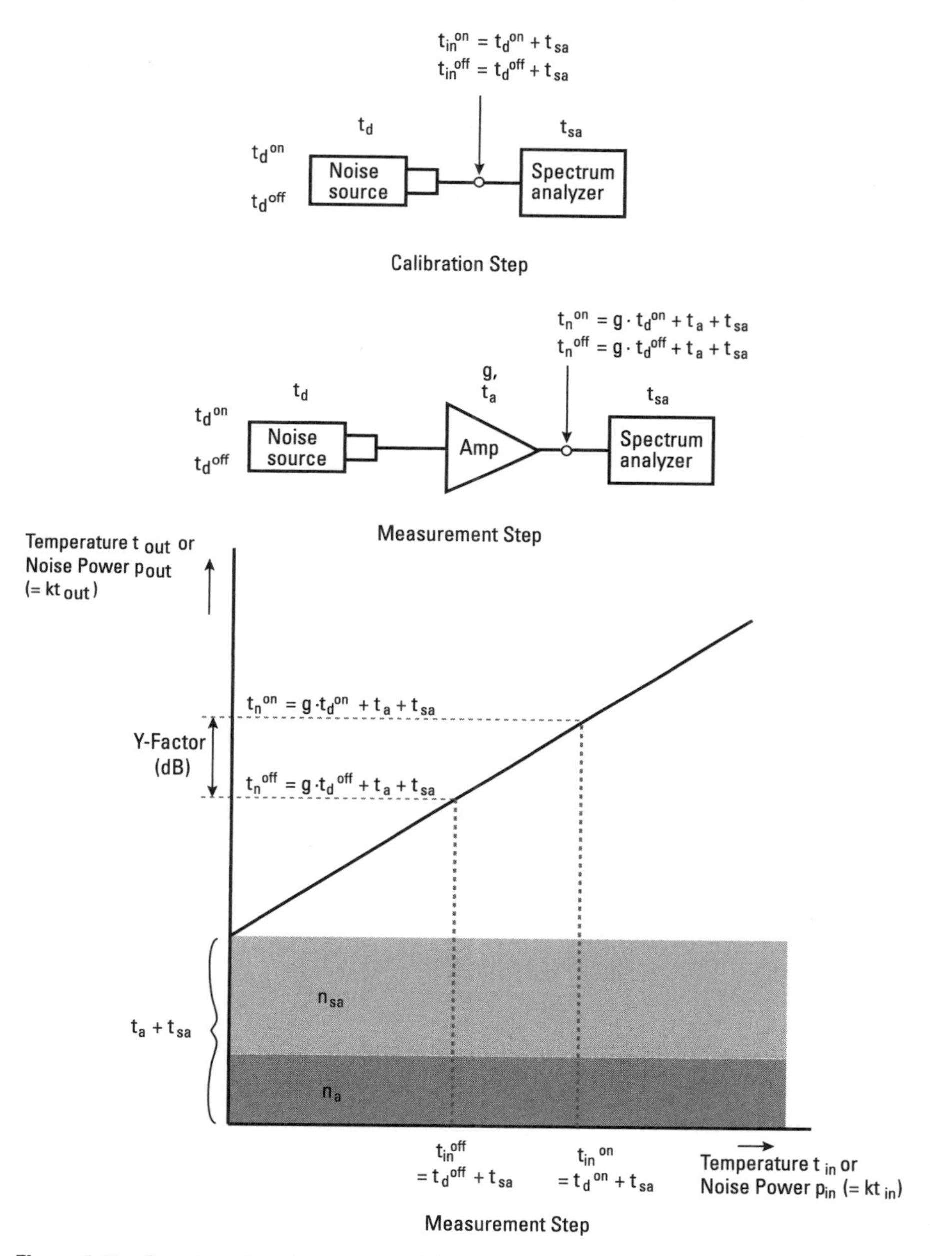

Figure 5.19 Complete description of the Y-factor method.

noise figure. This made the analysis easier to understand. However, the true beauty of the Y-factor method is that you can measure the noise figure of a device without resorting to any absolute power measurements. This is a significant benefit because it removes the need for absolute power calibration; you only need an instrument that can accurately measure relative differences in power levels. This is an important point and the derivation is rarely presented in noise figure literature. We start with the noise factor (5.16). We note that T_{DUT} can be represented using (5.20) (essentially the calibration stage from Figure 5.19):

$$t_{dut} = \frac{\left(t_d^{on} - yt_d^{off}\right)}{(y-1)} \tag{5.22}$$

Substituting (5.22) into (5.16) yields:

$$f = \frac{t_d^{on} - yt_d^{off}}{(y-1)t_0} + 1 \tag{5.23}$$

Now rewrite (5.23) with a common denominator:

$$f = \frac{\left(t_d^{on} - yt_d^{off}\right) + (y-1)t_0}{(y-1)t_0} \tag{5.24}$$

Add and subtract a common factor of $t_d^{off} - t_d^{off}$ (= 0) in the numerator:

$$f = \frac{\left(t_d^{on} - yt_d^{off}\right) + (y-1)t_0 + \left(t_d^{off} - t_d^{off}\right)}{(y-1)t_0} \tag{5.25}$$

Regroup terms:

$$f = \frac{\left(t_d^{on} - t_d^{off}\right) + (y-1)t_0 + \left(t_d^{off} - yt_d^{off}\right)}{(y-1)t_0} \tag{5.26}$$

Simplify:

$$f = \frac{\left(t_d^{on} - t_d^{off}\right)}{(y-1)t_0} + 1 + \frac{t_d^{off}(1-y)}{(y-1)t_0} \tag{5.27}$$

Recognizing the expression for excess noise ratio (5.17) contained in the first term and simplifying the last term yields:

$$f = \frac{enr}{(y-1)} + 1 - \frac{t_d^{off}}{t_o} \tag{5.28}$$

If we assume that $t_d^{off} = t_0$, we obtain a simplified expression for noise factor:

$$f = \frac{enr}{(y-1)} \tag{5.29}$$

Now express this as a noise figure by taking the logarithm of both sides (recognizing that division in linear equations becomes subtraction with logarithms):

$$F(dB) = ENR(dB) - 10\log_{10}(y-1) \tag{5.30}$$

To use (5.30), please note that the measured Y-factor (in decibels) must first be converted to linear form:

$$y(linear) = 10^{\left(\frac{Y(dB)}{10}\right)} \tag{5.31}$$

5.4.2.2 The Two-Step Y-Factor Measurement Process

Because any Y-factor measurement will be corrupted by the noise of the measuring instrument itself, we must break the process into two steps: a calibration step and a DUT measurement step. Only in this way can we remove the contribution of the measuring instrument from our device's noise figure. The tool for accomplishing this task is the cascaded noise figure equation [2]:

$$f_{1n} = f_1 + \frac{f_2 - 1}{g_1} + \frac{f_3 - 1}{g_1 \cdot g_2} + \ldots + \frac{f_n - 1}{g_1 \cdots g(n-1)} \tag{5.32}$$

where f_1 and g_1 are the first stage noise factor and linear gain, respectively, f_n is the noise factor of the nth stage, and f_{1n} is the noise figure of the entire cascade (stage 1 through stage n).

With respect to Figure 5.19, there are only two stages, with the amplifier representing Stage 1 and the spectrum analyzer representing Stage 2, and the cascade refers to stage 1 followed by stage 2. We can obtain an equivalent representation in terms of noise temperature by substituting (5.16) into (5.32):

$$t_{12} = t_1 + \frac{t_2}{g_1} \tag{5.33}$$

Calibration Step

In this step, we connect the noise diode directly to the input of the spectrum analyzer (Figure 5.20, "Calibration Step") and measure the diode's on and off noise power ratio (t_{in}^{on}/t_{in}^{off}), yielding the Y-factor (in decibels) for the second stage (spectrum analyzer). Since we know the diode ENR (provided by the manufacturer), we can use (5.29) or (5.30) to calculate the noise factor for the second stage, f_2.

Measurement Step

Next, we insert the amplifier between the noise diode and the spectrum analyzer (Figure 5.20, "Measurement Step") and measure the diode's on and off noise power ratio (t_n^{on}/t_n^{off}), yielding the Y-factor (in decibels) for the cascaded assembly of amplifier plus spectrum analyzer. However, the cascaded noise factor equation also requires us to know the gain of the amplifier, and requires that we know the absolute power values corresponding to the noise temperatures

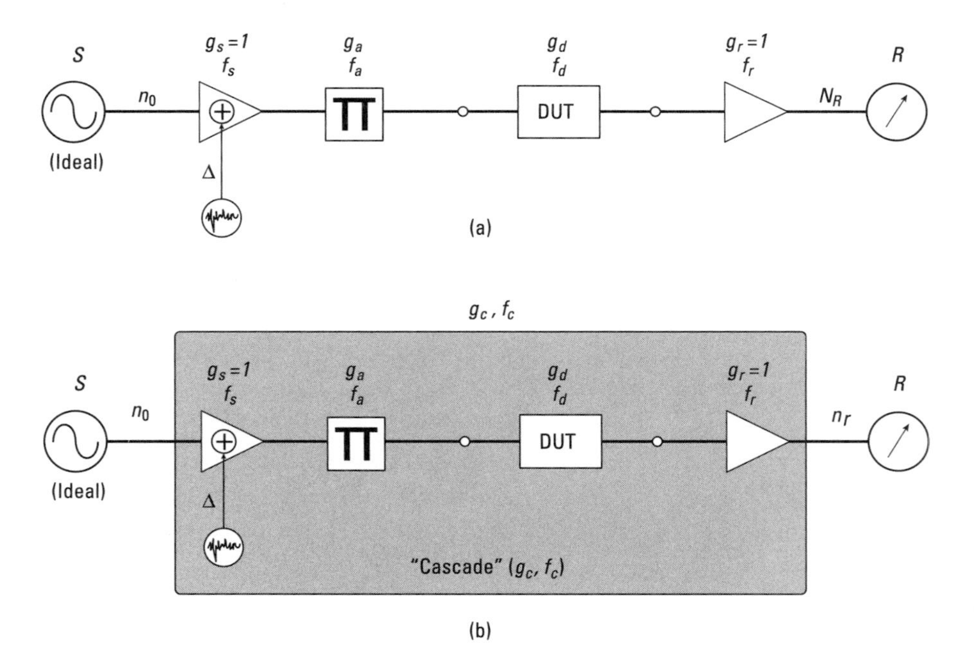

Figure 5.20 (a) The system block diagram for SNR method. (b) A simplified block diagram for SNR method derivation.

t_n^{on} and t_n^{off}. (Curiously, we do not need the power values from the calibration stage, because we can simply calculate t_d^{on} and t_d^{off} from the ENR equation and convert to power via Boltzmann's constant.) You can see this if you examine and simplify the gain calculation from Figure 5.20. In the first step, we show that the slope is actually equal to the amplifier gain (g):

$$slope = \frac{\Delta y}{\Delta x} = \frac{t_n^{on} - t_n^{off}}{t_{in}^{on} - t_{in}^{off}} =$$
$$\frac{\left(g \cdot t_d^{on} + t_a + t_{sa}\right) - \left(g \cdot t_d^{off} + t_a + t_{sa}\right)}{\left(t_d^{on} + t_{sa}\right) - \left(t_d^{off} + t_{sa}\right)} = \frac{g\left(t_d^{on} - t_d^{off}\right)}{\left(t_d^{on} - t_d^{off}\right)} = g \tag{5.34}$$

We also recognize that since we are actually measuring noise power (and not noise temperature) with the spectrum analyzer, we need to convert these noise powers in the numerator (p_n^{on} and p_n^{off}) to noise temperatures by dividing them by k (Boltzmann's constant):

$$g_1 = \frac{\Delta y}{\Delta x} = \frac{p_n^{on} - p_n^{off}}{k\left(t_{in}^{on} - t_{in}^{off}\right)} \tag{5.35}$$

In (5.33), we call this gain g_1 to acknowledge that it is the gain of the first stage (referencing the cascaded noise factor equation).

Now, we can calculate the noise temperature of the amplifier, t_1 by rearranging (5.33):

$$t_1 = t_{12} - \frac{t_2}{g_1} \tag{5.36}$$

Once we have calculated the amplifier's noise temperature, we can convert to noise factor or noise figure using (5.16) or (5.15), respectively.

Rule-of-Thumb Guidelines for Successful Noise Figure Measurements

In order to make noise figure measurements using the Y-factor method, there must be an observable difference in the noise power between the diode off and on states. If your spectrum analyzer or measurement receiver cannot discern a difference, a meaningful noise figure cannot be calculated. To aid in this quest, the following three guidelines will help ensure successful measurements. They address observable deltas between noise power measurements in the calibration step, the measurement step, and between the calibration and measurement step.

- *Guideline 1 (Calibration step):* Ensure a delta between diode on and off states of at least 3 dB for calibration results. This requires a noise source with an ENR at least 3 dB greater than the noise figure of the receiver. Mathematically,

$$ENR(dB) > NF_{RX}(dB) + 3dB \tag{5.37}$$

If this guideline is not met, consider using a noise diode with higher ENR or connecting a low-noise preamp to the input of the measurement receiver.

- *Guideline 2 (Measurement step):* Ensure a delta between diode on and off states of at least 5 dB for measurement results. This requires a noise source with ENR at least 5 dB greater than the noise figure of the DUT. Mathematically,

$$ENR(dB) > NF_{DUT}(dB) + 5dB \tag{5.38}$$

If this guideline is not met, consider using a noise diode with higher ENR.

- *Guideline 3 (Difference between Calibration and Measurement steps):* Ensure that the noise figure of the DUT will be at least 1 dB greater than the noise figure of the measurement receiver alone. The noise entering the DUT will be boosted by the gain of the DUT, so the equation governing this relationship is

$$NF_{DUT}(dB) + G_{DUT}(dB) > NF_{RX}(dB) + 1dB \tag{5.39}$$

If this guideline is not met, consider connecting a low-noise preamp to the input of the measurement receiver.

Meeting all three guidelines will ensure stable and repeatable measurements. If one or more guidelines are not met (but are within a few decibels), reliable measurements can still be achieved, but will likely require more averaging in the calibration and measurement phases to achieve stable and repeatable noise figure results.

5.4.2.3 Noise Figure Difficulties for VNAs

These guidelines help to illustrate two common difficulties with the Y-factor method (in particular) and most noise figure methods (in general). First, stimulus levels (ENR) tend to be quite low and can make the measurements susceptible to extraneous noise or interfering signals. This is one reason why some laboratories choose to make noise figure measurements in RF-shielded chambers. Second, measurement performance is largely dictated by the noise contribution of the measurement receiver. That is why noise figure measurements on spectrum analyzers commonly use preamplifiers, which reduce their noise figure from the 18 to 30-dB range down to 5 to 7 dB. What about using a VNA for Y-factor measurements? The basic VNA receiver has a noise figure in the 20–30-dB range (similar to a spectrum analyzer), but has the additional disadvantage of a 10–13-dB coupler loss directly in front of the receiver, resulting in a noise figure in the 30–40-dB range. Even a casual inspection of the three guidelines presented above suggests a standard VNA will be hard-pressed to make noise figure measurements using the Y-factor method. To get around this problem, some VNA manufacturers provide a low noise receiver path used specifically for Y-factor noise figure measurements. Alternatively, new noise figure measurement approaches have been adopted specifically for use with VNAs.

5.4.3 Signal-to-Noise Method

In the signal-to-noise method, no noise diodes (or impedance tuners) are required. Instead, this method uses a CW signal from the VNA source and applies different detectors to measure the level of the signal and the signal plus noise power at the output of the DUT. A calibration step determines the signal and signal plus noise power present at the input of the DUT (along with the noise of the VNA receiver itself). Finally, since it is a VNA, it can independently measure the DUT gain, and use all this information to determine the noise figure of the DUT. The system block diagram is shown in Figure 5.20(a). Here, a CW signal, S, is injected into the system along with additional noise, modeled by an amplifier with gain of 1 and noise factor f_s. The next component in the cascade is an attenuator with gain g_a and noise factor f_a. This composite signal is applied to a DUT with gain g_d and noise factor f_d. Finally, the output of the DUT is applied to the nonideal receiver R, modeled as an amplifier with $g_r = 1$ and noise factor f_r.

As a first step, consider the simplified block diagram in Figure 5.20(b). Here, we place the DUT along with all nonidealities inside the box, so we can use the noise factor equation ((5.11) and (5.12)) to obtain:

$$f_c = \frac{s_{in}/n_{in}}{s_{out}/n_{out}} = \frac{\dfrac{s}{n_0}}{\dfrac{s \cdot g_c}{n_r}} = \frac{n_r}{n_0 \cdot g_c} - \frac{n_r}{n_0 \cdot g_a \cdot g_d} \tag{5.40}$$

Within the box, we can use the cascaded noise factor expression (5.33) to obtain:

$$f_{cascade} = f_s + \frac{f_a - 1}{g_s} + \frac{f_d - 1}{g_s \cdot g_a} + \frac{f_r - 1}{g_s \cdot g_a \cdot g_d} \tag{5.41}$$

Recognizing that $g_s = 1$ and that the noise factor of an attenuator is simply the reciprocal of the gain:

$$f_a = 1/g_a \tag{5.42}$$

we obtain:

$$f_{cascade} = f_s + \frac{1}{g_a} - 1 + \frac{f_d - 1}{g_a} + \frac{f_r - 1}{g_a \cdot g_d} \tag{5.43}$$

Recognizing that $f_{cascade}$ is equal to f_c of (5.40) (and simplifying) yields:

$$\frac{n_r}{n_0 \cdot g_a \cdot g_d} = f_s - 1 + \frac{f_d}{g_a} + \frac{f_r - 1}{g_a \cdot g_d} \tag{5.44}$$

Rearrange this expression to solve for the DUT noise factor, f_d:

$$f_d = \frac{n_r}{n_0 \cdot g_d} + g_a(1 - f_s) - \left(\frac{f_r - 1}{g_d}\right) \tag{5.45}$$

This expression tells us that the DUT noise factor is calculated from the measured noise power (n_r) measured DUT gain (g_d) and system parameters associated with the attenuation (g_a), receiver noise figure (f_r), and CW signal source noise figure (f_s). n_r and g_d are obtained during the measurement sweep. The three system parameters g_a, f_r, and f_s are obtained during the calibration process.

Similar rules of thumb apply to the signal-to-noise method as we observed in the Y-factor method. First and foremost, we must ensure that the noise contributed by the receiver does not obscure the noise created by the DUT. To quantify this, we start with (5.44), and solve for n_r.

$$n_r = n_0 \cdot g_a \cdot g_d \left(f_s - 1 + \frac{f_d}{g_a} + \frac{f_r - 1}{g_a \cdot g_d} \right) \tag{5.46}$$

Simplifying and adding/subtracting the term $n_o\, g_d$:

$$n_r = n_0 \cdot g_a \cdot g_d (f_s - 1) + n_0 \cdot g_d \cdot f_d + n_0 (f_r - 1) + (n_0 \cdot g_d - n_0 \cdot g_d) \quad (5.47)$$

Regrouping terms:

$$n_r = n_0 \cdot g_a \cdot g_d \cdot f_s + n_0 \cdot g_d (1 - g_a) + n_0 \cdot g_d (f_d - 1) + n_0 (f_r - 1) \quad (5.48)$$

Consider the noise at the output of the source n_s. This can be calculated from the classic noise figure expression given in (5.11):

$$f_s = \frac{s_{in}/n_{in}}{s_{out}/n_{out}} = \frac{\dfrac{s}{n_0}}{\dfrac{s \cdot g_s}{n_s}} = \frac{n_s}{n_0 \cdot g_s} \quad (5.49)$$

Solving (5.49) for n_s, recognizing that $g_s = 1$ and inserting into (5.48) gives:

$$n_r = n_s \cdot g_a \cdot g_d + n_0 \cdot g_d (1 - g_a) + n_0 \cdot g_d (f_d - 1) + n_0 (f_r - 1) \quad (5.50)$$

The first term represents the noise power contribution from the source. For large attenuation values ($g_a << 1$), the first term will be very small compared to the others and can be neglected. The second term tends towards $n_o\ g_d$. Rewriting, we get:

$$n_r = n_0 \cdot g_d + n_0 \cdot g_d (f_d - 1) + n_0 (f_r - 1) \quad (5.51)$$

Regrouping terms, we at last obtain:

$$n_r = n_0 \cdot g_d \cdot f_d + n_0 (f_r - 1) \quad (5.52)$$

The first term represents the contribution of the DUT to the measured receiver noise. The second term represents the noise generated by the receiver itself. To ensure that the noise contributed by the receiver does not obscure the noise created by the DUT, we can set:

$$g_d \cdot f_d > c(f_r - 1) \quad (5.53)$$

or

$$f_d > \frac{c(f_r - 1)}{g_d} \quad (5.54)$$

The DUT's noise factor must be greater than the noise factor of the measurement receiver divided by the DUT gain, multiplied by a rule-of-thumb constant, c. In typical applications, $c = 0.1$. In this equation, it is clear that it will be easier to make noise figure measurements on high-gain DUTs, or if the receiver noise factor is reduced.

Equation (5.54) can be further simplified, realizing that many devices have a noise factor close to 1, and the typical measurement receiver has a noise factor >> 1. Assuming $c = 0.1$,

$$g_d > \frac{f_r}{10} \tag{5.55}$$

or

$$F_{receiver}(dB) < G_{dut}(dB) + 10dB \tag{5.56}$$

So the receiver's noise figure should be less than 10 dB higher than the DUT's gain.

Table 5.5 shows this is a lofty (and generally unreachable) goal for typical devices (10–20-dB gain) when considering noise figure measurements between port 1 and port 2 of a ZVA. However, this table also reveals that noise figure measurements become much easier if the B16 direct receiver access option is used. Other possibilities include placing a low-noise preamp either at VNA port 2 or, more optimally, placing the low-noise preamp between the DUT output and the direct receiver access input.

5.4.3.1 Setup Considerations

Figure 5.21 shows four possible VNA port configurations for noise figure measurements. There are multiple downsides to using any port configuration but the native option (1). For example, setup (2) precludes full two-port S-parameter measurements because s_{22} measurements of the DUT are blocked by the preamp. Setup (3) bypasses port 2 of the test set altogether, so the port-2 generator and reference receiver cannot be accessed. Additionally, when using any configuration involving an external preamp or direct receiver access, care must be taken to avoid compressing either the preamp or the VNA receiver (or both). Table 5.3 shows the difference in receiver compression levels for the ZVA test port versus the direct access panel.

There are five parameter categories comprising a total of 13 different factors that impact noise figure measurement uncertainty:

- *ZVA source parameters:* Source match, noise figure, and attenuator gain;
- *ZVA receiver parameters:* Receiver match, noise figure;

Table 5.5
ZVA Receiver Noise Figure (dB) over Frequency

Device	Input	4 GHz	8 GHz	12 GHz	16 GHz	20 GHz	24 GHz	28 GHz	32 GHz	36 GHz	40 GHz	44 GHz	50 GHz
ZVA 8	Port	35	35	—	—	—	—	—	—	—	—	—	—
	B16	23	23	—	—	—	—	—	—	—	—	—	—
ZVA 24	Port	42	42	42	45	45	45	—	—	—	—	—	—
	B16	31	31	31	33	33	33	—	—	—	—	—	—
ZVA40	Port	35	35	35	35	35	35	48	48	48	50	—	—
	B16	28	28	28	28	28	28	38	38	38	38	—	—
ZVA50	Port	35	35	35	35	35	35	48	48	48	50	53	53
	B16	28	28	28	28	28	28	38	38	38	38	40	40

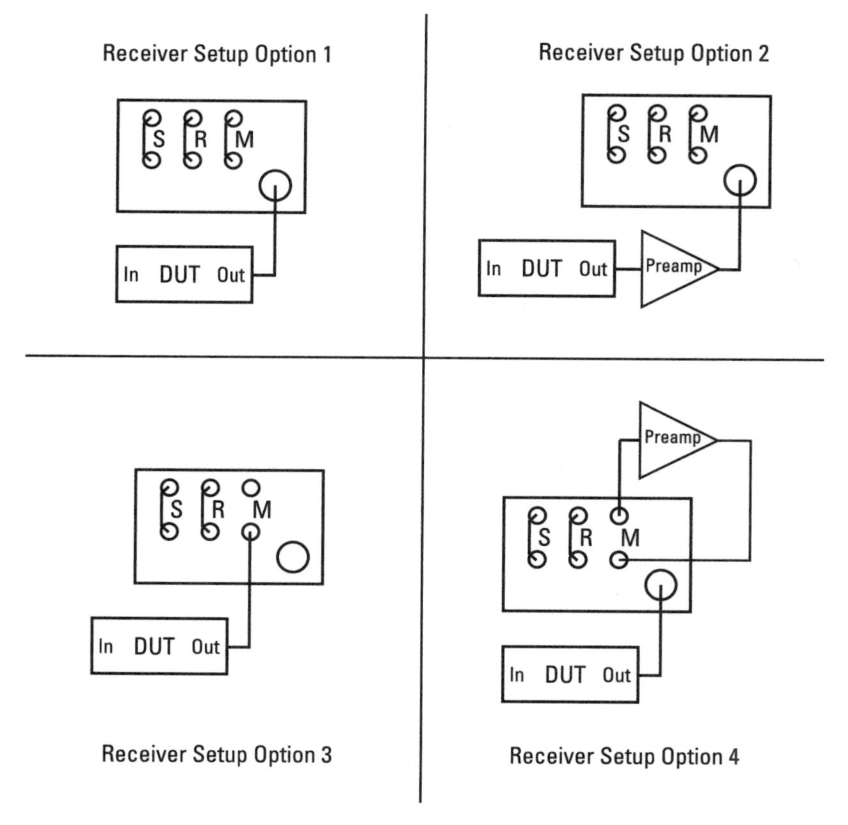

Figure 5.21 Port configurations for noise figure measurements.

- *DUT parameters:* DUT noise figure, gain, isolation, input reflection, and output reflection;

- *Power meter parameters:* Power meter match and accuracy;
- *Measurement temperature.*

Fortunately, there is a free software program available that calculates the impact of these different parameters.[10] Additionally, it provides typical default values for all settings, making the job considerably easier. To demonstrate the use of this tool, two setups are considered for a DUT having 18-dB gain and 3.5-dB noise figure in the 1–3-GHz range. The first setup uses port configuration 1 from Figure 5.20 with a ZVA40. Table 5.5 indicates the ZVA40 has a receiver noise figure of 35 dB in this frequency range. Applying these DUT and VNA values (and using default values for all other system and DUT parameters), we obtain the results shown in Figure 5.22, which results in a noise figure measurement uncertainty of +/−2.14 dB rms. The screenshot shows actual measurement results for this setup.

Clearly, there is room for improvement. If we introduce a preamp with 10-dB gain into the direct receiver access path (port configuration 4 from Figure 5.21), we effectively reduce the receiver NF by 10 dB, which reduces the uncertainty to 0.21 dB (Figure 5.23).

5.4.3.2 Noise Figure Measurement Procedure

The following noise figure measurement is based on an example from App Note 1EZ61, which uses an ultralow noise amplifier, the BFP740F from Infineon Technologies. This device is designed for use in the 1.4–2-GHz range, with gain of approximately 18 dB and 0.7 dB noise figure.

- *Preliminary Step 1:* Determine a test setup that is compatible with the DUT's gain and noise figure. In this frequency range, the ZVA40 has a port-2 receiver noise figure of 35 dB. Using (5.56), we see that the conventional measurement setup (port 1 to port 2, Figure 5.21, option 1) does not meet the noise figure guidelines for this DUT: 35 dB $\nless$ 18 dB + 10 dB (=28 dB).

  Instead, we elect setup option 4, which places the preamp in front of the measurement receiver via the direct receiver access panel. This setup allows a conventional connection between the DUT and ZVA port 2 so that a complete set of S-parameters can be measured in parallel with the noise figure. Alternatively, setup option 3 would have eliminated the

10. See App Note 1EZ61_2e "Noise Figure Measurement without a Noise Source on a Vector Network Analyzer" and the accompanying application "RohdeSchwarzZVAB-K30-NFErrorEstimationRev1_0", available from theRohde and Schwarz website, https://www.rohde-schwarz.com..

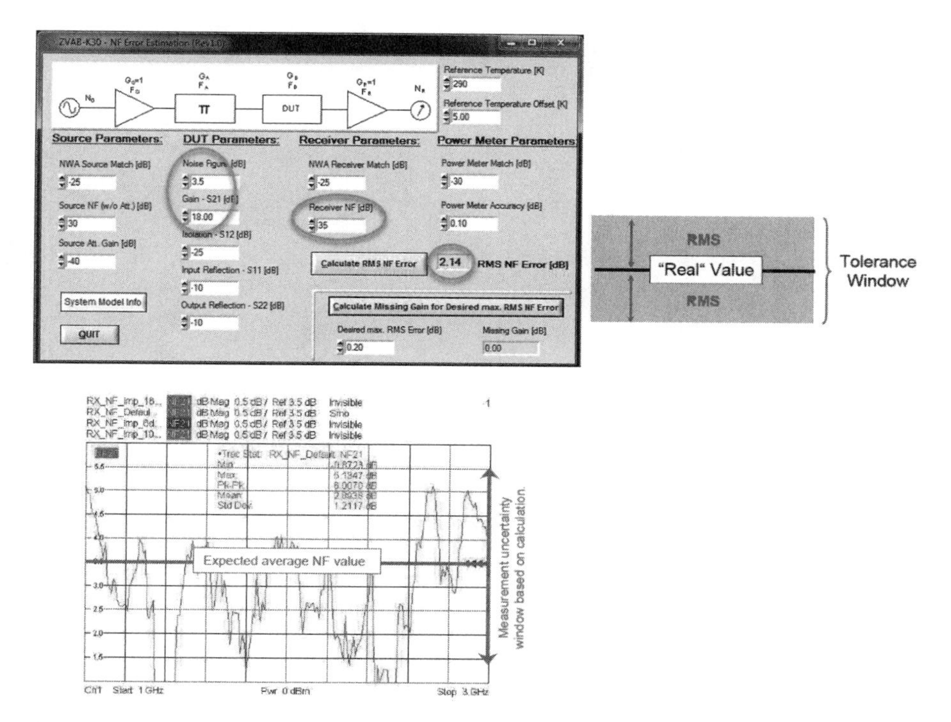

Figure 5.22 Noise figure uncertainty: example 1 using conventional connection to VNA port 2. As shown at right, the RMS NF error describes a 95% probability tolerance window around the real value.

need for a preamp, but would have also precluded the ability to make full two-port S-parameters in parallel with noise figure measurements.

With setup option 4, we reduce the receiver noise figure by the gain of the preamp. Using a preamp with 19-dB gain, (5.56) is now satisfied: 35 dB – 19 dB = 16 dB < 28 dB.

- *Preliminary Step 2:* Assess signal levels throughout the setup to ensure that the following three compression point requirements are met:
  - 1. DUT input power is at least 10 dB below its input compression point.
  - 2. Preamp input power is at least 10 dB below its input compression point.
  - 3. ZVA measurement receiver input power is at least 10 dB below its input compression point.

Figure 5.24 shows expected power levels for this setup, which uses the internal source attenuator option (0 to 70 dB in 10-dB steps). In (Figure 5.24(a)), a 30-dB attenuator is applied at port 1, and only two

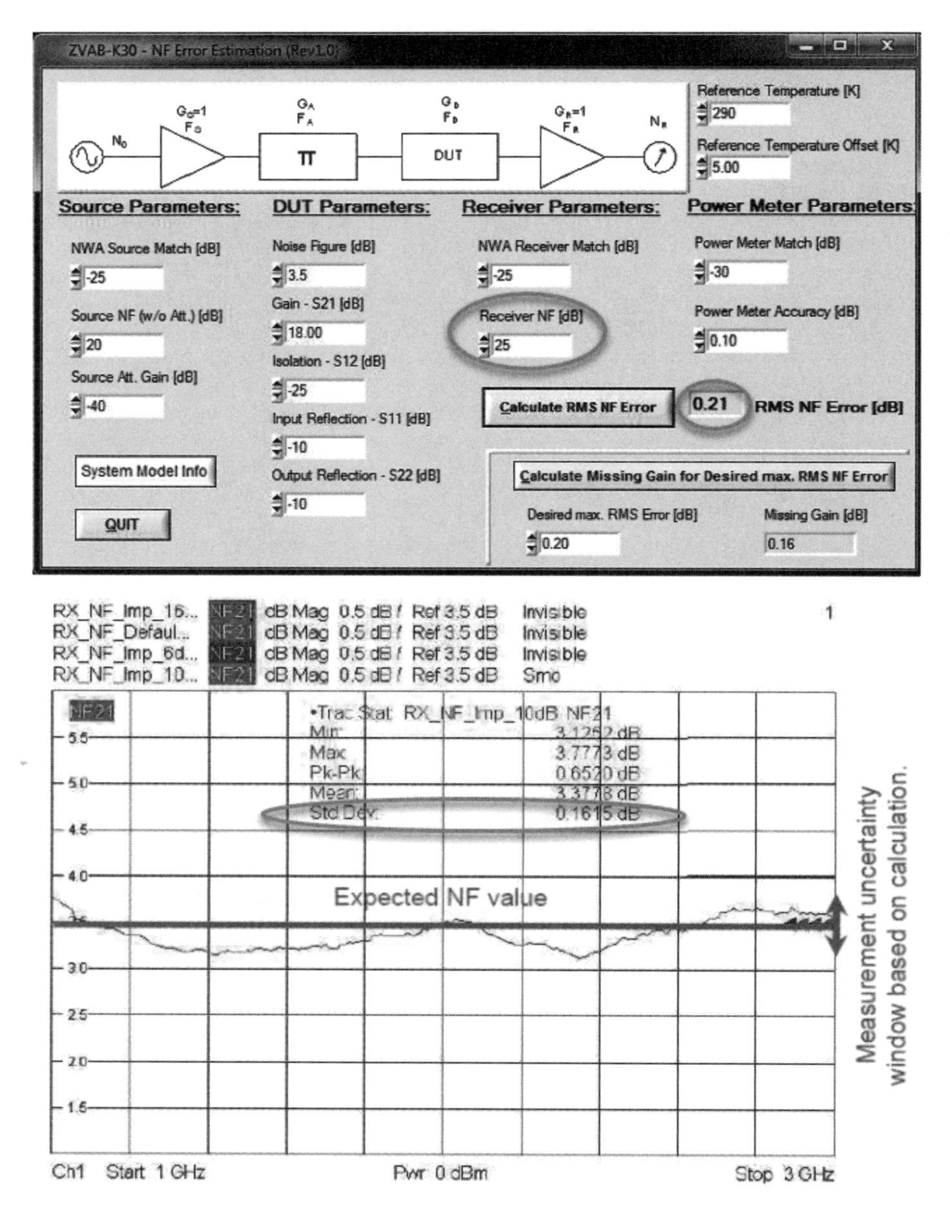

Figure 5.23 Noise figure uncertainty: example 2 using external preamp connected at direct receiver access panel.

of the three compression point requirements are met. By adding 20-dB additional attenuation to port 1, all three criteria are easily met (Figure 5.24(b)).

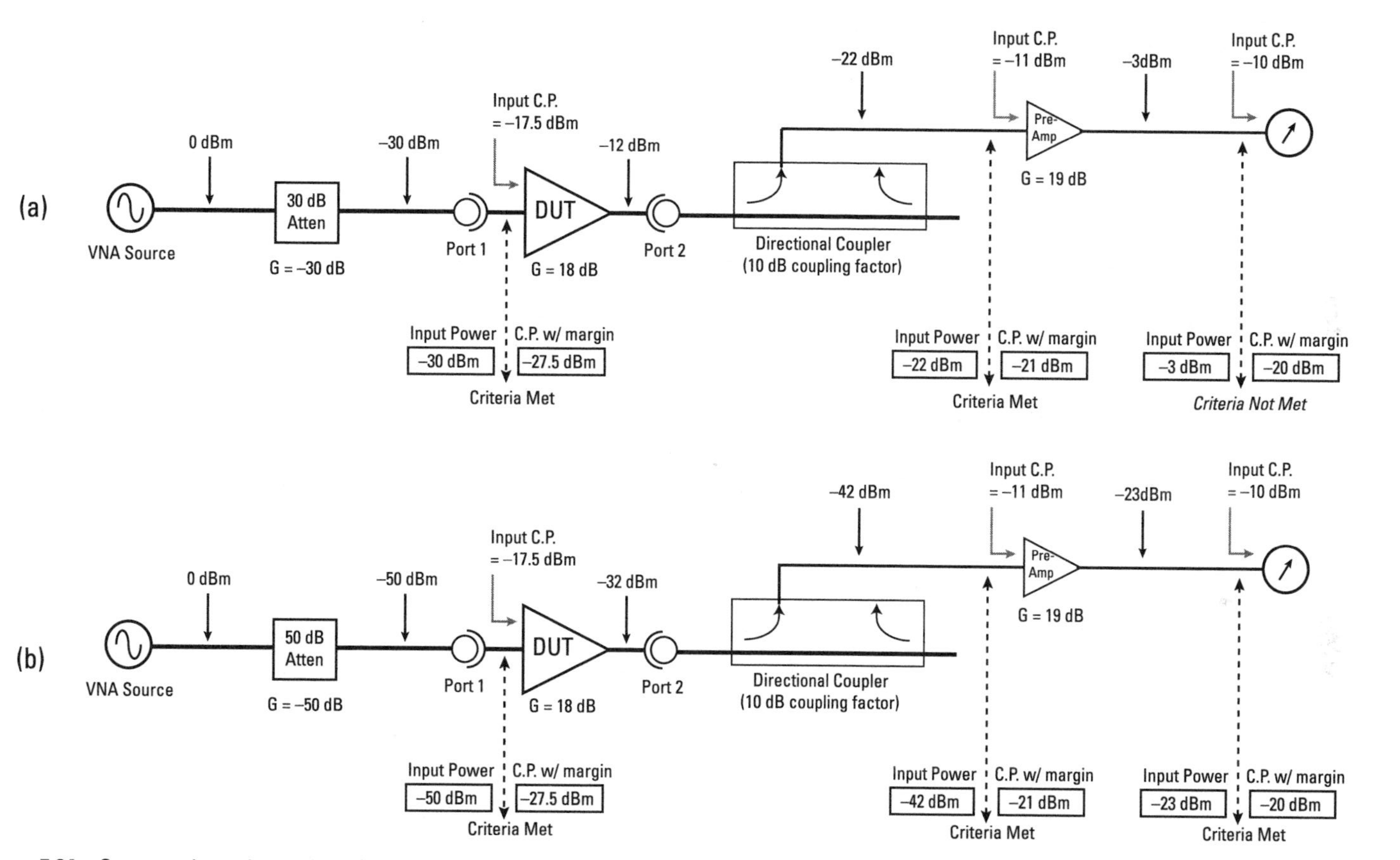

Figure 5.24 Compression point analysis for noise figure measurement. Configuration (a) meets CP criteria for DUT and preamp, but fails to meet CP criteria for VNA measurement receiver. Configuration (b) meets CP criteria for all three (DUT, preamp, and VNA receiver).

- *Preliminary Step 3:* Use the NF Error Estimation application to verify that the selected setup meets NF error requirements. Figure 5.25 shows the results, indicating an NF measurement uncertainty of less than 0.2 dB.
  - *NF Step 1:* Begin with a System > Preset on the ZVA. Configure start and stop frequencies and number of measurement points. From "Channel" > "Mode" > "Noise Figure Meas," select "Noise Figure Setup Guide." Starting at "1. DUT Port Connection," select the measurement ports. (We use default port settings for this example: ZVA Port 1 connected to DUT input and ZVA Port 2 connected to DUT output.)
  - *NF Step 2:* Connect a power sensor to the port-1 test cable/calibration plane and perform a source power calibration via the "2. Source Power Calibration" dialog. Use the default power value (0 dBm) to ensure a good SNR. For this application, only a reference receiver calibration is required, the source power does not need to be flat, it must simply be known (accurately).
  - *NF Step 3:* Adjust the ZVA source power to avoid compression and perform receiver power calibration using "3. Receiver Power Calibration." Here, we connect the ZVA port-1 to port-2 test cables together (no DUT). Since there is no DUT (and the DUT has a gain of 19 dB), we set the source attenuation to 30 dB and still easily meet the

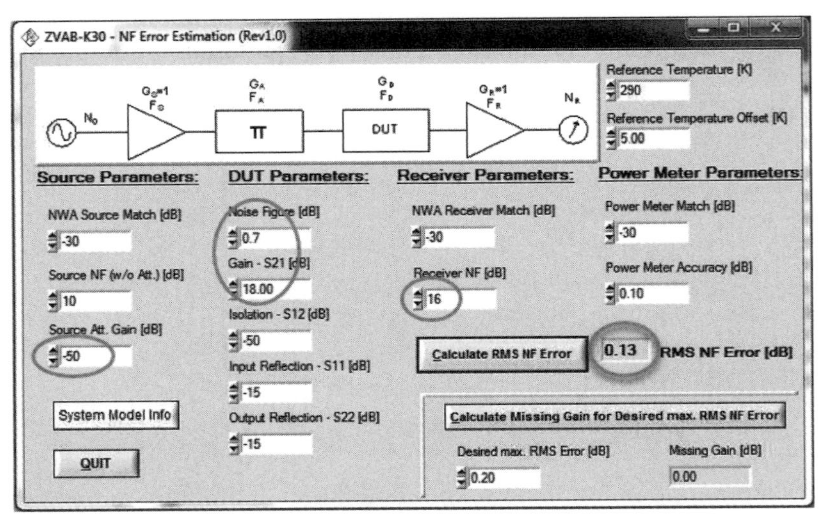

Figure 5.25 NF error estimation for ultralow noise LNA example.

compression criteria of Figure 5.24. Be sure to check the checkbox "Preamp or Direct Receiver Access Used."

- *NF Step 4:* Define the Noise Figure Measurement receiver configuration using "4. Define Noise Figure." In this example (and in most instances), default settings are sufficient. Click "OK."
- *NF Step 5:* Perform noise figure calibration using "5. Noise Figure Calibration." The procedure measures three VNA parameters (VNA receiver noise figure, source noise figure, and attenuator gain) and uses them for the final noise figure calculation. This is probably the most critical step (and most error-prone), due to the requirement to set the ZVA generator attenuator properly for two measurements. Refer to Figure 5.26. The first setting "Gen Atten for Src NCal:" should be the value from step 3 (receiver power calibration) and measures the excess noise of the ZVA source. The second setting "Gen Atten for DUT Meas:" is the attenuation value used when measuring the DUT. We determined this value (50 dB) from the system compression point analysis (Figure 5.24(b)). Use the default settings for all other parameters.

The next three calibration steps can be performed in any order, but care must be taken to ensure the proper test cable connections and terminations are used.

- *Rcv Noise Cal:* Make sure you terminate the port-2 test cable with a 50Ω match. Then click the left-most box to commence the calibration sweep. The completed sweep shows the ZVA's effective noise figure (with preamp).
- *Src Noise Cal:* For this calibration step, use a thru adapter to connect the port-1 and port-2 test cables. Then click the associated box to perform the source calibration sweep.
- *Attenuator Cal:* For this calibration step, use the same thru adapter as the step above to connect the port-1 and port-2 test cables. Then click the associated box to perform the attenuator calibration sweep. You will hear the mechanical attenuator change states (30 dB to 50 dB) during the calibration sweeps. The completed sweep shows the ZVA's attenuator value.

At the completion of the three steps, click "Apply."

- *NF Step 6 (optional):* We perform this optional step when we want to measure S-parameters in parallel with noise figure. Two calibration

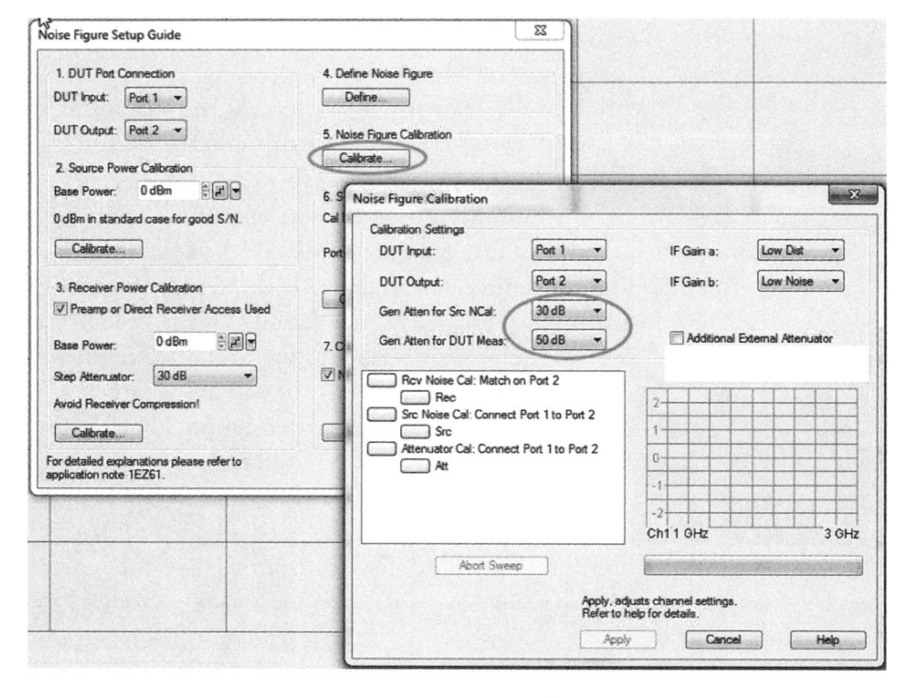

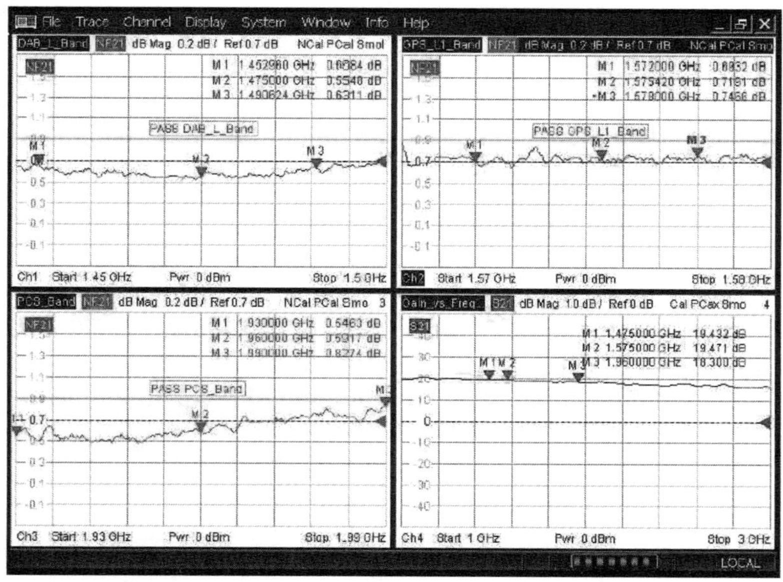

Figure 5.26 (a) The Noise Figure Calibration dialog. (b) Noise figure measurement results.

methods are supported: full two-port or one-path, two-port. (We would employ the latter method if gain measurements are desired but "receiver direct access" is used instead of ZVA port 2, making full two-port calibration impossible.) The port-2 power offset is available for full two-port calibration and is used when the ZVA source power setting is either too low (resulting in a poor SNR for port-2 calibration) or too high (no internal port-2 source attenuator), risking damage to a sensitive small-signal DUT. After selecting a calibration method, click "Calibrate" to commence the S-parameter calibration process. The resulting calibration dialog requires you to set the connector type and gender at the end of each test cable and select an appropriate test kit. One other setting that is unique to this mode is the use of the AVG detector. This is because wider bandwidths are used for the noise figure measurement, and so the same bandwidths are used for the S-parameter measurements as well. The AVG detector dramatically reduces trace noise compared with the "Normal" detector for wider IF bandwidths. Click "Next" and commence calibration using conventional "TOSM" calibration standards, and click "Apply" to complete the calibration process.

- *NF Step 7:* Calibration is complete; all that remains is to select the desired measurement. The available traces will depend on the S-parameter calibration (if any) selected in NF Step 6. Measurement results are shown in Figure 5.26(b).

5.4.4 Cold Source Method

In the Y-factor method of Figure 5.19, we saw that using two (measurement) points to determine a line ended up yielding noise figure (from the y-intercept) and gain (from the slope) of the line. The cold source method is a variation on this theme, recognizing that a line can also be determined from only one (measurement) point, as long as you also know the slope of the line. In this method, error-corrected device gain (s_{21}) provides the slope, while the measurement of a cold source (input of DUT terminated in a room-temperature impedance) positions the y-intercept to provide the DUT noise figure. Along the way, source matching (vector noise calibration) may be applied to improve the noise figure accuracy if the DUT source impedance differs from 50Ω. This approach requires additional hardware in the form of a low-noise receiver preamp and an ECal module dedicated for use as an impedance tuner. Alternatively, scalar noise calibration can be used to simplify the calibration process and speed up measurements at the expense of degraded measurement accuracy. The cold source technique is one of the noise figure methods offered by Keysight in their PNA-X VNA [3].

5.4.5 Other Concerns

Clearly, VNA-based noise figure measurements can be tricky and may be influenced by the thirteen factors mentioned in Section 5.4.3.1. However, there are two other factors that can make or break noise figure measurements: external interference and VNA receiver architecture. Interference is likely to show up when measuring an unshielded device, like an unpackaged amplifier on a wafer or printed circuit board. Interference typically shows up as a spike on the noise figure trace that may or may not be repeatable. This kind of interference is commonly observed in cell-phone bands (700–900 MHz, 1.8–2.1 GHz) and WLAN bands (2.4 GHz and 5.9 GHz). Other times, the interference may be broadband in nature, causing the noise figure of the DUT to be higher than it actually is. In both cases, the remedy is to test the device in an RF-shielded enclosure or to relocate the entire test setup to an RF-shielded room.

VNA receiver architecture is more of an insidious problem, because architectural details are rarely published. For example, the ZVA receivers operate in their fundamental mode up to 24 GHz. Above this frequency, they use harmonic mixing. The impact on S-parameters is minimal, usually showing up as a slight step increase in noise floor (or, conversely, a step reduction in dynamic range) above 24 GHz. The big problem is with noise figure measurements, where the noise from the fundamental receive band and the harmonic band both mix down to IF. The simple answer is to install a highpass filter when making noise figure measurements above 24 GHz on the ZVA. However, it certainly becomes challenging to make noise figure measurements straddling 24 GHz. Fortunately, advances in receiver architecture are making this issue moot, as the latest generation of VNAs from R&S (ZNA) use fundamental mixing up to 40 GHz. This eliminates the problem for the majority of popular aerospace and defense and commercial 5G millimeter-wave applications.

References

[1] Friis, H., "Noise Figures of Radio Receivers," *Proc. IRE*, Vol. 32, July 1944, pp. 419–422.

[2] Meer, D., "Noise Figures," *IEEE Transactions on Education*, Vol. 32, May 1989, pp. 66–72.

[3] "High-Accuracy Noise Figure Measurements Using the PNA-X Series Network Analyzer," 5990-5800EN, Keysight Technologies, August 3, 2014.

6

Measurements on Mixers and Frequency Converters

6.1 Introduction

One of the most important and fundamental RF tasks involves translating electromagnetically encoded information from one frequency band to another. In transmitting systems, information at lower intermediate frequencies (IF) is translated to microwave or millimeter-wave RF for subsequent over-the-air transmission. In receiving systems, the process is reversed. Information embedded in signals at very high RF is first downconverted to a lower IF band before digital processing is applied to extract the information. This frequency translation process is referred to as heterodyning (Greek hetero: different; dyne: power), and has been employed since the early days of radio.

The heart of any frequency-converting scheme is the mixer, which accomplishes heterodyning by exploiting the nonlinear voltage-current relationship in a semiconductor device. Active or passive semiconductors are arranged in (generally) symmetrical configurations and may employ anywhere from a single device to eight or more devices in elaborate bridge or ring structures. When two (or more) signals are applied, for example, an RF signal and a local oscillator (LO) signal, the result is nonlinear multiplication in the time domain. This gives rise to new sum and difference terms in the frequency domain, such as $nf_{LO}+/-mf_{RF}$, where m and n are integers. The selected architecture determines which products of this mixing process are enhanced and which are suppressed. Generally, only a single mixing product is desirable (e.g., $f_{LO}+f_{RF}$ or $f_{LO}-f_{RF}$), and a big part of the design challenge is to suppress or eliminate the unwanted products.

Engineers are drawn to VNAs for characterizing mixers, just as they would any other electronic component, but there are few instruments naturally less suited to frequency converting measurements than VNAs. Consider that a VNA's native language is scattering parameters, measured by injecting a single (reference) signal at one DUT port, and examining the energy emanating from every port. This requires a set of receivers that are coherent to each other (no frequency or phase drift), and they must be tuned to the injected tone frequency to get meaningful results. By forming ratios from different receiver measurements, amplitude and phase differences can be obtained. If any of these conditions are violated, S-parameter measurements cannot be made.

Mixer applications violate all these prerequisites:

1. Multiple tones are required, for example, for RF and LO. This violates the single-tone excitation requirement.
2. Different frequencies are injected at different ports. This violates the "receiver frequency same as generator frequency" requirement.
3. Different frequencies are received at each port. This invalidates the receiver coherency requirement.

If VNAs were restricted to the native operation alone, this would be a very short chapter. Fortunately, thanks to some innovative operational tweaks, the impressive receiver linearity, independent generator capability and sweep synchronization of VNAs can be leveraged for mixer measurements. Until recently, this has involved a number of compromises:

1. Phase measurements were excluded from native operation. If we set all the receivers (coherency principle) to the frequency of the VNA generator at port 1, the LO generator's phase at VNA port 3 appears as a rapidly spinning phasor (relative to port 1), preventing a meaningful phase measurement. A similar situation exists for the IF signal at port 2.
2. Power calibration was (and is) still required. Since full two-tone error correction is only valid when all ports are operating at the same frequency, the ratio measurements of classic S-parameter theory (along with their associated error-correction methods) must be jettisoned. By applying power calibration at each port to characterize generator levels and calibrate receiver signal levels (in absolute power, dBm), ratios

of these corrected scalar wave quantities will yield accurate gain/loss measurements.

3. Measurement time increases because gain/loss measurements must be made at two frequencies before computing a ratio result (e.g., first measure the RF at port 1 and then change the receivers to measure the IF at port 2). Contrast this with a conventional two-port amplifier application, where input and output frequencies are measured concurrently.

For typical VNA measurements, one mixer port will be swept (e.g., the RF port), one will be defined as fixed (e.g., LO port), and the third (IF port) will be automatically configured for sweeping the correct output range, either at $F_{RF} + F_{LO}$, $F_{RF} - F_{LO}$, or $F_{LO} - F_{RF}$. Depending on the defined frequency relationships, the converted output frequency range will be inverted or noninverted. Figure 6.1 illustrates the possibilities. At Figure 6.1(a), a low-frequency swept IF input is mixed with a high-frequency fixed LO, leading to an output containing a noninverted upper sideband and an inverted lower sideband (note: a single sideband (SSB) mixer will suppress one of these sidebands). The upper and lower sidebands are each separated by the value of the IF. At Figure 6.1(b), a swept IF input is mixed with a closely spaced fixed LO, leading to an inverted, low-frequency difference term ($F_{LO} - F_{RF}$) and a noninverted sum term ($F_{LO} + F_{RF}$) at nearly twice the LO frequency. The third possibility is shown in Figure 6.1(c), where a low-frequency fixed LO is mixed with a much higher swept IF. The difference ($F_{RF} - F_{LO}$) and sum ($F_{RF} + F_{LO}$) terms are both noninverted.

In this chapter, we briefly address the primary mixer architectures and mention the advantages and disadvantages of each. This exercise provides important insights concerning mixer parameters (and key performance metrics). We then discuss each datasheet parameter, and delve into associated VNA-based scalar measurement approaches. Subsequent sections introduce more complicated (and messy) vector techniques that allow relative or absolute phase to be obtained despite DUTs having different input and output frequencies. We also describe the latest VNA synthesizer technology and show how it drastically simplifies vector mixer measurements. Finally, we discuss a powerful two-tone method for measuring group delay on frequency converting devices with inaccessible LOs.

6.2 Mixer Architecture

As mentioned in the introduction, one of the biggest challenges associated with mixers is optimizing desired mixing products while suppressing unwanted products over the frequency range and power levels of interest (while maintain-

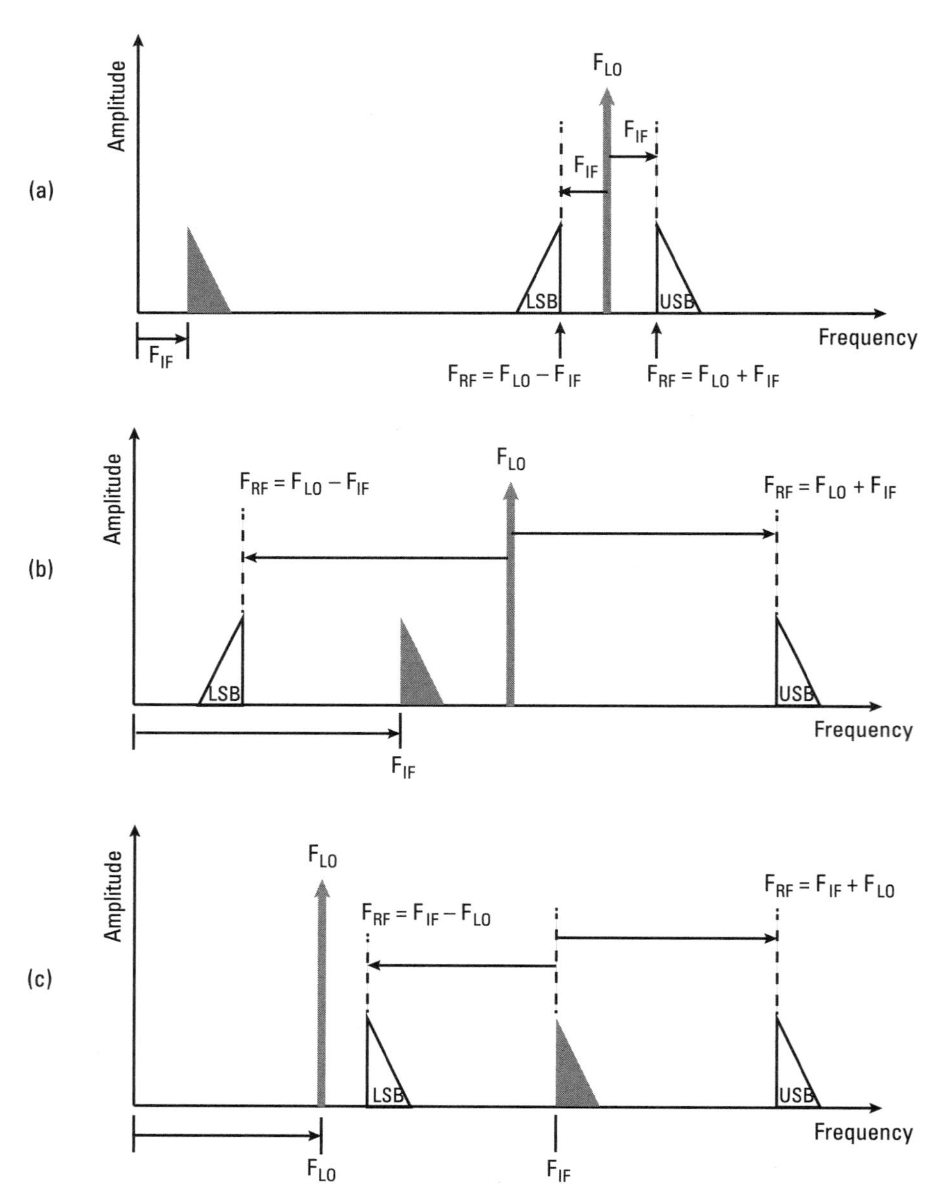

Figure 6.1 Mixer frequency conversion relationships.

ing cost objectives). Selecting a mixer architecture is the first step in this process, as architecture largely dictates the power levels (and LO drive levels), IMD suppression, and port isolation achievable.

Mixer architectures fall into four categories: single-ended, single-balanced, double-balanced, and triple balanced. While we refer to diodes as the

RF switching devices, some mixer architectures employ passive or active CMOS transistors or FETs instead. Generally, higher-performance wideband mixers tend to favor diodes, while high-volume, low-cost, narrowband designs gravitate towards transistors.

Single-ended mixers usually have a single diode. A shunt tank circuit may be used to isolate the output from the input. Figure 6.2(a) shows a schematic diagram of a single-ended downconverting mixer. The output contains not only the IF but also the LO, RF, and image (LO + RF), as well as higher-order IMD products.

The single-balanced mixer uses either two or four diodes in a balanced arrangement (Figure 6.2(b)). Circuit symmetry is exploited to ensure that the LO voltage does not appear at the IF output (though the RF signal is not suppressed at the output). Generally, the LO power is much greater than RF signal in a mixer, so suppressing the LO at the output is important to avoid interaction with IF components (e.g., saturating wideband IF amplifiers or reflecting LO energy off IF filters and back into the mixer IF port).

A doubly-balanced mixer (DBM) uses four diodes in either a ring (Figure 6.2(c)) or star configuration with two baluns or two hybrid junctions that isolate both the RF and LO voltages from the output. During half of the LO cycle, two of the diodes are biased on (low-resistance state) and the other two are biased off (high-resistance state). During the next half of the LO cycle, the diodes switch states. This symmetrical switching combined with careful circuit balancing achieves high isolation between the IF port and the RF and LO ports.

A triple-balanced mixer (TBM) is also known as a doubly double balanced mixer and is constructed by doubling the components of the double-balanced mixer, while also adding an additional coupler or balun for the IF port (Figure 6.2(d)). This design is useful when extremely high spurious suppression is required, such as when the mixer is used with overlapping RF, LO, and IF bands. The required LO drive power is also doubled, and conversion loss tends to be higher than with the DBM due to additional circuit losses.

Other, more specialized designs take advantage of these basic mixer building blocks for extended functions. For example, a balanced image-rejection mixer (or single sideband (SSB) modulator/demodulator) can be constructed from two mixers, a 0° power splitter and two 90° hybrid couplers (Figure 6.2(e)). The power splitter provides the same IF signal into both mixers, while the LO is split with 90° phase offset between the mixers. The RF output of the mixers are recombined via a second 90° hybrid with one sideband (e.g., LO + IF) reaching the RF port while the other sideband (LO – IF) is dissipated by the termination on the other hybrid port. To switch sidebands at the RF port, simply switch the connections at either the output of the LO hybrid, the input of the RF hybrid, or the output of the RF hybrid.

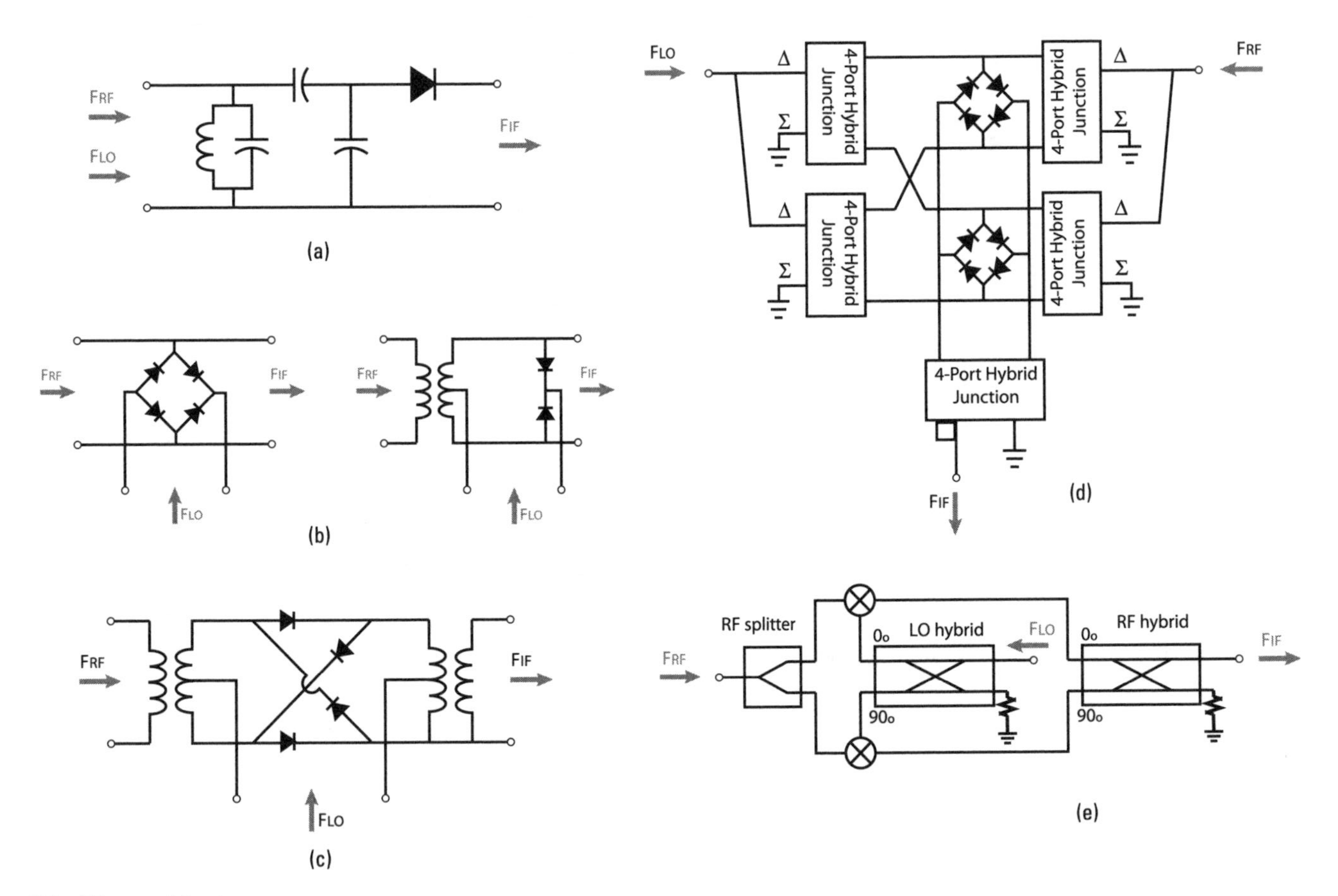

Figure 6.2 Mixer architectures.

Within these architecture discussions, we have already introduced important mixer characteristics like port isolation and IMD suppression. In the next section, we will expand this list, examining each parameter in greater detail and discussing ways of measuring these parameters with a VNA.

6.2.1 Conversion Loss

Conversion loss in a mixer is defined as the difference in power between the input signal and the output signal. For the purposes of this section, we will assume the input signal is provided to the RF port and the output signal is obtained from the IF port.

$$\mathit{Conversion\ Loss}(C.L.) = P_{RF} - P_{IF} \tag{6.1}$$

Like many mixer parameters, conversion loss must be specified at a particular LO drive level, and for a particular arrangement of RF, LO, and IF.

As it turns out, conversion loss is a benchmark mixer metric that correlates closely with other metrics like isolation and the 1-dB compression point. Depending on architecture, the conversion loss will vary between 4.5 and 9 dB for a passive diode mixer. Transmission line losses, balun mismatch, diode series resistance, and mixer imbalance can all contribute to increased conversion loss. In general, wider bandwidth mixers tend to have greater conversion loss because of the difficulty involved in maintaining circuit balance over the entire operating range.

Conversion loss is often the first mixer measurement attempted with a VNA. Since input and output frequencies are different, generator and receiver sweep synchronization is crucial to success. Additionally, power calibration (including flatness calibration) is required for the RF and LO generators, and then one of these calibrated generators is used to power calibrate the IF measurement receiver. Further, the power levels used for the RF excitation signal and LO drive can have a strong impact on conversion loss results. Use the recommended LO drive level, and keep the RF excitation 10 to 20 dB below the expected 1-dB compression point for accurate results.

6.2.2 Isolation

Isolation is associated with signal leakage from one port to another. For this measurement, there is no frequency conversion involved: all measurements are made at the generator (input) frequency. Isolation is approximately reciprocal and tends to decrease with increasing frequency due to transformer imbalance, lead inductance, and capacitive unbalance between diodes. VNAs can measure isolation using S-parameters since there is no frequency conversion required.

However, the mixer should be measured in a configuration that conforms to the expected application environment: the RF input level should be the same, and the appropriate LO drive power must be applied. As with conversion loss, the VNA generator(s) used for stimulating the RF and LO ports should be power calibrated to ensure proper excitation. In terms of S-parameters, "high isolation" means the same as "high loss" when performing a through measurement (e.g., S_{21}, where port 1 is the input (RF) port and port 2 is the output (IF) port). Isolation is also approximately reciprocal: dBMag(S_{21}) ~ dBMag(S_{12}).

6.2.2.1 LO-to-RF (L-R) Isolation

Isolation between the local oscillator port and the RF port is important for downconverting front ends because LO energy can leak into the RF path and possibly be radiated, causing interference to other circuitry in the system or to other colocated users of the radio spectrum. Typical L-R values range from 25 to 35 dB. Normally, only LO isolation is specified, not RF isolation, since RF signal power is generally much lower than LO drive levels.

6.2.2.2 RF-to-IF (R-I) Isolation

R-I isolation is not usually an issue because the RF signal power is much smaller that the LO drive power, and the RF and IF are sufficiently well separated to prevent problems with saturation of low-frequency IF amplifiers. However, R-I is a good mixer metric; high isolation suggests a well-balanced design, and conversion loss therefore tends to be low. Poor R-I leads to higher conversion loss and lack of conversion loss flatness. Typical values are in the 25–35-dB range.

6.2.2.3 LO-to-IF (L-I) Isolation

L-I isolation is typically the worst of the three types of isolation. This can be a problem when the LO is close to the IF, because the LO energy can saturate a high-gain IF amplifier. Also, poor L-I isolation can cause conversion loss flatness problems. Typical values of L-I isolation are in the 20–30-dB range.

6.2.3 1-dB Compression Point

A mixer's 1-dB compression point compares the linearity between the RF input and IF output. Conversion loss is determined at low RF input levels. As the RF input level increases, the mixer output begins to compress. As the mixer is driven into compression, its RF input power competes with the LO power so that diode switching efficiency is impacted. When the RF is within 3 dB of the 1-dB compression point, the mixer may begin to behave unpredictably (including the levels of other intermod products). Eventually a point is reached where the input/output loss is 1 dB greater than the conversion loss. This is the 1-dB compression point and is generally expressed relative to the input RF power level

A 1-dB compression point measurement is configured similar to a conversion loss measurement, with proper power calibration of the RF and LO generators, as well as the IF measurement receiver. The 1-dB compression point is then characterized by placing the VNA in a power sweep mode and measuring the compression at a single input frequency. A rule of thumb for mixers is that the 1-dB compression point will be 4 to 7 dB below the minimum recommended LO drive level. So for a +7 dBm mixer that has a LO power range of 4 to 10 dBm, a conservative estimate for the compression point would be +4 dBm – 7 dB = –3 dBm. Since we want to start the power sweep in the mixer's linear range, we subtract 10 to 15 dB from this number, and set an initial start value of –18 dBm. For the stop value, we want to work our way up to the 1 dB compression point so as not to overload (or damage) the mixer. We set an initial stop value of –6 dBm. We can then observe the gain compression (b_2/a_1) with a vertical scale of 0.5 dB/div. The mixer's conversion loss (approximately 6 dB, nominally) is typically placed between the midpoint and the top of the vertical scale, so that the gain compression can be easily observed. Then we increase the power sweep's ending value in half-decibel steps until we observe slightly more than 1 dB of gain compression. By invoking the VNA's compression point marker function, the instrument will automatically determining the input and output power levels associated with the 1-dB compression point (see Section 5.1.2.2).

Other variations are also possible. For example, compression point changes with LO drive, so another interesting measurement involves keeping the RF level constant and sweeping the LO power to characterize the mixer's LO drive level sensitivity. These measurements can then be repeated at other important frequencies (e.g., RF_{low}, RF_{mid}, RF_{hi})

During a compression point measurement, it is important to ensure that the VNA's receiver is not compressing. This can easily be checked by placing more attenuation in front of the VNA receiver (a built-in receiver step attenuator is extremely handy for this purpose). If the shape of the compression curve changes when you add attenuation, the VNA receiver is likely compressing. Add attenuation until the shape no longer changes, at which point you can be sure that you are only measuring DUT compression. If the 1-dB compression point of your mixer ends up being too low for your application, you might have to consider a different mixer model or architecture, which may in turn require a higher LO drive level.

6.2.4 VSWR

VSWR is a curious mixer metric. Some consider it meaningless, because a mixer that is perfectly matched at an input port will still emit mixing products at other frequencies from the same port, and the levels of all frequency components will be highly dependent on the terminations applied at all ports in the mixer's final application. Then there is the problem of reciprocal mixing: Consider a downconverter with F_{rf} and F_{lo}, with the desired output at $F_{if} = F_{lo} - F_{rf}$. The mixer produces intermodulation components at $nF_{LO} + mF_{RF}$. Consider the component at $2F_{LO} - F_{RF}$, which, due to limited L-R isolation, exits the RF port and is reflected back by a narrow bandpass filter. It reenters the RF port, and is downconverted to $(2F_{LO} - F_{RF}) - F_{LO} = F_{lo} - F_{rf}$. This is at the same frequency as the desired IF signal. However, the path length between the mixer and filter will vary as a function of frequency, leading to vector combination of the desired and reciprocal mixing product that will add constructively or destructively. This will cause ripple across the operating band due entirely to reciprocal mixing, not VSWR. (This is a perfect example of why it is a good idea to add 3, 6, or even 10-dB pads to bare mixers to control the effects of unwanted reflections at mixer ports.)

Aside from reciprocal mixing, broadband mixers may exhibit different VSWR versus frequency characteristics due to circuit resonances. Also, diode impedance in influenced by LO power changes, and the input impedance of the various ports is load-dependent, even though the ports may be isolated from each other, and this load dependency gets worse at higher frequencies as isolation decreases.

VSWR can easily be measured by any VNA in conventional S-parameter mode, with the caveat that all ports should be terminated in 50Ω and/or driven by application-specific power levels. LO drive should always be applied when making VSWR measurements at any port.

6.2.5 Noise Figure

Noise figure is a mixer parameter that is gaining in popularity. A mixer's noise figure should be around a half-decibel higher than its conversion loss. Many performance-grade VNAs support frequency-converting noise figure measurements with dialogs that simplify the task of assigning frequency and power levels to the VNA ports as well as guiding the user through the unique calibration and configuration procedures.

6.2.6 Dynamic Range

Dynamic range is the signal level range over which a mixer provides useful operation. It is bounded by noise figure at the lower end, and by the 1-dB compression point at the upper end. This may be an important metric for mixers that are used in an RF front-end receiver application. Conversely, if the mixer is slated for use in an instrumentation or modulator/upconverter application where a relatively high and constant input level is expected, dynamic range probably does not matter.

6.2.7 Single-Tone IMD

Single-tone IMD is important for gaging the levels of all nF_{LO} +/− mF_{RF} intermodulation products. The expected intermod frequencies are known, and measurements made at those frequencies are used to characterize mixer performance. This measurement was traditionally carried out by a spectrum analyzer, but can be performed more easily and quickly by a VNA using scalar power and flatness calibration for all generators (RF, LO) and power calibration of all receivers (IF). The VNA can be configured to sweep only the receiver while maintaining fixed generator frequencies. Additionally, since the mixer products can be calculated, a segmented sweep table may be used to measure only these IMD frequencies, further minimizing sweep time.

As in the case of conversion loss and 1-dB compression point, power sweeps can be configured using different LO or RF power sweep ranges to see how these changing levels impact critical IMD products.

6.2.8 Multitone IMD

Multitone IMD in a mixer can vary dramatically depending on the mixer type and technology. As a rule of thumb, the third-order intercept (TOI) for two-tone IMD will be 15 dB above the 1-dB compression point for diode mixers, and 10 dB above the 1-dB compression point for FET mixers. A slightly different rule of thumb is based on the recommended LO drive level. For DBMs and TBMs, the TOI will be a few decibels above the recommended LO drive level.

A VNA performs two-tone IMD measurements for mixers in a manner similar to amplifiers (see Chapter 5). The slight deviation involves the need for three generators: two for the stimulus tones, and one to provide the LO drive power. Some four-port VNAs offer up to four independent internal generators for these purposes. Alternatively, most VNAs can be configured to control an external generator (amplitude and frequency), which is very useful for situations involving a swept LO configuration. However, sweep time may be impacted due to the additional delay required for sending commands to the external generator and waiting for it to settle (in both frequency and amplitude) at each measurement point.

6.2.9 DC Polarity

DC polarity is associated with mixers that have an IF port response down to DC. DC polarity defines the polarity of the IF output voltage when the mixer is used as a phase detector (RF and LO ports driven with equal frequencies and equal phase (0° difference)). Some performance-class VNAs have a defined coherence mode where the amplitude and phase of the internal generators can be set to specific offsets relative to any of the other generators. By utilizing multiport S-parameter calibration, the reference plane for these measurements is moved to the end of the test cables (calibration plane), and resulting amplitude and phase accuracy is in the range of 0.1 dB and 1°, respectively. These amplitude and phase relationships can be maintained over a frequency sweep. By using a built-in voltmeter capability in the VNA, the DC polarity characteristics of a mixer can be characterized as a function of frequency, amplitude, power, phase, amplitude imbalance, or phase imbalance.

6.2.10 DC Offset

DC offset in a mixer's IF port provides a measure of mixer imbalance. A perfectly balanced mixer will have zero offset. DC offset also defines the IF output voltage range when the mixer is used as a phase detector. For DC offset measurements, the signal is applied only to the LO port, with the RF port terminated in 50Ω and a voltmeter attached to the IF port. (The IF port obviously must be DC coupled.) As with the DC polarity measurements mentioned previously, the VNA's internal DC voltmeter can be used to characterize mixer performance as a function of frequency and/or power level.

Curiously, DC current flowing through the IF port can impact isolation between the LO and RF ports. This characteristic can be exploited for voltage-controlled attenuator applications, where the RF and LO ports serve as the attenuator ports, and the attenuation value is varied by applying a small DC current to the IF port. With no current, maximum attenuation is achieved (on the order of 50 dB). As current increases, attenuation decreases down to several decibels (but care must be taken to avoid excessive current into the IF port to prevent damage to the mixer diodes).

6.3 Scalar Mixer Measurements

Scalar mixer measurements take full advantage of the tight sweep synchronization between VNA generators and receivers. Since many modern VNAs provide the ability to set source and receiver sweep ranges independently, it is very easy to tap this capability to make meaningful mixer measurements. In this section we describe the calibration and measurement steps required to configure a VNA (R&S ZNB) for making scalar mixer measurements. This procedure delivers important mixer parameters including mixer conversion loss (or gain), RF port isolation, RF port match, IF port match, leakage from the LO to the RF port, and feedthrough from the LO to the IF port.

6.3.1 Scalar Mixer Setup (ZNB)

The first step is to define the RF, LO, and IF relationships. Consider a downconverter application. The RF input is between 3.0 and 3.5 GHz, and insertion loss measurements are required every 2.5 MHz (201-point sweep). If the IF is fixed at 100 MHz, then the LO generator must sweep in step with the RF generator. Two choices are possible. A high-side LO will sweep from 3.1 to 3.6 GHz. A low-side LO will sweep from 2.9 to 3.4 GHz. Either of these LO frequency sweeps injected into the mixer will produce a 100-MHz difference term (LO-RF or RF-LO, respectively) at the mixer's IF port (as well as a sum term at RF+LO, and many other mixing products). Since the VNA receiver is fixed at the IF (100 MHz), it will measure this difference term and ignore or reject the other mixing products.

Instead of a fixed IF, consider the case where we want to establish a fixed LO and let the IF vary between 500 MHz and 1 GHz (this is a common scenario involving block downconverters). Choosing a fixed LO frequency of 2.5 GHz would produce a difference term within our desired IF range. Alternatively, if we placed the LO at 4.0 GHz, we would require the IF receiver to sweep in the reverse direction (1.0 GHz to 500 MHz).

Upconverter applications tend to focus on the sum rather than the difference term. For example, we might choose an IF range of 500 MHz to 1 GHz, and upconvert this to a RF range of 3.0 to 3.5 GHz. Again, we would place a fixed LO at 2.5 GHz, but this time the VNA source/generator would sweep 500 MHz to 1 GHz, and the VNA receiver would sweep 3.0 to 3.5 GHz.

The one subtle departure from this nomenclature is that, for configuring a frequency conversion measurement in a VNA, the DUT input port is always considered RF and the DUT output port is IF, regardless of whether the DUT is used as an upconverter or a downconverter. In the following example, we will consider a downconverting application (RF range 3.0 to 3.5 GHz, LO fixed at 2.5 GHz, IF range 0.5 to 1.0 GHz).

- *Setup step 1:* Configure the RF sweep range. The VNA uses the RF range as the basis for all subsequent generator (LO) and receiver (IF) settings. Start from an instrument preset. Set the start frequency to 3.0 GHz and the stop frequency to 3.5 GHz. Use the default number of points (201), yielding 2.5-MHz steps.
- *Setup step 2:* Configure LO and IF ranges via the Mixer Measurement dialog. Entering this dialog with the RF range already set simplifies the process: Channel > Channel config > "Mixer Mode" tab > "Mixer Meas Wizard..." Under "Conversion," set Mixer 1 to "IF = RF – LO (Down, USB)." Under "Fixed," select "LO" and set the LO frequency to 2.5 GHz. In the block diagram, select "Port 3" for the LO. Click the "Pow-

er" tab and set the LO power to +7 dBm. (The default value of −10 dBm is fine for the RF and IF ports.)

- *Setup step 3:* Select desired mixer measurements. The user's selection will determine the type(s) of calibration that will be required and also creates the channels necessary for the different generator and receiver frequency combinations. Click "Next" and select the required measurements. In this case, we select all six: conversion gain (S_{21}), RF isolation (S_{21}), RF reflection (S_{11}), LO leakage (S_{13}), IF reflection (S_{22}), and LO feedthrough (S_{23}).

6.3.2 Scalar Mixer Calibration Steps

Proper calibration is essential for scalar mixer measurements. The process is very similar to the combined S-parameter and power calibration described in Chapter 5 for active device applications. The primary difference is in the multiple frequency lists, since calibration will be required at the RF, LO, and IF ranges (a total of 13 calibration steps plus three (optional) for source leveling). Fortunately, the Mixer Measurement Wizard takes care of these details (along with the channel configurations).

- Click "Next" and select "Continue with SMARTerCal" (either "Manual" or "Cal Unit" if you have an autocal). Press "Finish" to close the Mixer Wizard and commence the calibration process.
- On the next dialog, if using a manual calibration kit, select the correct kit and test cable gender from the pull-down menus. Also, be sure to select the correct calibration kit in this menu. You should also see your power sensor under "Power Meter" (e.g., Pmtr 1). Pick the port you would like to use for power measurements or accept the default (Port 1). This will ensure that all measurement receivers associated with the generators are power calibrated. You should also click the checkbox for "Source Flatness" to calibrate the output level of the port-1 generator as well. Then click "OK."
- Commence S-parameter calibration in the next dialog, starting with port 1. Attach the appropriate calibration standards (OSM), as prompted by the wizard, and then "Start Cal Sweep" after attaching each of the standards. Continue for the remaining ports, as determined by your choice of measurements (Section 6.3.1) and ports.

- Step 10 in the ZNB calibration dialog asks you to attach the power sensor to the port you selected for power measurements. Do so and click "Start Cal Sweep."
- The remaining steps prompt you to connect an unknown thru between the different ports to complete the S-parameter calibration and scalar power calibration process.
- When a sufficient number of unknown thru connections have been made, an information dialog appears stating: "A sufficient number of two-port standards has been measured. Additional ones, however, increase accuracy." (This message only appears if you are calibrating more than two ports.) You can either click "Apply" now or continue with the remaining calibration steps and then click "Apply."
- The next dialog is used to achieve source flatness over the entire measurement frequency band. Connect the DUT per the block diagram and, if using an external LO source, ensure the LO drive is applied. On the right "Flatness Cal" menu, set the "Max Iterations" to 10 and the "Tolerance" to 0.1 dB. Click "Start Cal Sweep." At the end of the process, there will be green checkmarks beside each port in the block diagram. Click "Apply" at the bottom of the dialog.

6.3.3 Scalar Mixer Measurement Steps

After completing the calibration steps, connect the mixer. You might want to drag the different traces to different display windows (for individual amplitude scaling). Note that the S_{21} designation associated with conversion loss is not a true S-parameter, but simply a ratio measurement associated with the wave quantities at port 1 (RF) and port 2 (IF). Conversely, the s_{21} measurement associated with RF isolation is a true two-port, error-corrected S-parameter measurement. It characterizes the amount of unconverted RF-band energy leaking from the RF port to the IF port. This is also true for s_{13} (LO leakage) and s_{23} (LO feedthrough). The RF and IF reflection measurements take place exclusively at the RF and IF, respectively.

Other required measurements may be added and configured individually. If the desired input-output frequency relationships already exist in a channel, simply click on any existing trace in that channel and add a new trace. If you need a new frequency relationship, select a channel/trace close to what you need, and click "New Ch + Tr" from the top of the screen. For example, if I want to measure the LO return loss, I will need a new channel. In this case, select either of the reflection measurements (s_{11} RF reflection or s_{22} IF reflection) and then click “New Ch + Tr.” In the pop-up dialog, select S_{33}. A new trace and channel are created. Since the LO frequency is fixed, the measurement is not too exciting. Suppose I want to see how the LO match changes with LO power. I can change the sweep type to a power sweep. The steps are provided here:

- Channel > Sweep> “Sweep Type” tab > “Power”. You will immediately get a “Frequency Out of Range” error, because by default the “Power” sweep sets the Fb value (e.g., RF) to 1 GHz. (Keep in mind we originally defined an RF sweep range from 3.0 to 3.5 GHz.) So the first order of business is to change Fb to the middle of the mixer’s RF range (Channel > “Stimulus” tab > “CW Frequency”: Set the “CW Frequency” to 3.25 GHz.)
- Next, set the power sweep range: Channel > “Stimulus” tab > “Start Power…” set to −15 dBm. Channel > “Stimulus” tab > “Stop Power…” set to +7 dBm. Channel > Sweep > “Sweep Params” tab > set “Number of Points” to 89 (0.25-dB steps).
- Finally, assign the power sweep variable to the LO port: Channel > Channel Config > “Mixer Mode” tab > “Mixer Meas…” Select the “Power” tab and for the RF port, change the “Type” from “Base Pwr” to “Fixed” and set the level to −10 dBm. Similarly, for the LO port, change the “Type” from “Fixed” to “Base Pwr” and note that the sweep range is the same as the stimulus range we set above. Finally, for the IF port, change the “Type” from “Base Pwr” to “Fixed” and set the level to −10 dBm.

In addition to all the other mixer parameters, the LO reflection is now measured as a function of power. Figure 6.3 shows the resulting measurements.

6.4 Mixer Phase Measurements with a VNA

In the introduction, we made the claim that there were few instruments naturally less suited to frequency converting measurements than the VNA. The key word here is *naturally*. If additional hardware is added to a traditional VNA, vector mixer measurements become possible. However, the setup complexity grows in proportion to the phase measurement capability. We will explore several different setups in this section and describe the capabilities and limitations of each.

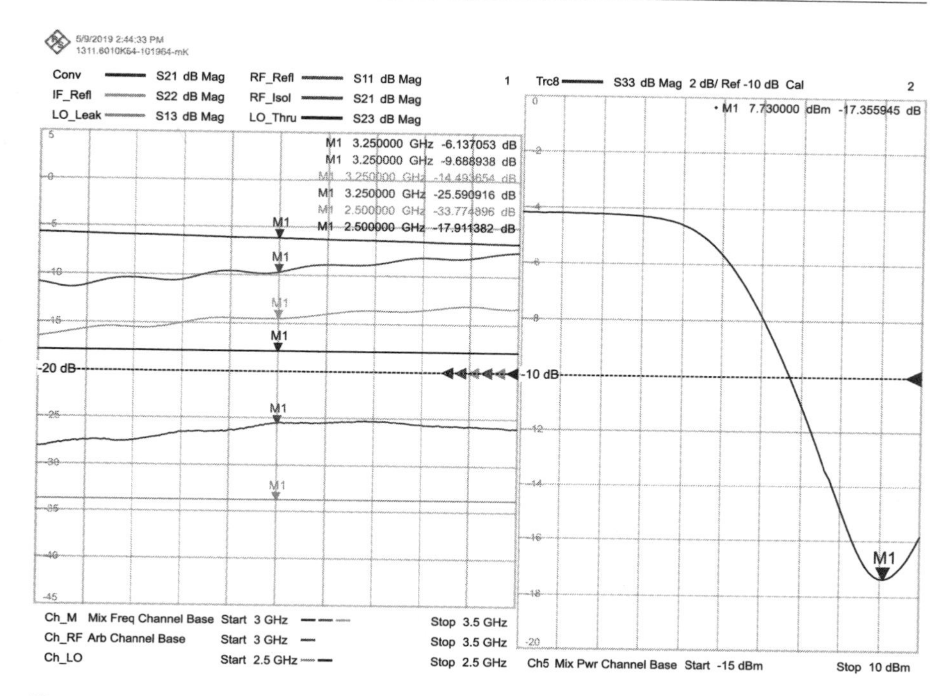

Figure 6.3 Mixer measurements, including LO port return loss versus power sweep.

6.4.1 Reference Mixer Method

By adding a reference mixer to the VNA test set, relative phase measurements can be made on a frequency-converting DUT. However, the resulting phase values are always relative to a reference mixer. This is perfect for applications comparing the relative phase performance of two or more devices. Figure 6.4 shows the basic technique for a four-port ZNB and a two-port ZVA.

Here, the port-1 generator and port-2 receivers operate at different frequencies. By adding a mixer into the reference path, we can obtain a meaningful and repeatable thru phase reference at the IF. This phase reference is intimately associated with the propagation delay through the reference mixer, which is not known and cannot be measured with this setup. However, it is stable and repeatable, which is all that matters. Different DUTs can be inserted into the MEAS path, and their phase is measured relative to the reference mixer. Generally, the phase of a golden DUT is obtained relative to the reference mixer over some frequency range, and the response is normalized to zero. Then any other mixer inserted into the measurement path will show phase deviations from this normalized value.

This method also relies upon a common LO source for both the reference mixer and the DUT. This LO is provided by a second internal VNA generator

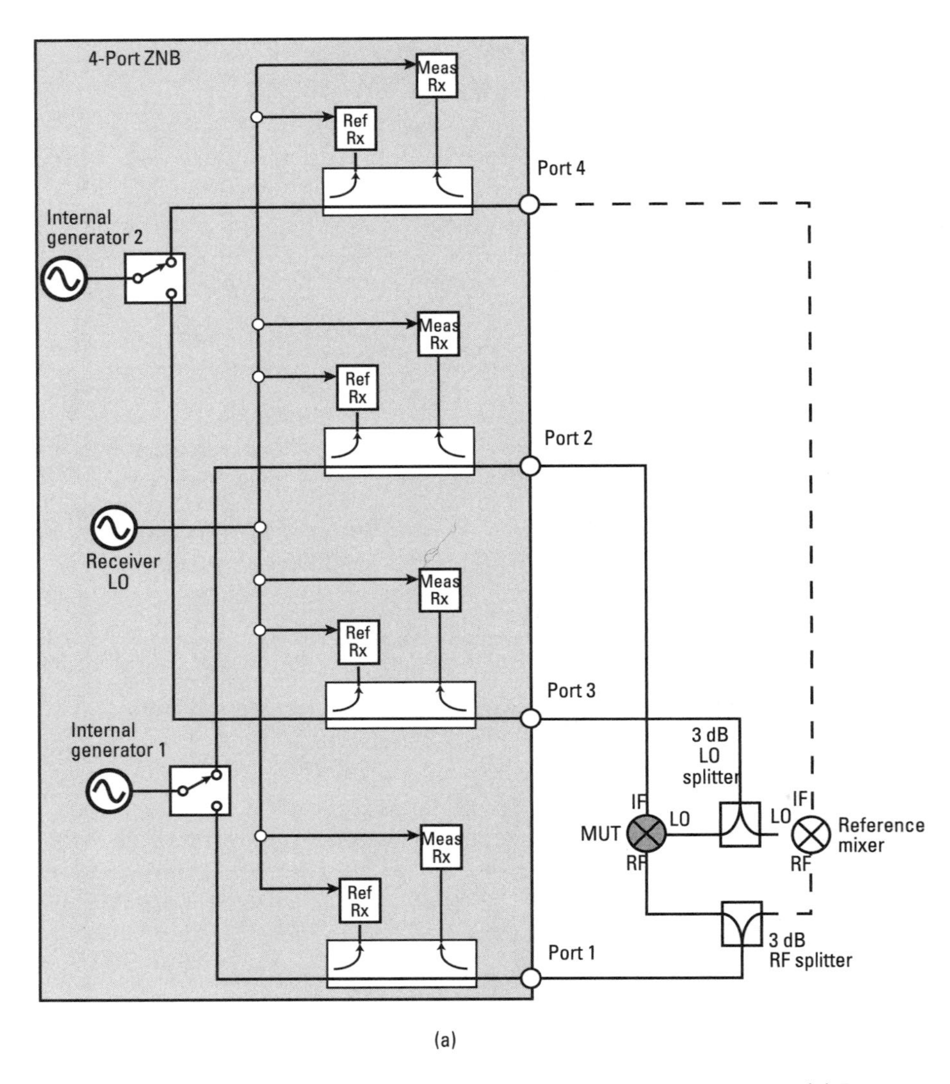

Figure 6.4 Setups for phase measurements using the relative mixer method. (a) Four-port ZNB. (b) Two-port ZVA with direct generator/receiver access.

or an external signal generator. By supplying this LO to both the reference path and the DUT measurement path, any LO frequency or phase drift will be common to both paths, and this common-mode drift eliminated when the phase difference (MEAS – REF) is calculated.

One shortcoming of this approach is that vector error correction of the through path is not possible; only trace math or normalization may be applied. Reflections (and hence phase ripple) may be tamed by applying attenuators (3

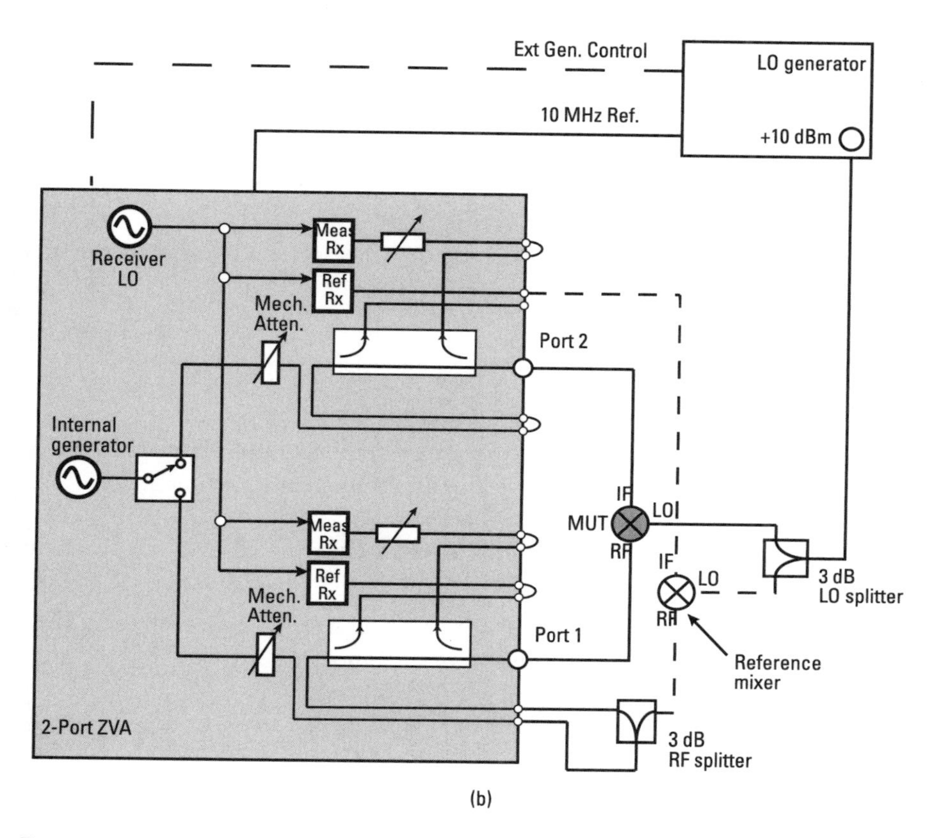

Figure 6.4 (continued)

to 6 dB typically) to the ends of the test cables (and at the input and output ports of the reference mixer).

6.4.1.1 Relative Phase Measurements with the ZNB

Figure 6.4(a) shows an example setup of the reference mixer method with a ZNB. Here, a four-port instrument is used with a second internal generator providing the common LO drive at port 3 for both the reference mixer and DUT via an external power splitter. Follow the same setup procedure as for a scalar mixer measurement (Section 6.3). A two-port instrument such as a ZVA with direct receiver access can also be used for this measurement as long as an external signal generator with sufficient drive power is available to provide the necessary LO drive for both mixers via a power splitter (Figure 6.4(b)).

After performing the S-parameter and power calibration (Section 6.3.2) and selecting desired measurements (Section 6.3.3) per the Mixer Measurement Wizard, an additional trace may be added for these phase measurements. Since port 4 is the reference mixer's IF port, configure the new trace for a ratio measurement. (For these next configuration steps, both the reference mixer and a golden mixer must be attached.)

- Select (highlight) the “Conv S21” trace and then click on “New Trace” from the menu at the top of the screen.
- Select any S-parameter that appears (e.g., S_{21}).
- Select Trace > Meas > “Ratios” tab > “More Ratios…” and set the numerator to “b2” and the denominator to “b4.” Then, click “OK” to close the dialog.
- Select Trace > Format and choose “Phase.”

Next, we need to normalize this phase trace to zero. (Here the golden DUT will provide the phase reference for all subsequent measurements.)

- Select Trace > Trace Config > “Mem Math” tab, then click “Data to New Mem.”
- Click the “Trace Math” button (to activate it) and deselect the checkbox in front of “Show Mem” (to hide the memory trace).

Now you are ready to make comparative phase measurements. Remove the golden mixer and insert another mixer that you wish to evaluate.

6.4.2 Vector-Corrected Mixer Method

The vector-corrected mixer method is considerably more sophisticated than the reference mixer method. Besides increasing the number of reference mixers required, this method also imposes a fundamental change in the VNA’s operation. Specifically, it requires the VNA generator to sweep the mixer’s RF range in the forward direction and its IF range in the reverse direction (Figure 6.5). By mounting an additional two reference mixers directly in front of the VNA reference and measurement receivers, all four receivers can make coherent phase measurements in the same frequency range, at the same time.

This approach has some other benefits: the reference mixers do not have to be specifically matched to each other. Their only requirement is that they must cover the same frequency range (RF, LO, and IF) as the DUT. However, precautions are still required. For example, filters are recommended on the output of the DUT and reference mixers to pass the desired frequency range (e.g., RF – LO) and exclude the image frequency range (RF + LO). Otherwise, this image band can remix with the LO to produce an artifact (error signal appearing as ripple) at the RF.

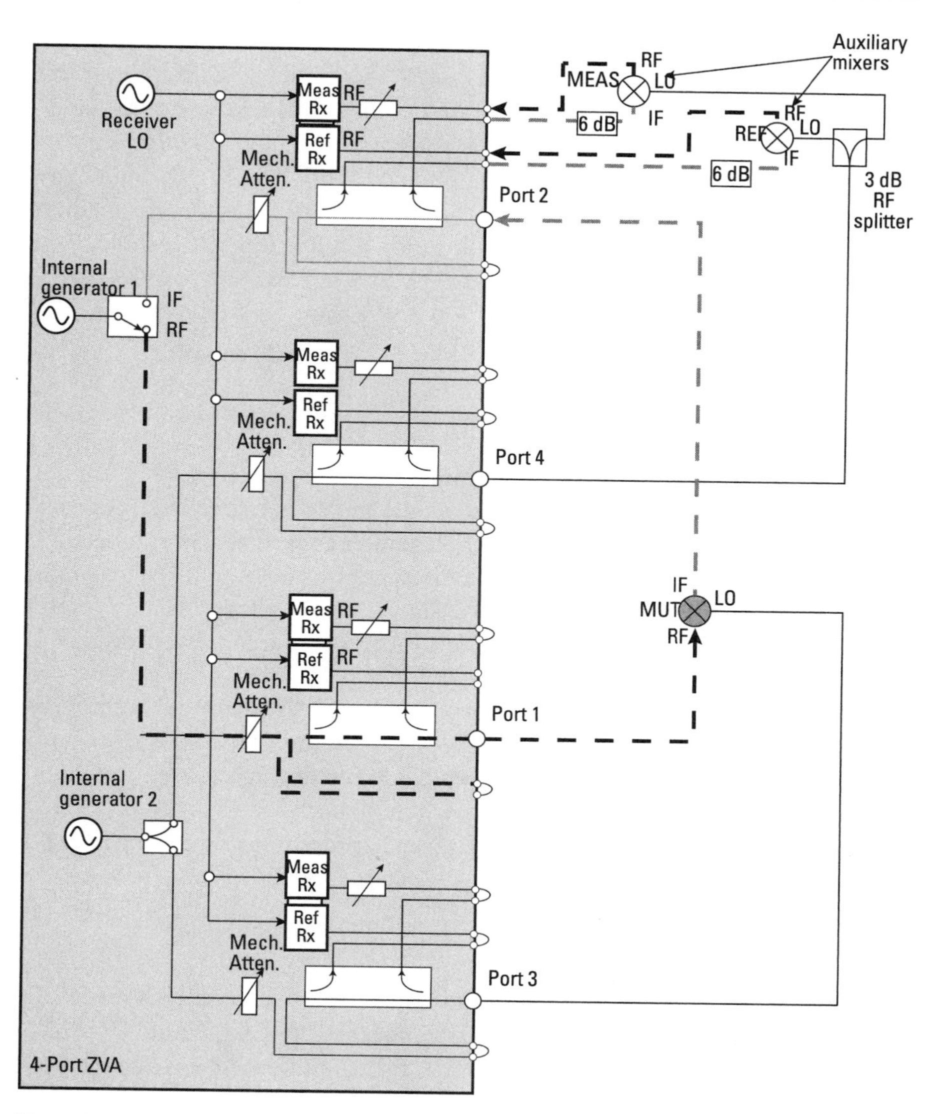

Figure 6.5 Frequency relationships with the vector corrected mixer method. When the internal generator switch is in the RF position, it provides a sweep of the DUT's RF at port 1. When the switch is in the IF position, it provides a sweep of the DUT's IF at port 2. In either case, all the receivers operate at the same frequency range (RF range).

Power calibration is optional, but generally advised for accurate generator levels and receiver power measurements. It may be performed before S-parameter calibration to establish proper RF, IF, and LO injection levels. (It is generally wise to ensure that the reference mixers receive the correct LO power before S-parameter calibration.) S-parameter calibration follows and takes advantage of the UOSM calibration method, where the unknown thru is assumed to be a reciprocal device. (See Section 2.4.4.5.) As previously mentioned, "reciprocal" implies $S_{21} = S_{12}$ (in both magnitude and phase). While this is easy to achieve for many passive structures (same-sex adapters, cables, filters), truly reciprocal mixers do not exist, but they get close enough (within certain operating ranges) and this is often sufficient. A manual calibration kit provides the open, short, and match standards at both ports, and then a reciprocal mixer is inserted for the unknown thru standard, with proper LO drive applied to this mixer during the calibration process. At the completion of S-parameter calibration, a DUT may be inserted and two-port S-parameters may be measured. The step-by-step procedure is provided in the next section.

6.4.2.1 External Generator Control with the ZVA

For the vector mixer measurement example, we use a two-port ZVA with external generator/receiver access (looping plugs) and an external signal generator to provide the required fixed LO signal. Although a mixer that uses a fixed LO does not require a communications link between the ZVA and the generator, there are many instances where the ZVA must control the external generator. (Examples include LO power sweeps and/or mixer configurations with fixed IF and swept LO.)

Below are the steps required to configure a ZVA to control an external generator. This same process may be used for providing an additional generator for two-tone IMD measurements on frequency-converting devices.

- System > System Config > (use arrows at top of dialog until you get to the "External Generators" tab).
- Click on "External Generators" tab and in the lower diagram, click "Add Other…"
- Create a descriptive name for the generator and then select the physical interface that you will use to communicate with the generator (GPIB or Ethernet). For the former, select "GPIB0" and provide the appropriate GPIB address for the generator in the adjacent text box. Although the ZVA has a GPIB connector on the rear panel, this is only for remote control of the ZVA from an external PC. This built-in GPIB connector cannot be used for controlling other instruments. Instead, employ a USB-to-GPIB adapter, such as the GPIB-USB-HS Interface Adapter from National Instruments. Appropriate driver software must also be installed directly on the ZVA to support the device. One other tip is to use one of the front-panel USB jacks for the USB-to-GPIB adapter. The front-panel jacks provide higher current than the rear-panel jacks. (Use the rear-panel USB connectors for keyboard, mice, and memory sticks,

and save the front-panel USB connectors for things like autocal units, power sensors, and the USB-to-GPIB adapter.)

For Ethernet communications, select VXI-11 and enter the IP address of the generator in the adjacent text box. If both the ZVA and external generator are connected to a router, they should both be set to "DHCP" so that the router assigns the IP addresses. For the ZVA, you can set this from Windows > Control Panel > Network and Internet > View network status and tasks > Change adapter settings. Then click on the "Local Area Connection" for the connected Ethernet line (there are two independent connectors on the back of the ZVA), and double-click "Internet Protocol Version 4 (TCP/IPv4)" and click "Obtain an IP address automatically." If you are only connecting the ZVA to the generator through an IP switch, or are using a direct cable connection from the ZVA to the generator, select "Use the following IP address" and type in a conventional address. (I prefer simpler addresses like "10.0.0.2.") Assign a similar address to the external generator (keep the first three parts of the address the same, and set the last digit to something different, like "10.0.0.5"). If the subnet mask is not applied automatically, use "255.0.0.0" for this example. (Other IP addresses, like "192.168.2.20," will use a subnet with the first three fields set to 255, for example, "255.255.255.0.") Then click "OK" to close each dialog. (In this example, I will assume the IP address of the external generator is 10.0.0.5, so I will type this IP address in the text box.)

- Select an appropriate driver from the Driver pull-down menu. There are quite a few predefined generators. Even if you cannot find an exact match, select a driver that is a similar model (more on this below).
- After selecting the driver, click on "Identify Type." If everything is connected properly, it should respond to the "*IDN?" command that was sent to the instrument, and show the response string next to the word "Type:"
- For "10 MHz Ref." typically select "External" if you intend to use the ZVA's 10-MHz reference as master and the generator as slave. (Make sure you provide the appropriate BNC cable between the ZVA 10-MHz reference OUT and the generator's 10 MHz reference IN.) Alternatively, select "Internal" if you intend to use the generator's 10-MHz reference as master and the ZVA as slave.
- Finally, the Fast Sweep option uses TTL hardware trigger/handshake signals between the ZVA and generator for faster sweeps. Without this option, the standard way of communicating with an external generator is to send a GPIB command for each frequency point as it is needed, and

wait for confirmation that the generator has completed this task before the ZVA makes a measurement at that frequency. Sweeps performed under fast sweep conditions are generally 10 and 50 times faster. However, the fast sweep mode is only supported by a handful of R&S instruments.

There are some cases where the desired generator is not included in the ZVA driver list. If a remote-control manual for the generator is available, then a custom driver can be easily created. Figure 6.6 shows an excerpt from a driver file. It is completely ASCII-based, so it can be created (or an existing driver can be modified) by substituting the instrument's SCPI syntax in the appropriate lines.

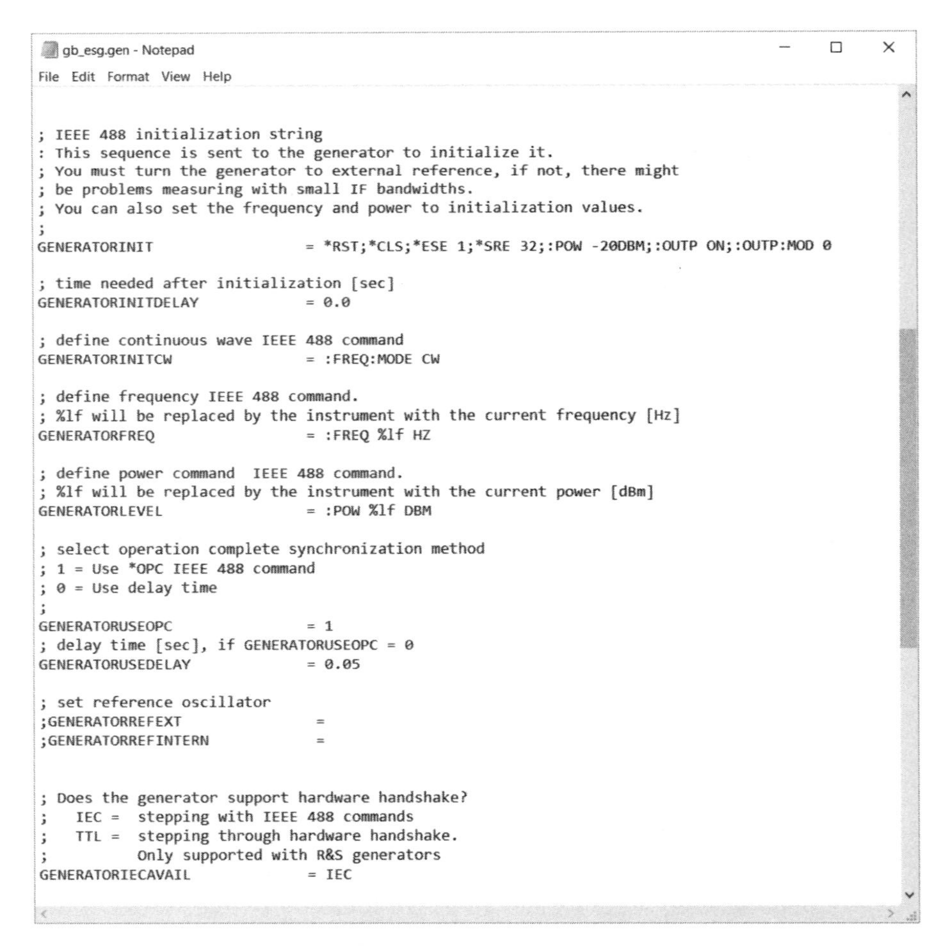


```
; IEEE 488 initialization string
: This sequence is sent to the generator to initialize it.
; You must turn the generator to external reference, if not, there might
; be problems measuring with small IF bandwidths.
; You can also set the frequency and power to initialization values.
;
GENERATORINIT                 = *RST;*CLS;*ESE 1;*SRE 32;:POW -20DBM;:OUTP ON;:OUTP:MOD 0

; time needed after initialization [sec]
GENERATORINITDELAY            = 0.0

; define continuous wave IEEE 488 command
GENERATORINITCW               = :FREQ:MODE CW

; define frequency IEEE 488 command.
; %lf will be replaced by the instrument with the current frequency [Hz]
GENERATORFREQ                 = :FREQ %lf HZ

; define power command  IEEE 488 command.
; %lf will be replaced by the instrument with the current power [dBm]
GENERATORLEVEL                = :POW %lf DBM

; select operation complete synchronization method
; 1 = Use *OPC IEEE 488 command
; 0 = Use delay time
;
GENERATORUSEOPC               = 1
; delay time [sec], if GENERATORUSEOPC = 0
GENERATORUSEDELAY             = 0.05

; set reference oscillator
;GENERATORREFEXT              =
;GENERATORREFINTERN           =


; Does the generator support hardware handshake?
;   IEC =  stepping with IEEE 488 commands
;   TTL =  stepping through hardware handshake.
;          Only supported with R&S generators
GENERATORIECAVAIL             = IEC
```

Figure 6.6 Signal generator driver file excerpt.

For example, to change the syntax for initializing the generator, replace the SCPI commands after "GENERATORINIT =" with the appropriate initialization commands for your generator. Then save your modified file using a unique filename. After saving this file, you must completely exit the System Configuration > "External Generators" dialog. To use your new driver, go to the System > System Configuration > "External Generators" tab and select the new driver.[1]

6.4.2.2 Absolute Phase Measurements with the ZVA

First, connect the auxiliary mixers and other components as shown in Figure 6.5. Then follow the steps below to configure the ZVA for vector mixer measurements.

- Start from a System > Preset.
- Set the RF: Channel > Stimulus > Start 3.0 GHz, Channel > Stimulus > Stop 3.5 GHz.
- Set the number of sweep points: Channel > Sweep > Number of Points… > 201 (2.5-MHz steps).
- Set the base power level (Pb) to −10 dBm: Channel > Power Bandwidth Average > Power… > −10 dBm.
- Select Channel > Mode > Vector Mixer Meas… > Define Vector Mixer Meas…
  - Leave the "Aux. Mixer" set to "Port 2."
  - Under LO, for the mixer under test (MUT) and auxiliary mixers, select "None." (Or, if you have configured your external generator for remote control by the ZVA, select your generator from the pull-down menu.)
  - Under Frequencies, set the proper relationship for the application. (Here, our RF range is 3.0 to 3.5 GHz, and the LO is fixed at 2.5 GHz. The IF sweeps from 500 MHz to 1 GHz.)
  - Set the "Conversion" to "IF = RF – LO (Down, USB)."
- Click "OK." The next step will be to set the power levels. The external LO generator must provide appropriate power levels for the two auxil-

1. In the ZVA and ZNB, the generator files are stored in the "Resources\ExtDev" folder of the program directory and have a ".gen" file extension. On a Windows 7 machine, the complete ZVA path is: "C:\Program Files (x86)\Rohde&Schwarz\Network Analyzer\resources\extdev". On a ZNB, the complete path (for a Windows7 machine) is: "C:\Program Files\Rohde-Schwarz\Vector Network Analyzer\Resources\ExtDev". (The external generator configuration dialog is similar for the ZNB, and may be accessed from System > System Config… > "External Devices" tab > "Generators…" button.)

iary mixers and the DUT. If all three mixers require an LO drive of +7 dBm, and a four-way power splitter is connected to the output of the generator, this means the generator must be able to source greater than +13 dBm. (The fourth output of the power splitter is terminated in 50Ω.)

- If the external generator is being remote-controlled by the ZVA, click the "Set Powers" button, and set the LO to "Fixed" (Set the power level to +7 dBm. This is the power level established at the end of the test cable.) If the external (LO) generator is not being remote-controlled, set its frequency to 2.5 GHz and set the power level using a power meter. Adjust the power until +7 dBm is achieved.
- Click "OK" to close the "Define Powers" dialog, and "OK" to close the "Define Vector Mixer Measurement" dialog.

Power Calibration

Connect an appropriate (preferably USB) power sensor to a front-panel USB jack and click "Mixer Power Cal." This will open the "Vector Mixer Measurement Power Calibration" dialog. If you are remote-controlling the LO generator, there will be three entries in the dialog for RF, IF, and LO levels. Otherwise, there will only be two entries (RF and IF).

- For more accurate source power levels, click on "Modify Settings…" and change the maximum number of readings to 10 and the tolerance to 0.1 dB. Leave all other settings at default values.
- Follow the dialog wording closely. For example, the first entry "Port 1 (RF): Connect Pmtr to Port 1" tells you to connect the power meter to the end of the test cable that feeds the RF port of the mixer. Then click that entry to begin the calibration.
- Repeat for the other port(s) and then click "Close."

Mixer Calibration

If you intend to use an autocalibration unit, connect it to one of the front-panel USB jacks. (Otherwise, you will use a manual calibration kit.)

- Click "Mixer Cal" and select either "Manual Cal" or "Calibration Unit." Follow the dialog steps.

- For the unknown thru step, connect a reciprocal mixer (including proper LO drive) and commence the measurement.
- In the next dialog ("Unknown Thru Characteristics"), simply click "Apply" unless using a dispersive media such as a waveguide. For the latter case, click "Dispersive." For a dispersive device, the delay time is frequency dependent and the analyzer determines two solutions for the transmission phase at the start of the calibration sweep. These two solutions differ by 180°, so the correct solution must be selected from a drop-down list. (The analyzer also assumes the transmission phase differs by less than 90° for any two consecutive sweep points.) If you are not sure which solution is correct, pick one and verify the measured result for reasonableness. If you made the wrong choice, use Channel > Calibration > Repeat Previous Cal… to select the alternate value.

At the completion of power calibration and mixer calibration, you may insert the DUT and commence any two-port S-parameter mixer measurements. However, you cannot perform isolation or leakage measurements (e.g., RF to IF feedthrough) with this setup, because the auxiliary mixers prevent it. You first have to remove these mixers and configure a conventional two-port S-parameter measurement. Then you can perform a conventional S-parameter calibration and (if desired) a power calibration. If remote-controlling an external generator, you will also have to configure the LO separately:

- Channel > Mode > Port Config to open the Port Configuration dialog
- Under the "Frequency" column, click the "…" in the "Gen 1" row and click the "0 Hz" on the left side of the dialog, and insert "2.5 GHz" into the right-side text box and then "OK." When you get the popup dialog box "Set Arbitrary Mode active?" click "Yes."
- Similarly, under the "Power" column, click "0 dBm" on the left side of the dialog, and in the adjacent text box for "Port Power Offset," insert an offset value (relative to 0 dBm) that will provide the appropriate LO drive to the DUT. Then click "OK."
- Click "Gen" next to "Gen 1" to activate the external LO generator.
- Click "OK" to close the Port Configuration dialog.

6.4.3 Simply Better Vector Mixer Measurements

The ZNA introduced a new synthesizer architecture that has revolutionized vector mixer measurements. Gone are reference-path mixers and auxiliary components. Instead, the ZNA's new architecture enables true signal coherency, even across RF, LO, and IF ranges. Previously, fractional *N* synthesizers provided precise frequency steps for both generators and LOs. Each time the synthesizer changed to a new frequency, it would have to reacquire phase lock at a new (and random) phase value. So if a generator toggled between two frequencies f_1 and f_2, and the receiver was parked continuously at f_1, then each time the synthesizer returned to f_1, it would have a different phase value relative to the receiver LO. This is demonstrated in Figure 6.7.

The ZNA architecture uses a DDS synthesizer as the clock reference for an integer *N* synthesizer. The DDS synthesizer smoothly and continuously accumulates phase as it transitions from one frequency to the next, so the integer-*N* synthesizer, once initially locked, stays locked throughout a frequency sweep. Since all synthesizers in the ZNA share a common clock and an identical synthesizer architecture, the DDS synthesizers can all be programmed to start at the same phase value (across their different operating frequencies) for truly coherent, repeatable performance. In Figure 6.8(a), we see three harmonically related signals (RF, LO, and IF) that are coherent because they all reach a positive peak value when the lowest-frequency signal (IF) reaches its positive peak value. If we examine these three signals on three channels of an oscilloscope and

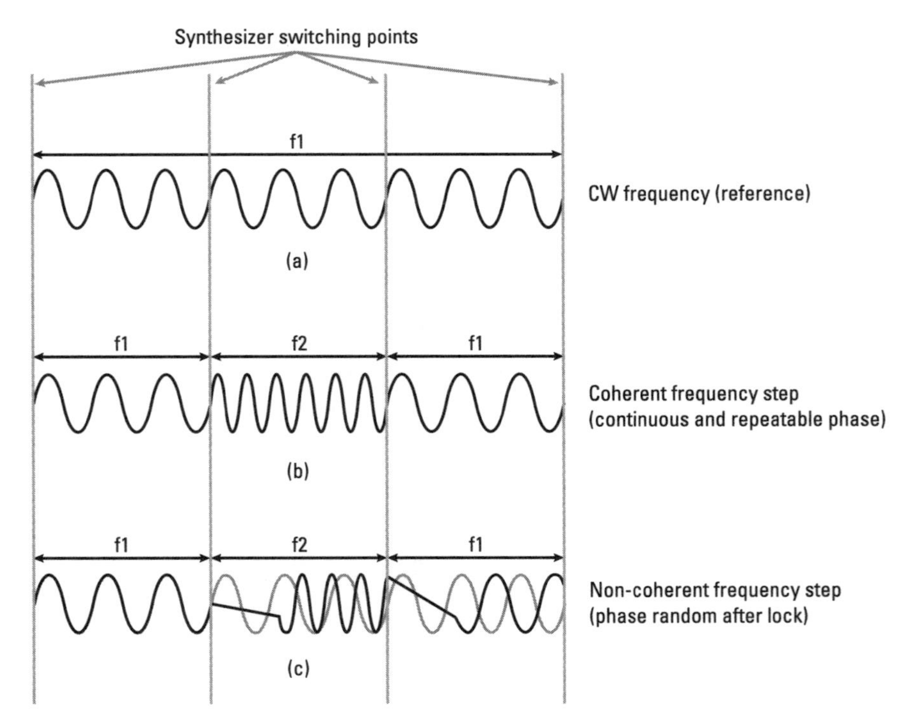

Figure 6.7 The problem with fractional-N synthesizers for mixer measurements: random phase angles produced when returning to a frequency.

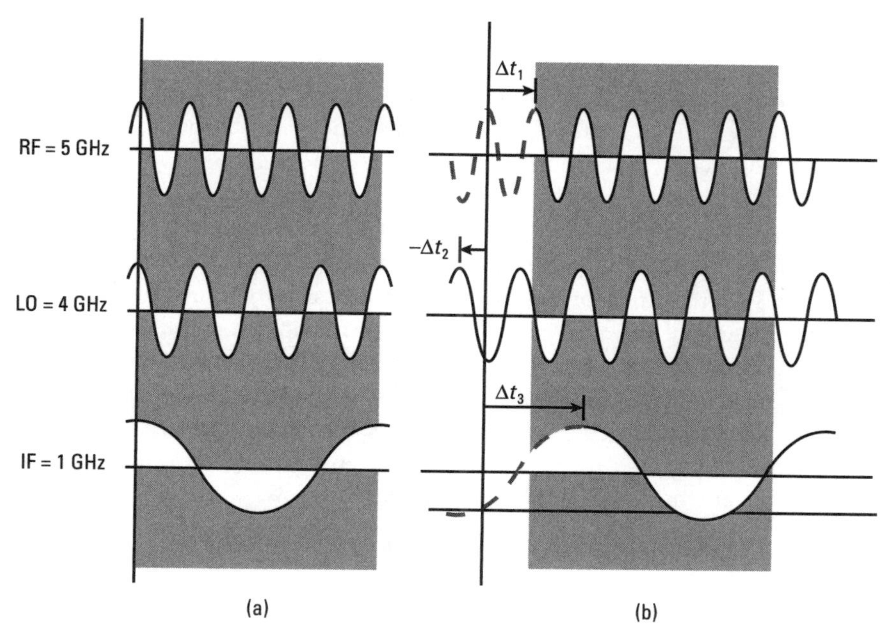

Figure 6.8 Example demonstrating phase relationship between coherent generators at three frequencies (RF, LO, IF). (a) Signal coherence demonstrated with 0 phase delay between LO, RF, and IF. (b) Signals are still coherent after adding electrical delay (e.g., different line lengths) between LO, RF, and IF.

apply an external trigger synchronized to the DDS synthesizer's start trigger, we see this exact same phase relationship every time a trigger event occurs. If we add different length cables to each generator and repeat the experiment, we will obtain a situation such as Figure 6.8(b), where the phase relationships between the three signals are different, but remain stationary with each trigger. However, if we tried this experiment with fractional-*N* synthesizers, the phase would be random every time the synthesizers reacquired their respective frequencies.

The receiver LO, which uses the same DDS synthesizer architecture, is applied to all reference and measurement receivers simultaneously. It too can be programmed to start at a specific phase value like the generator synthesizers, leading to coherent and repeatable phase values at each frequency step in a sweep, regardless of whether the LO is tuned to the RF, LO, IF, or any other frequency.

In standard operation, the ZNA receiver LO sweeps the RF to measure the phase at port 1, and then takes a second sweep across the IF to measure the phase at port 2. However, the ZNA has a provision for a second receiver LO. When this is used, the second receiver LO can be measuring the IF at port 2 while the original receiver LO measures the RF at port 1. This cuts overall mixer measurement time in half. In addition, since there are no auxiliary mixers required, it is possible to combine vector mixer frequency conversion measurements with conventional two-port S-parameter mixer measurements, such as RF to IF leakage (S_{21}).

The ZNA mixer method requires S-parameter calibration and uses the same UOSM vector-correction method as presented in Section 6.4.2. Amplitude, phase, and delay reciprocity are required by the UOSM calibration procedure, because these are a priori assumptions of the underlying correction algorithm. In reality, the only truly reciprocal parameter is likely to be the delay, which is determined primarily by the physical dimensions and layout of the mixer (electrical speed of propagation). For maximum amplitude accuracy, you may consider adding a new channel for a scalar mixer measurement (using SmarterCal for both S-parameter and power calibration). This will accurately reveal any differences in conversion gain/loss between the forward dBMag(S_{21}) and reverse dBMag(S_{12}) measurements.) The steps for calibrating and making a mixer measurement with the ZNA are presented below.

6.4.3.1 Absolute Phase Measurements with the ZNA

Setting up the ZNA for a mixer measurement is easy when using the Mixer Presetting Wizard. A mixer measurement can be configured in four steps, as shown here. Start from a System > Preset. Then launch the mixer measurement wizard by clicking Channel > Channel Config > Mixer Mode > "Mixer Meas Wizard…"

- *Step 1:* Define the mixer configuration.
  - On the block diagram, ensure that the RF input is set to "Port 1" and the IF output is set to "Port 2." Set the LO to "Port 3."
  - Under the "Frequency" tab, set the "Base Freq" parameter to the RF port, and set the start frequency to 3 GHz and stop frequency to 3.5 GHz. Set the Fixed parameter to LO, and set the LO frequency to 2.5 GHz. Set the "Auto" parameter to IF. Set the "Conversion" to "IF = RF – LO (Down, USB)."
  - Under the "Power" tab, set the RF port to "Base Pwr" and set the Start/CW level to −10 dBm. Set the LO port to Fixed and the level to +7 dBm. Set the IF port to "Base Pwr" (−10 dBm).
- *Step 2:* Click "Next" at the bottom of the dialog and select the mixer parameters to be measured (Table 6.1). Depending on your selection, the required channels and traces are set up automatically. Note that LO parameters can only be measured if a VNA port was selected to provide LO drive (currently only possible with a four-port R&S ZNA).
- *Step 3:* Click "Next" at the bottom to complete the mixer measurement configuration. From this final dialog, you can leave the wizard dialog

Table 6.1
Automatically Created Channels and Traces

Measurement Parameter	Channel Name	Trace Name	Notes
Conversion gain S<i><r>	Ch_M	Conv	
RF reflection S<r><r>	Ch_M	RF_Refl	
IF reflection S<i><i>	Ch_M	IF_Refl	
RF isolation S<i><r>	Ch_RF	RF_Isol	No frequency conversion
LO[m] leakage S<r><o>	Ch_LO	LO_Leak	[m] = mixer number if
LO[m] feedthrough S<i><o>	Ch_LO	LO_Thru	2 mixers are measured

r = RF test port #, i = IF test port #, o = LO test port #

immediately (without performing calibration), or continue with calibration.

- *Step 4:* Perform a SmarterCal (combined S-parameter calibration and power calibration). Just select the appropriate S-parameter calibration kit (auto or manual calibration unit). The calibration procedure guides you through the required S-parameter calibration standards (including attaching a reference mixer for the reciprocal thru calibration standard). It also guides power calibration using an appropriately configured power meter or USB-based power sensor. A source flatness calibration can be automatically included as the final calibration step.

At the completion of this four-step process, save the setup as a new recall set. This will save both the ZNA configuration settings and the calibration data, in case something happens to disturb the setup (e.g., power mains glitch, operator error). After this, a test mixer may be inserted and all previously selected measurements will be performed automatically.

6.5 Two-Tone Group Delay Measurements

Without LO access, a DUT's LO frequency will inevitably drift with respect to the VNA generators and receivers employed in a group delay measurement. This frequency drift will manifest itself as a linearly increasing or decreasing phase shift with time. This error can be momentarily reduced or cancelled by manual or automatic means, but the frequency error will likely return due to thermal drift or other effects. The problem can become considerably challenging when measuring converters using multiple downconversion stages.

The mathematical representation of group delay was first introduced in Section 3.3.2 for linear devices, where the input and output frequencies were the same. The two-tone method is somewhat similar, except that two signals are simultaneously injected at the DUT input and the phase difference between them is measured at both the DUT input and output. The two-tone method requires that the frequency spacing of the tones remains within the analog IF bandwidth of the VNA receivers. The tones are sampled by the port 1 reference receiver at the RF and digitized by an analog-to-digital converter (ADC). Once digitized, all subsequent operations take place in the mathematical realm of 1s and 0s using digital signal processing (DSP) techniques: The tones are digitally filtered and downconverted to baseband using parallel digital downconverters (see the inset at the top of Figure 6.9).

The VNA provides a master clock for all generator synthesizers as well as for the LO synthesizer (used for all VNA receivers). It also provides a clock source for the digital signal processing chain to maintain system-wide signal coherence. This results in a constant phase difference between the two tones.

When the receiver LO synthesizer retunes from the RF (input) frequency to the IF (output) frequency of the DUT, it reacquires phase lock. The absolute phase value at which relock occurs will vary randomly from frequency to frequency and from sweep to sweep due to the fractional-*N* synthesizer architecture. However, once locked, analog downconversion takes place and the phase difference between the two tones is preserved. By applying coherent digital downconversion to each tone, the phase difference at the DUT input and output can be measured, and subsequent division by the tone frequency spacing allows the group delay to be calculated.

The group delay equations for the two-tone method are a bit different from the single-tone method shown in Figure 3.6(a). For example, in the two-tone method the Δf is provided by the tone spacing (Figure 3.6(b)). It is worth examining the details of these equations to better understand why the two-tone method is impervious to LO drift, instability, or even FM (provided that both tones remain within the VNA receiver's IF bandwidth the entire time.)

The frequency of the first RF tone is mathematically represented by:

$$v_1^{RF}(t) = \cos\left(\omega_1 t + \phi_{1,in}\right) \tag{6.2}$$

Similarly, the second tone is represented by:

$$v_2^{RF}(t) = \cos\left(\omega_2 t + \phi_{2,in}\right) \tag{6.3}$$

The LO is represented by:

$$v^{LO}(t) = \cos\left(\omega_{LO} t + \phi_{LO}\right) \tag{6.4}$$

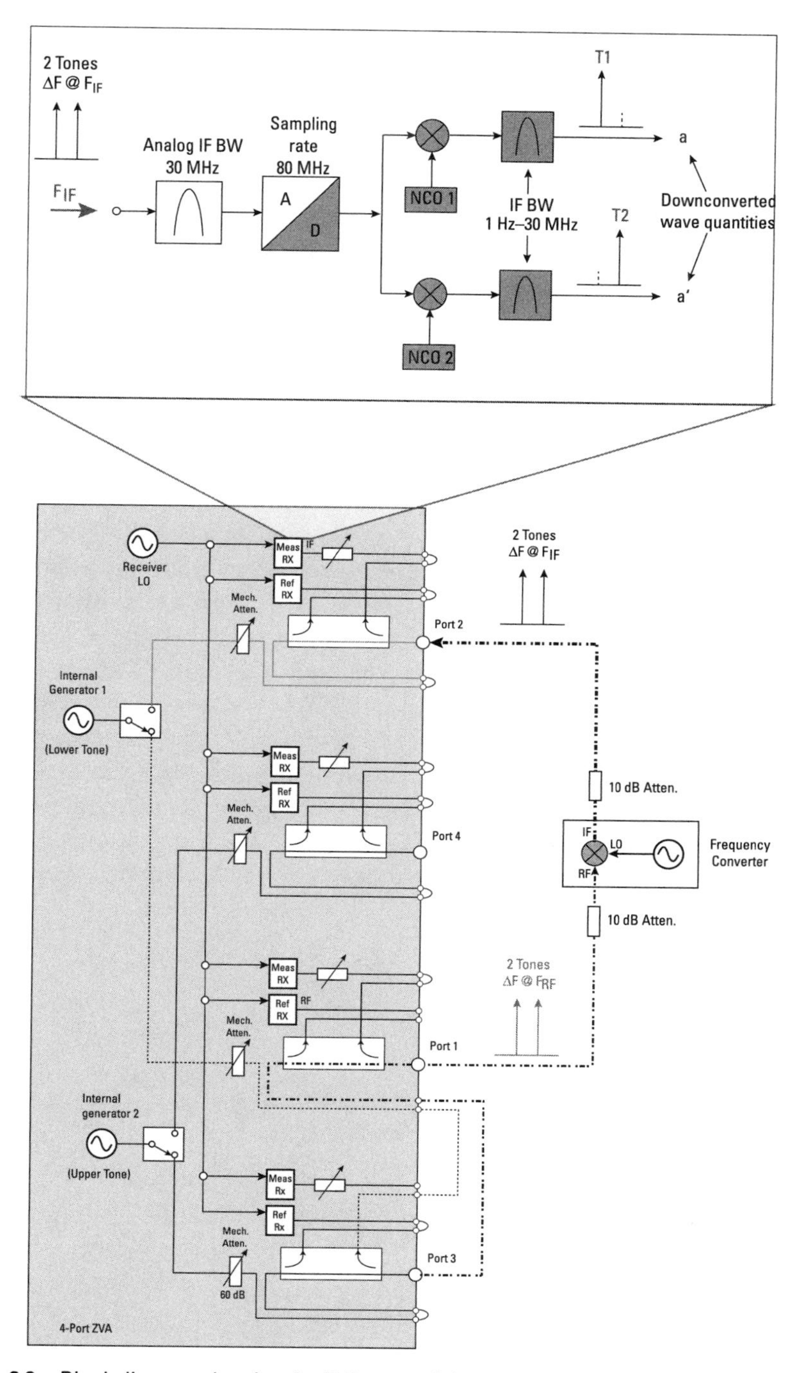

Figure 6.9 Block diagram showing the ZVA group delay configuration using internal combining for two-tone generation. The insert at the top shows the receiver processing for the two tones.

Assuming a downconverting mixer, IF = RF – LO. This means that at the output of the mixer, each tone will be downconverted to IF:

$$v_1^{IF}(t) = \cos\left(\left(\omega_1 - \omega_{LO}\right)t + \left(\phi_{1,out} - \phi_{LO}\right)\right) \tag{6.5}$$

$$v_2^{IF}(t) = \cos\left(\left(\omega_2 - \omega_{LO}\right)t + \left(\phi_{2,out} - \phi_{LO}\right)\right) \tag{6.6}$$

Notice that both tones have been impacted equally by the LO frequency and phase. Coherent digital downconversion of these tones to baseband eliminates the frequency terms, leaving only the phase terms.

The difference in phase between the two output tones will be:

$$\Delta\phi_{out} = \left(\phi_{2,out} - \phi_{LO}\right) - \left(\phi_{1,out} - \phi_{LO}\right) = \left(\phi_{2,out} - \phi_{1,out}\right) \tag{6.7}$$

The LO phase component disappears completely from this equation.

To compute the group delay, we start with the basic group delay formula:

$$\tau_g = \frac{-1}{360°}\left(\frac{\Delta_\varphi}{\Delta f}\right) \tag{6.8}$$

where $\Delta\varphi$ = change of phase in degrees, Δf = change of frequency in hertz, and τ_g = group delay in seconds.

Equation (6.7) provides the $\Delta\varphi$ term:

$$\Delta\varphi = \left(\Delta\phi_{out} - \Delta\phi_{in}\right) = \left(\phi_{2,out} - \phi_{1,out}\right) - \left(\phi_{2,in} - \phi_{1,in}\right) \tag{6.9}$$

Since the tone spacing is given by:

$$\Delta f = f_2 - f_1 = \frac{\left(\omega_2 - \omega_1\right)}{2\pi} \tag{6.10}$$

we can rewrite the group delay equation as:

$$\tau_g = \frac{-1}{360°}\left(\frac{\left(\phi_{2,out} - \phi_{1,out}\right) - \left(\phi_{2,in} - \phi_{1,in}\right)}{f_2 - f_1}\right) \tag{6.11}$$

6.5.1 Group Delay Measurements with the ZVA

To set up group delay measurements on the ZVA, two-tone stimulus is required. With a four-port ZVA, this can be accomplished in two ways:

1. Use an external power combiner to combine ports 1 and 3 together.
2. Use the B9 Cable Set to combine the signals internally using the port-3 coupler.

The setup for the second method is shown in Figure 6.9.

After making the required connections, the ZVA Mixer Delay measurement dialog provides all the steps necessary to configure and calibrate the instrument. These steps are presented below. Here, we again assume a mixer with input (RF) range of 3.0 to 3.5 GHz and a fixed (inaccessible) LO of 2.5 GHz.

- Start from an instrument preset: System > Preset
- Set the RF: Channel > Stimulus > Start 3.0 GHz, Channel > Stimulus > Stop 3.5 GHz.
- Set the number of sweep points: Channel > Sweep > Number of Points… 201 (2.5-MHz steps).
- Set the base power level (Pb) to −10 dBm: Channel > Power Bandwidth Average > Power… −10 dBm.
- Select Channel > Mode > Mixer Delay Meas > Define Mixer Delay Meas…
- Select "Port 3" for the upper tone.
- Use the default tone spacing: Aperture = 1 MHz.
- Accept the default "Measurement Setup." (Note: The ZVA does not have an actual internal power combiner. We are accomplishing power combining using the port-3 directional coupler.)
- Use the default Meas BW = 10 kHz.
- Set "Selectivity" to "High."
- Click "Define Mixer Measurement" and in the block diagram, change the LO source from "None" to "Embedded."
- Click "Set Frequencies" and change the LO frequency to 2.5 GHz. The default "Conversion" is fine, for example, "IF = RF – LO (Down, USB)." Click "OK."
- Click "Set Powers" and ensure that both RF and IF are set to "Base Power" and a value of −10 dBm. (The LO entries do not matter, since

we are assuming an inaccessible LO embedded inside the DUT.) Click "OK" to close the "Define Powers" dialog.

- Since this two-tone method is a radical departure from conventional S-parameter measurements, the only available correction is scalar only, implying power calibration, not S-parameter calibration. Use high-quality attenuators at the ends of the test cables (if necessary) to reduce ripple due to reflections between the VNA ports and the DUT ports. These attenuators should exhibit a very good match (better than 25 dB) and attenuation values of between 3 dB and 6 dB, typically. Click "OK" to close the "Define Mixer Measurement" dialog.
- Click "OK" again to close the "Define Mixer Delay Measurement without LO Access" dialog.

Calibration for the mixer delay measurement is limited to normalizing the delay response using a "Golden Mixer" or a reference frequency-converting device. Ideally, the device should have flat, well-characterized group delay. There are two calibration steps. First, load the data associated with the delay of the calibration mixer, and then perform the calibration sweep.

- Select "Cal Mixer Delay Meas."
- If the group delay is known, it may be entered as either a constant value or a table. For tabular data, click "Variable Delay" and click the "Load…" button to select an appropriate file. A tabular mixer delay must be stored in ASCII file format with a ".csv" extension. The ZVA provides two examples in the default directory "C:\Program Files\Rohde&Schwarz\Network Analyzer\Calibratio\MixerDelay". This file format is compatible with the ZVA's Trace Export function "Trace > Trace Funct > Trace Data > Export Formatted Data…ASCII Files (*.csv)" If the delay information is collected using a conventional S-parameter reflection measurement, then the delay values represent a round-trip delay. In this case, click the "Divide Calibration Data by 2" box on the dialog. Often, the delay of the calibration device (a bare mixer) may not be known, but will be essentially constant and much less than the DUT (if the DUT is a complex frequency converter containing many RF components (especially filters)). Simply assuming that the mixer has a constant delay of 0s (default) is suitable when only the variation in the group delay over the device's frequency range is of interest (not absolute delay).
- Click "Take Cal Sweep." At the conclusion of this sweep, you may save the calibration sweep data (cal Step 3 "Cal Data Manager" is optional).

- Click "Close" to dismiss the "Mixer Delay Meas Calibration" dialog.

You may now replace the calibration device with your DUT and commence delay measurements. Several other quantities can be derived from this basic measurement as well:

- Trace > Meas > Mixer Meas > Delay Derivative shows how much the mixer delay changes (delay slope) at each sweep point as the sweep variable (frequency) changes.
- Trace > Meas > Mixer Meas > Mixer Phase is the integral of the mixer delay over frequency, multiplied by −360°. This yields the transmission phase of the mixer relative to an initial phase of 0° at the first sweep point.

With group delay measurements, there is a natural trade-off between group delay resolution and trace noise. By increasing the aperture (e.g., spacing between tones), the trace noise will be reduced, but the overall group delay response will tend to be smoothed, leading to inability to discern sometimes important small phase details. (This follows from the definition of group delay. As the denominator (e.g., frequency spacing) gets larger, the quotient tends to get smaller.)

There is another limitation to increasing the aperture. For devices showing significant group delay, care must be taken to avoid a phase change of more than 180° between the two tones. Otherwise, aliasing occurs (Figure 6.10). Fortunately, the maximum aperture can be calculated based on the maximum group delay of the DUT. We start with the basic group delay formula (6.8) and rearrange it to solve for the phase difference:

$$\Delta\varphi = \tau_g \cdot 360° \cdot \Delta f \tag{6.12}$$

Assuming the maximum phase difference can be no greater than −180°, we obtain:

$$\Delta f = 0.5/\tau_g \tag{6.13}$$

As an example, assume the DUT has a maximum delay of 100 ns. This means that the maximum aperture will be limited to 5 MHz. Keep in mind that this limitation applies only to the aperture size (tone spacing). The actual measurement step size (e.g., the amount by which the two tones are incremented (together) at each point in the frequency sweep) is typically much smaller, on the order of kilohertz.

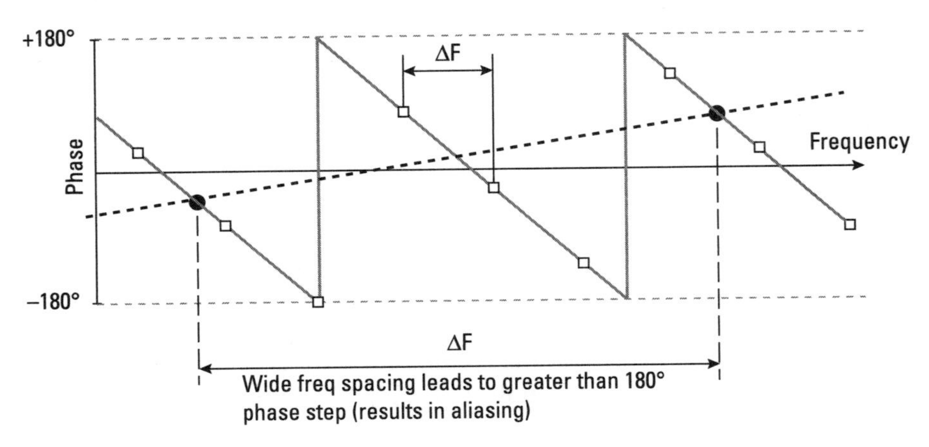

Figure 6.10 Group delay aliasing effects due to excessive frequency spacing.

If other conventional scalar measurements are desired (such as conversion gain/loss), they must be configured separately in a new channel following the scalar mixer measurements procedure of Section 6.3.

References

[1] Bednorz, T., "Group Delay and Phase Measurement on Frequency Converters," App Note 1EZ61_1E, Rohde & Schwarz, August 27, 2012.

[2] Davis, W., *Microwave Semiconductor Circuit Design*, New York: Van Nostrand Reinhold, 1984, pp. 253–262.

[3] Henderson, B., "Predicting Intermodulation Suppression in Double-Balanced Mixers," *Watkins-Johnson Company Technical Notes*, Vol. 10, No. 4, July/August 1983.

[4] Marki, F., and C. Marki, "Mixer Basics Primer," Marki Microwave, Morgan Hill, CA, 2010.

[5] "Frequency Mixers: Frequently Asked Questions About Mixers," Mini-Circuits, AN00-011, 2017.

[5] "How to Select a Mixer," Mini-Circuits, AN00-010, 2017.

[6] "Low Cost, Triple Balanced, LTCC Mixer," Mini-Circuits, AN00-002, 2017.

[7] "Mixers: Terms Defined, and Measuring Performance," Mini-Circuits, AN00-009, 2017.

[8] "Novel Passive FET Mixers Provide Superior Dynamic Range," Mini-Circuits, AN00-003, 2017.

[9] "Selecting the Right Mixer for your Application," Mini-Circuits, AN00-014, 2017.

7

Pulse Measurements

7.1 Pulse Measurement Introduction

There are a number of RF components specifically designed for pulsed operation. Radar systems (for both automotive and military applications) immediately come to mind. Other examples include components for time-division duplex (TDD) communications systems, which send data in short bursts or pulses. These systems began in earnest with time-domain multiple access (TDMA) in 2G mobile telephones,[1] and are still found today in the latest 5G-NR technology standard. Additionally, there are classes of devices that simply cannot tolerate continuous-wave (CW) RF excitation. For example, it is becoming popular to evaluate component performance via on-wafer measurements, where devices lacking proper heat-sinking must be tested with pulsed excitation to avoid thermal destruction.

Pulse operation is frequently accomplished by switching the RF signal on and off in a controlled manner, applying a CW RF signal and switching a control voltage on and off, or by a combination of these methods. Modern IQ modulation systems generate the pulsed waveform at baseband and modulate it onto the RF carrier using vector modulation.

The modern VNA uses constant-amplitude pulses to evaluate the performance of pulse-mode devices in a manner similar to conventional CW RF devices, since in both cases the objective is to measure the complex ratio of reference, reflected, and transmitted energy. Synchronization between the

1. GSM (Global System for Mobile communications), first introduced in 1991.

pulse generator and the VNA may or may not be necessary, depending on the measurement technique employed. Other unique challenges posed by pulse measurements include pulse desensitization (which impacts the measurement's dynamic range) and pulse power calibration. These topics will be covered in this chapter. However, before discussing specific measurement approaches, it is important to examine pulses in both the time and frequency domain to appreciate how characteristics of one domain affect the other.

7.1.1 Time-Domain Representation

In the time domain, application of periodic on-off modulation to an RF carrier produces a pulse train. This train consists of individual pulses that are characterized by two fundamental parameters: pulse width (τ) and pulse period (T). Pulse period is sometimes referred to as pulse repetition interval (PRI). You may also encounter pulse repetition frequency (PRF), which is simply 1/. Figure 7.1 shows a pulse train as observed on a spectrum analyzer. The pulse duty cycle (d.c.) is derived from the pulse width and period, and is expressed as:

$$d.c. = \tau / T \tag{7.1}$$

Both pulse width and pulse period have units of time, and the PRF is given in units of frequency. The duty cycle is usually expressed as a percentage. In Figure 7.1, each pulse has a width of 10 μs and a period of 100 μs, resulting

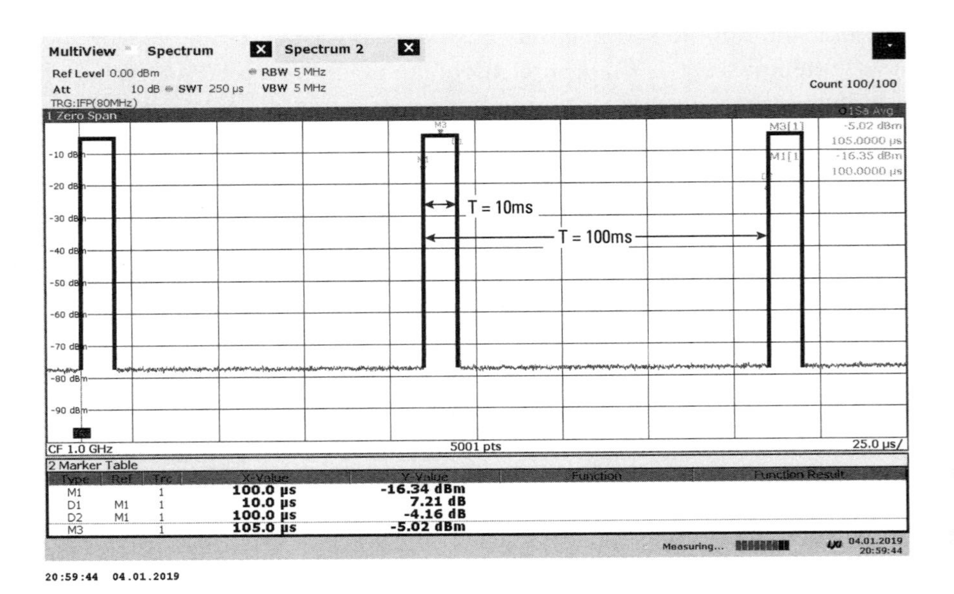

Figure 7.1 Pulse train viewed in the time domain.

in a 10% duty cycle.This means that the RF is only on for 10% of the time. If the pulse width decreases and/or the pulse period increases, the duty cycle will become smaller.

7.1.2 Frequency-Domain Representation

In the frequency domain, application of periodic on-off modulation causes the RF energy to be spread across a band of frequencies as shown in Figure 7.2. More specifically, for a repeating pulse stream, the resulting modulation spectrum consists of a large number of discrete CW frequency components mirrored on either side of a central CW tone. The amplitude envelope has a $\sin(x)/(x)$ roll-off with well-defined nulls occurring at offset frequencies that are the reciprocal of the pulse width (τ). Additionally, the closely spaced spectral lines are separated by exactly the reciprocal of the pulse period (T). This reciprocal relationship is maintained regardless of the duty cycle. As the pulse period gets longer, the adjacent spectral lines crowd closer together (and vice versa). Also, as the pulse width gets wider, the location of the spectral nulls move in closer to the central tone with fewer spectral lines between the nulls.

Both Figures 7.1 and 7.2 show measurements of the same RF pulse with a peak amplitude of −5.0 dBm. In the time domain (Figure 7.1), marker 1 clearly shows that the pulse peak amplitude is −5.0 dBm, but the average power (as measured with a power sensor or a spectrum analyzer in the time-domain

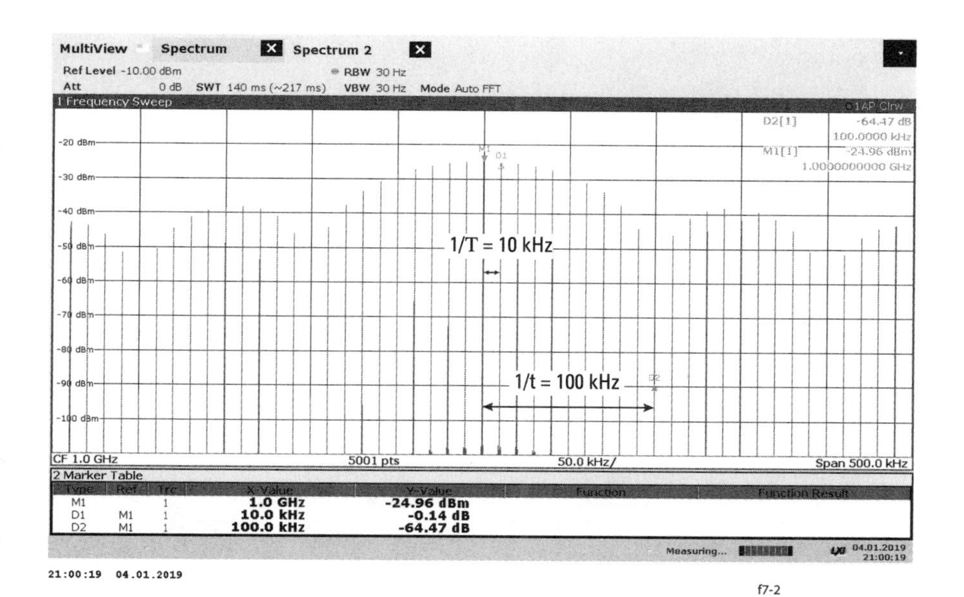

Figure 7.2 Same pulse train viewed in the frequency domain.

power mode) is −15 dBm (Figure 7.3). This is because the pulse is only on for one-tenth of the pulse period. This represents pulse desensitization, and for a wideband detector, is given by:

$$Desense\left(Wideband\right) = 10\log_{10}\left(d.c.\right) \tag{7.2}$$

Similarly, in the frequency domain, since the pulse power is dispersed across a wide frequency range, we must integrate the energy of all the spectral components, taking both amplitude and phase into consideration to arrive at the average power. Conversely, if we know how much power is contained in any particular spectral component (relative to the unmodulated carrier), we can use this information to compute the peak pulse power. Fortunately, this relationship is well known and calculated using the central tone as a reference:

$$Desense\left(central\ tone, Narrowband\right) = 20\log_{10}\left(d.c.\right) \tag{7.3}$$

In Figure 7.2, the marker table at the bottom of the screenshot displays the magnitude of the center tone as approximately −25 dBm. The desensitization for this 10% duty cycle pulse is calculated from (7.3) as −20 dB, leading to the conclusion that the peak pulse power is −25 dBm (measured) + 20 dB (desensitization correction factor) = −5.0 dBm. This agrees with the peak pulse power measured in Figure 7.1.

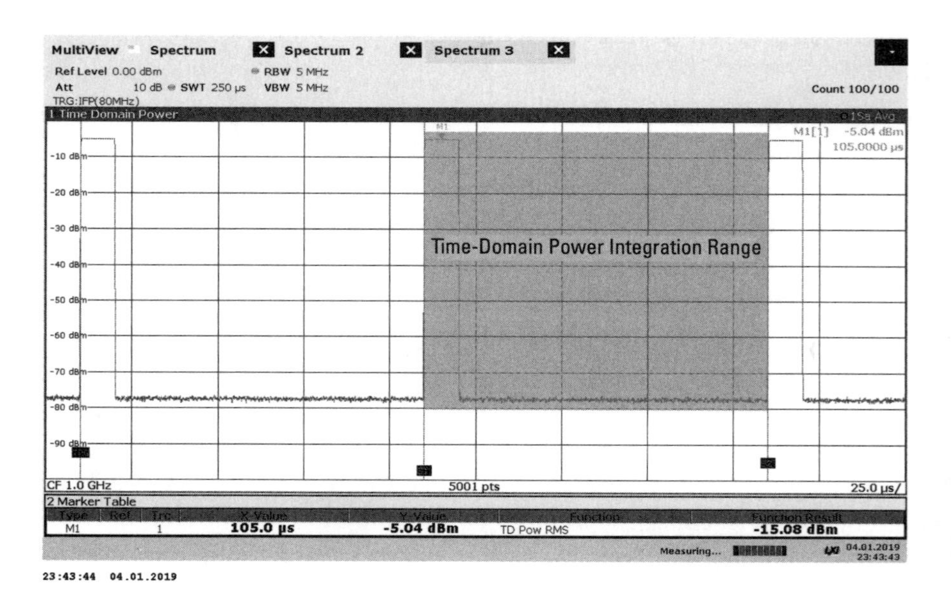

Figure 7.3 Pulse desensitization in the time domain.

7.2 VNA Pulsed Measurement Techniques

There are several different ways of making pulse measurements with a VNA. They differ according to measurement simplicity, speed, dynamic range, and pulse resolution capability. Table 7.1 presents trade-offs associated with the most popular methods. We will examine each in the order presented in the table.

With all these methods, it is important to understand that modern VNAs, like most state-of-the-art radio designs, employ digital filtering. Digital filters, like their analog counterparts, share a similar time-domain behavior in that narrower filters require more time to make a measurement. In the analog world, we speak of narrowband filters requiring more settling time than wideband filters. In the digital domain, we speak in terms of sampling time and note that narrowband digital filters are longer than their wideband counterparts, requiring more IQ samples to meet periodic sampling requirements for discrete-time signal processing.[2] This is an important trait to remember as we delve into the different pulse measurement techniques.

Table 7.1
Comparison of Pulse Measurement Techniques

Pulse Technique	Simplicity	Speed	Dynamic Range	Pulse Resolution Capability	Notes
Pulse Averaging	+++	– –	–	– – –	No pulse synchronization required
Point-in-Pulse (Wideband)	– –	+/–	+/–	+	Speed and dynamic range depend on duty cycle and VNA performance
Point-in-Pulse (Narrowband/ Pulse Gating)	– – –	– –	–	++	Complicated receiver gating; significant desensitization; requires optional hardware and software
Pulse Profile (Wideband)	++	+++	+	+++	Simple synchronization and high performance; requires optional software
Pulse Profile (Narrowband)	– – –	– –	–	++	Complicated receiver gating; significant desensitization; requires optional hardware and software

2. See Chapter 4, "Sampling of Continuous-Time Signals," in A. V. Oppenheim and R. W. Schafer, *Discrete-Time Signal Processing*, 3rd ed., Englewood Cliffs, NJ: Prentice Hall, 2010.

7.2.1 Pulse Averaging

With pulse averaging, we focus only on the central spectral line. As counter-intuitive as it may seem, this line and all the other spectral lines are actually continuous in the time domain. From (7.3), we know that only its amplitude is impacted by the pulse duty cycle. Because this center spectral line is always there (as long as RF pulses are being sent), it does not matter when the measurement is made. This means no pulse synchronization is necessary.

The VNA's signal processing treats a pulse-average measurement the same as it would a conventional S-parameter measurement using the VNA's built-in CW source. The only difference is that a narrowband IF filter is required to isolate the center spectral line from all the others. For example, if the spacing between adjacent spectral lines ($1/T$) is 10 kHz, the IF filter should be considerably narrower (e.g., 1 kHz) to filter out adjacent spectral lines. Most importantly, a 1-kHz digital filter may have a sample time of nearly 1 ms, meaning that it will require at least 10 pulses (1-ms filter sample time/(100 μs /pulse)) to satisfy sample time requirements. If you try to make a measurement with a pulse burst of fewer than 10 pulses, it will have the effect of lowering the amplitude of the desired tone and consequently increase trace noise (due to lower SNR). Even using a continuous pulse stream, (7.3) shows us that the pulse average method will exhibit amplitude desensitization according to the pulse duty cycle. In this case, the noise floor will be unchanged from a conventional s-parameter measurement, but the amplitude of the reference, reflected, and transmitted signals will be impacted by this desensitization factor. Pulse desensitization can be compensated by using a narrower IF BW filter, which reduces the noise floor without impacting the center tone amplitude. The trade-off is measurement time, since a narrower filter requires a longer sampling time.

Two-port, error-corrected S-parameter measurements will still require sweeps in both forward and reverse directions, but in the case of amplifiers requiring pulse excitation, the reverse-direction sweeps may often be performed with unmodulated RF (no pulse modulator required in the reverse direction).

7.2.2 Point-in-Pulse

While the pulse average method requires many pulses to provide a macro or averaged view of the DUT behavior (amplitude and phase) as a function of frequency or power, we sometimes want to zoom in to a portion of a pulse to see how the DUT responds during that particular snapshot in time. Figure 7.4 shows many important pulse characteristics that may be impacted by a DUT. In addition to these amplitude effects, the phase may change within a pulse due to thermal or energy storage effects. For these investigations, the point-in-pulse method provides resolution inside a pulse to gain insights into DUT temporal behavior. This requires tight synchronization between the pulse generator and

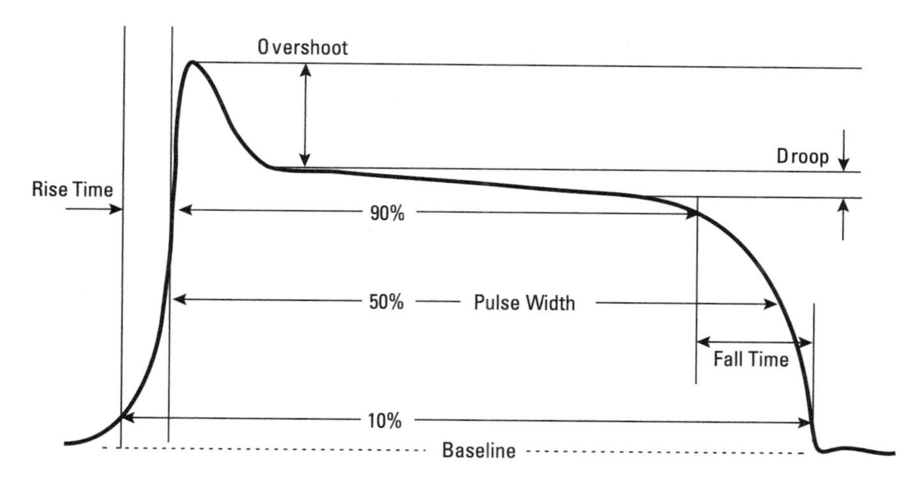

Figure 7.4 Pulse temporal characteristics.

the VNA receivers so that a particular point within the pulse may be examined while sweeping frequency, power, or time.

7.2.2.1 Wideband Point-in-Pulse

For the wideband point-in-pulse method, wide frequency-domain IF filters are used because they have short time-domain settling-time (or sampling time) requirements. For example, Figure 7.5 shows the impact of filter bandwidth on pulse resolution for a point-in-pulse measurement. A 1-MHz filter achieves 2-μs resolution, which represents 20% of the pulse width. Conversely, a 5-MHz filter achieves 0.5-μs resolution, which represents only 5% of the pulse width. Clearly, the 5-MHz filter would do a better job of isolating and characterizing, for example, amplitude and phase overshoot at the beginning of the pulse (if we wanted to trigger in that region).

However, better pulse resolution comes at a price. Whereas the previous pulse average method suffers reduced SNR due to pulse desensitization of the center spectral tone, the wideband point-in-pulse method suffers reduced SNR due to the use of wider IF filters, which allow more noise into the passband. Referring again to Figure 7.2, the amplitude of the spectral lines beyond the first null (at $1/t$) are quite low, so using a wider bandwidth may capture a bit more signal energy, but it also captures much more thermal noise. This noise may not seem very high, but when it is integrated across the entire filter bandwidth, it becomes appreciable. Mathematically, the thermal noise increases by $10\log_{10}$(ratio of bandwidths). So, increasing the bandwidth from 1 MHz to 5 MHz improves pulse resolution by a factor of 4 (0.5 μs versus 2.0 μs), but imposes a penalty of a 7-dB less dynamic range (= $10\log_{10}$ (5 MHz/1 MHz)). Unlike the pulsed average method, we cannot recover this dynamic range by

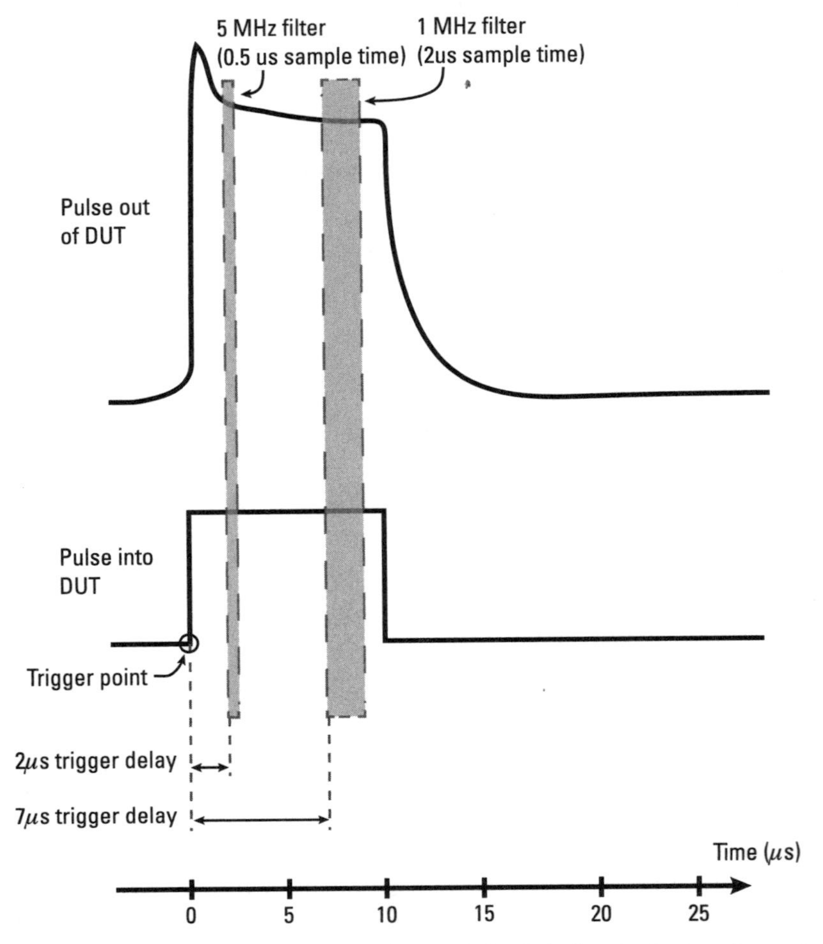

Figure 7.5 Impact of filter bandwidth on pulse resolution (wideband point-in-pulse method).

reducing the IF bandwidth, because in doing so we would lose our pulse resolution. The only option is to employ trace averaging over multiple pulses to reduce the trace noise at the cost of increased measurement time. If you have sufficient SNR due to the pulse characteristics (duty cycle) and/or sufficient signal levels into the VNA receivers, you will not need to apply pulse averaging.

Theoretically, a VNA using the wideband point-in-pulse method can perform a two-port calibrated S_{21} measurement at a single frequency or power level using only a single pulse, but this requires two conditions. First, the pulse must be long enough for two (forward and reverse sweep) partial measurements. Second, the VNA must be fast enough (switching from the forward sweep to the reverse sweep) to accommodate these measurements (Figure 7.6(a)). However, even here, there are additional caveats. Measurements will be obtained at two

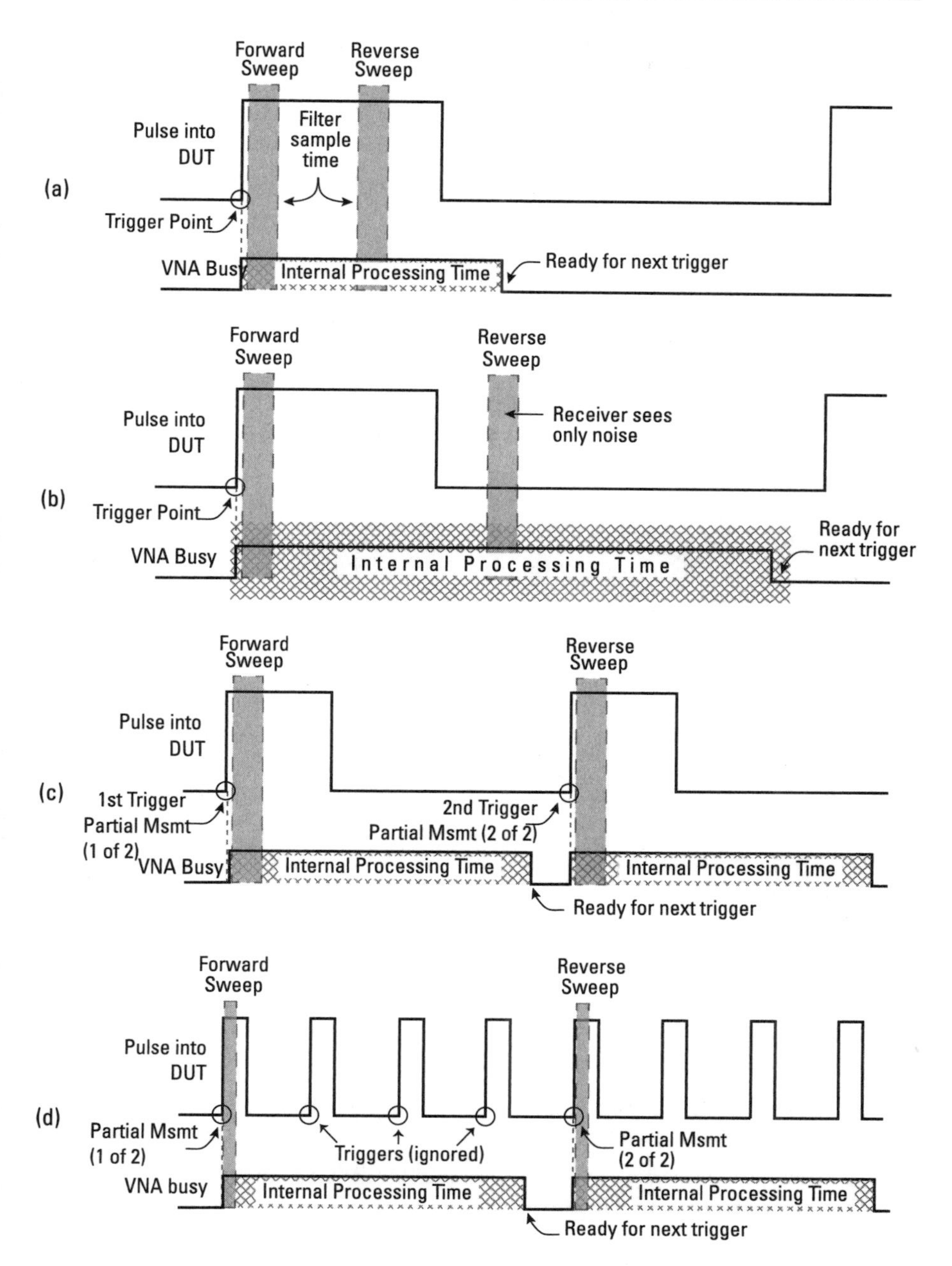

Figure 7.6 Point-to-pulse synchronization details. (a) Both forward and reverse measurements made during a single pulse, but at different times within the pulse. (b) Forward measurement properly triggered, but reverse measurement occurs during pulse off-time, resulting in erroneous s-parameter results. (c) Partial sweep mode allows triggering of forward and reverse measurements during subsequent pulses. (d) VNA internal processing time may result in a number of unmeasured pulses while switching between forward and reverse sweep.

different times within the pulse. If the DUT's response to the pulse is well behaved over the duration of the pulse, a good measurement will result. More often, the situation depicted in Figure 7.6(b) occurs. A measurement is comfortably triggered within the pulse on-time for the forward sweep, but by the time the analyzer has reconfigured for the reverse sweep, the VNA receivers see only noise associated with the pulse off time. In this case, S_{11} and S_{21} measurements will be degraded, and S_{22} and S_{12} will display only random noise. There are several ways to circumvent this problem. Perhaps the simplest method is to use a one-path, two-port calibration instead of a full two-port calibration. This type of calibration requires only a single sweep in the forward direction. If the DUT has good reverse isolation, the resulting S_{21} measurement will be as good as one made with full two-port calibration. Alternatively, if full two-port calibration is required, the two required partial measurements can be made using two distinct pulses: one for the forward sweep, and one for the reverse sweep. To ensure that these two partial measurements are made at the same point in consecutive pulses, the trigger mode is changed to partial measurement. Then the two partial measurements are combined by the VNA to calculate the full two-port S-parameters for a given trace point. Figure 7.6(c) shows how this works: The analyzer prepares for a new sweep and waits for the first trigger. When it receives the trigger, it makes the first (partial) measurement (forward direction only) and then reconfigures itself for the reverse direction and waits for the next pulse and trigger signal. When the next trigger occurs, the VNA makes the second (partial) measurement (reverse direction) and then calculates the associated two-port S-parameters from these two partial measurements, displays the results as a new trace point, increments its generator to the next frequency point, and prepares for the next partial measurement in the forward direction.

If the DUT's RF characteristics are time-invariant (meaning that they do not change over many pulses), then all the forward partial measurements can be performed for a frequency sweep, followed by all the reverse partial measurements. To do this, one must activate the alternate sweeps mode. This mode provides a modest time advantage over the conventional sweep mode.

Depending on the pulse characteristics, the VNA's internal data processing may not be fast enough to keep up with the pulse repetition rate. In such cases, a number of pulses will be missed before the VNA becomes armed and ready for its next partial measurement (Figure 7.6(d)). Depending on the DUT characteristics, this may or may not be important. If the DUT is already susceptible to thermal destruction, sending multiple pulses that will not be measured is foolish. Better to determine the minimum time between pulses that the VNA needs for its internal processing, and adapt the pulse duty cycle accordingly. If the duty cycle cannot be changed due to DUT requirements, consider a VNA with faster internal processing.

7.2.2.2 Narrowband Point in Pulse

The narrowband point-in-pulse method employs special hardware gating that restricts the receivers to seeing only a small fraction of each pulse. With this technique, the hardware gating is applied to the receiver's RF or IF stage. By synchronizing this gating with the beginning of each pulse, and using a specific gate delay and aperture, measurements can be made at a particular instant within the pulse, as shown graphically in Figure 7.7.

The narrowband point-in-pulse technique actually simplifies the job of the VNA receivers, which only need to recover the center spectral line of this gated pulse. Because the receiver gating circuitry is located in front of the VNA receivers, it allows them to operate in the pulse average mode, meaning they use a narrow IF filter (with consequently long acquisition time) to acquire energy over many pulse periods. Once the IF filter's sample time requirements are met, amplitude and phase information is recovered and used to calculate and display the associated trace data point. After this, the RF is incremented (for a frequency sweep), and the process is repeated for the next data point. Other variations of narrowband point-in-pulse measurements include power sweeps (where the pulse power is incremented in small steps at a particular RF), and time sweeps where pulse power and RF are held constant, and the point-in-pulse measurement shows long-term drift effects (e.g., impact of thermal heating on the DUT's amplitude/phase response).

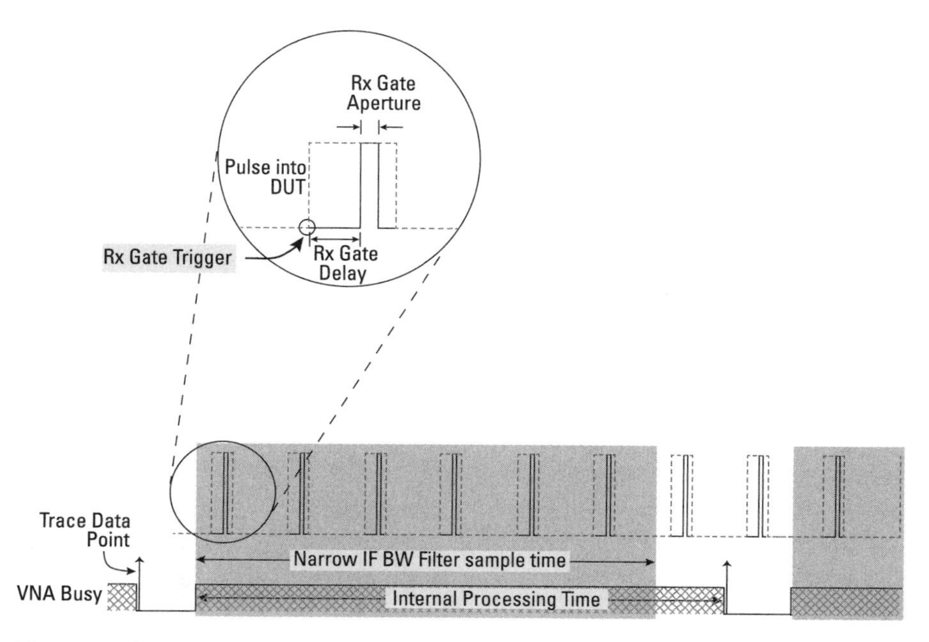

Figure 7.7 Narrowband point-in-pulse timing.

7.2.3 Pulse Profile

Often, it is useful to see how a DUT's amplitude and phase varies across a single pulse, such as the one shown in Figure 7.4. For this kind of measurement, the RF and pulse power are kept constant, and either wideband or narrowband techniques are used to obtain a pulse's time-domain profile. These techniques are discussed below.

7.2.3.1 Wideband Pulse Profiling

As we have seen from wideband point-in-pulse measurements, VNA internal data processing overhead renders the instrument blind to a significant portion of the pulse period, and limits the number of measurements that can be made during a single pulse. Because of this, Rohde & Schwarz (R&S) introduced an innovative wideband mode that fundamentally changes how a VNA operates for pulse profile measurements. Figure 7.8 shows a block diagram of this new approach, which collects samples from an ADC operating at 80 MHz. This ADC provides samples every 12.5 ns, and then at the end of the time-domain acquisition/sweep, it processes all the samples as a single block, applying system error correction (for S-parameter measurements) and updating the VNA screen with the results. As a result, this method delivers a seamless profile of a single pulse in a manner that is indistinguishable from a conventional VNA sweep. It can accommodate sweeps up to 25 ms long and offers IF bandwidths up to 30 MHz wide for examining short-duration pulses. Other benefits include compensation for DUT group delay, so that ratio measurements (e.g., S_{21}) on pulses shorter than the group delay of the DUT are possible.

A trigger signal at the beginning of the pulse is all that is required for synchronization in the pulse profile mode. The only drawback is the same one

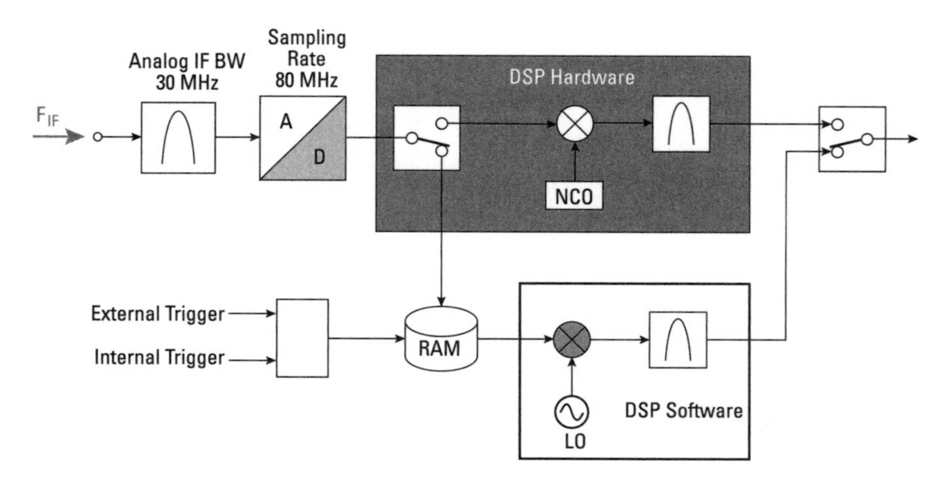

Figure 7.8 Block diagram of R&S pulse profile mode.

that afflicts all wideband point-in-pulse measurements. The wide IF bandwidth required for providing high pulse temporal resolution also introduces considerable noise, so achieving high SNR for a noise-free trace may be a problem. However, trace averaging can be applied for pulse profile measurements to reduce trace noise, with improvements on the order of $10\log_{10}$ (number of trace averages).

7.2.3.2 Narrowband Pulse Profiling

Narrowband pulse profiling uses the same basic approach as the narrowband point-in-pulse technique. Additional hardware is used to provide pulse gating for the receiver, and narrow IF filters measure only the central spectral component of the pulse (like in the nonsynchronous pulse-average mode). The narrow IF filtering also helps to preserve SNR, which is required because this additional pulse gating imposes an even lower pulse duty cycle on the receivers and consequently more pulse desensitization.

Instead of incrementing the frequency (or power) after each measurement point as in the narrowband point-in-pulse method, the narrowband pulse profile technique sequentially increments the gate delay by the amount of the gate aperture until the entire pulse is measured. However, hundreds or thousands of individual pulses may be needed to reconstruct a single pulse using this method. Additionally, power calibration depends on the pulse gating aperture. Changing this aperture invalidates the power calibration.

7.3 Instrument Setups and Power Calibration Approaches for Pulse Measurements

For discussion purposes, we will assume a high-power DUT using external high-power test set components. If full two-port S-parameter measurements (S_{11}, S_{12}, S_{21}, S_{22}) are required, use the setup shown in Figure 5.6, but place a pulse modulator between the power combiner and the DUT input directional coupler. If the reverse path also requires pulsed RF, add a second pulse modulator before or after the port 2 driver amp. Both modulators should be driven by the same pulse generator.

If only S_{11} and/or S_{21} is required, a one-path, two-port calibration may be performed. This has several advantages: only a single pulse modulator is required (for VNA source port 1), and only a single measurement in the forward direction is needed for S-parameter error correction. To keep things simple, we will assume a test set configuration appropriate to a one-path, two-port measurement. For example, Figure 7.9 shows several setups for the ZVA and ZNB. Neither of these instruments has built-in pulse modulators, but the ZVA does have built-in pulse generators. In Figure 7.9(a) the ZVA uses direct

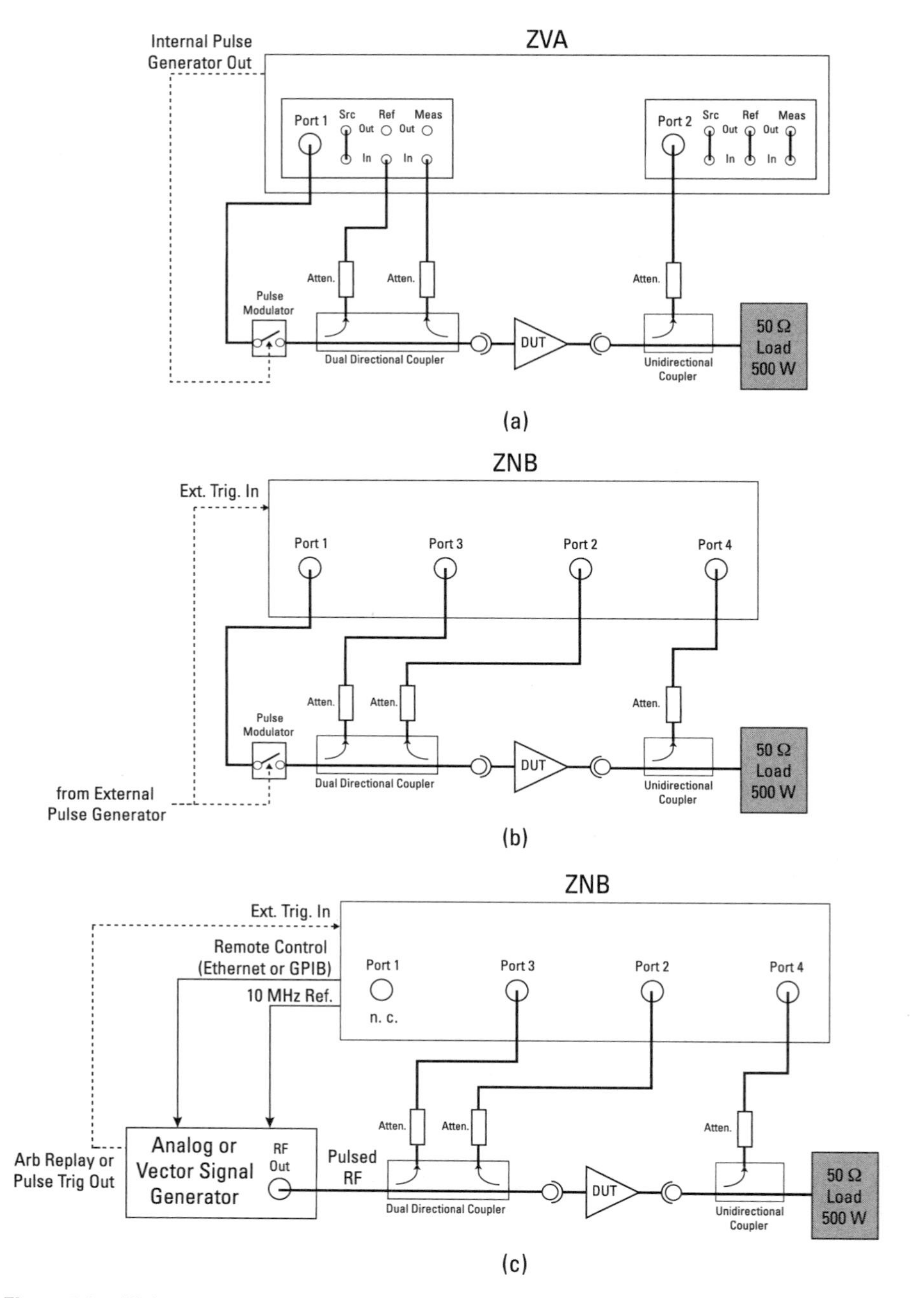

Figure 7.9 High-power pulse setups for ZVA and ZNB.

receiver/generator access ports to interface with external high-power components. The pulse modulator function can be supplied by the user or provided by the ZVAX-TRM configurable test extension unit. The setup in Figure 7.9(b) depicts a four-port ZNB using port reconfiguration to allow external components to be used for the same one-path, two-port measurement. For the ZNB, both the pulse generator and pulse modulator must be supplied by the user. Depending on the frequency range, power level, and switching speed requirements, simple single-pole double-throw (SPDT) electronic RF switches can serve as good pulse modulators. Some VNAs (such as the R&S ZNA and the Keysight PNA-X) have pulse generators and pulse modulators available as built-in hardware options. Another setup is shown at in Figure 7.9(c) where an external signal generator is used to supply the pulsed RF signal. Here, the VNA is configured for remote control of the external generator. A shared 10-MHz reference is also crucial for ensuring optimum performance.

If you are using your own pulse generator and pulse modulator (as opposed to built-in VNA hardware or an external signal generator), it would be wise to first verify the performance of your pulse generator/modulator system with a multichannel oscilloscope and signal generator. Apply the pulse trigger from the pulse generator to the oscilloscope “Ext Trig” input and ensure that the oscilloscope is synchronized. Next, apply the pulse generator output to the oscilloscope’s Channel 1 input. Route the RF out of a signal generator to the RF in of the pulse modulator. Connect the RF out of the pulse modulator to the oscilloscope’s Channel 2 input using a coaxial cable (set the Channel 2 input impedance to 50Ω, if available). Set the RF to a frequency that is no more than half the oscilloscope bandwidth (and within the operating range of the pulse modulator), and set the signal level to a relatively high level (0 or +10 dBm). The shape of the RF pulse should be observable on the oscilloscope trace.[3]

Configure a 1-ms-wide pulse with 10% duty cycle. Once you can see the pulse waveform on Channel 1 and the RF shape on Channel 2, begin reducing the pulse width and duty cycle to meet the DUT measurement requirements. The RF envelope should be distinct, without appreciable delay with respect to either the trigger signal or the rising edge of the pulse waveform. (If there is any appreciable pulse delay (more than 1% of the pulse width), note it and use this as a trigger delay for the VNA measurement.) Once you observe a good waveform and corresponding proper RF envelope, you are ready to configure the VNA by following the steps below. (These instructions are for the ZNB, but will be quite similar for the ZVA or ZNA.)

Chapter 5 emphasized the point that high-power measurements require special attention to all test set components to ensure that power levels are not exceeded. This is especially true for the VNA. Many components internal to the VNA are sensitive to peak voltage, not average power (i.e., heating). Therefore,

3. Alternatively, feed the RF output of the modulator into an RF (diode) detector. The output of the detector will show the pulse envelope.

peak pulse power levels should not exceed the CW power levels listed in the VNA specifications.

Another problem with high-power pulse measurements involves instrumentation amplifiers that may be required to drive a high-power DUT. These amplifiers may have different gain characteristics in pulse operation than CW operation. When using such an amplifier, it is a good idea to perform both calibration and measurement under pulsed operating conditions.

7.3.1 Pulse Averaging Measurement Setup

Configuring a narrowband pulse average measurement is very straightforward. Configure the VNA as it would be for a conventional swept measurement. As a minimum, one should have traces for monitoring the DUT input power (a_1 source port 1), DUT output power (b_2 source port 1), and gain (S_{21}). For the narrowband pulse averaging method, set the VNA BW to a value that is a factor of 10 less than the pulse PRF, to ensure that only the center tone of the pulse waveform is measured. Position a pulse modulator between the VNA's port 1 source and the DUT, as shown in Figure 7.9. Apply a thru at the DUT calibration plane, and you should see the pulse average measurement on the VNA traces. S_{21} should be approximately zero (depending on losses associated with the external test set components), and a_1 and b_2 should show the impact of the narrowband pulse desensitization. Specifically, (7.3) indicates that for 10% pulse duty cycle, both a_1 and b_2 should show 20-dB desensitization.

7.3.1.1 S-Parameter Calibration

S-parameter calibration is performed the same as for a conventional swept measurement with a CW RF source. From Channel > Cal > "Start Cal" tab > "Start...(Manual)" button, select the one-path, two-port calibration type, and "Next." Choose the DUT connector type and gender, along with your calibration kit model, and commence the calibration. At the completion of the calibration, click "Apply." It would be a good idea to save this calibration using the calibration manager: Cal > "Use Cal" tab > Cal Manager ... (Click the "Add" button to add this calibration to the calibration pool.) Rename the calibration in the pool as desired.

7.3.1.2 Power Calibration

Power calibration is the trickiest part of the entire pulse measurement process. This is because an average power sensor always operates in CW mode and thus sees pulse desensitization of $10\log_{10}$ (duty cycle), while the narrowband IF filters cause the VNA receivers to see desensitization of $20\log_{10}$ (duty cycle). Fortunately, this measurement discrepancy between detectors amounts to nothing more than additional signal attenuation in front of VNA receiver that is removed via the power calibration process.

Our initial focus will be on the desensitization associated with the power sensor, because it always operates in average power mode for VNA power calibration. Average power is less than peak power by the pulse desensitization factor. Another way of looking at this is to consider that this desensitization is nothing more than simple attenuation, reducing the power sensor reading from the true peak pulse power. So we apply a correction factor to compensate for the signal attenuation and thus recover the peak power.

1. Ensure that an appropriate power sensor is connected to the VNA. Use either a diode sensor (e.g., NRP18S) or a thermal sensor (NRP40T). Then, from Channel > Cal > "Start Cal" tab > "Power Cal…" set the maximum iterations to 10 and the tolerance to 0.1 dB.
2. Click on the "Transm. Coefficients." Assuming a 10-μs pulse width and 100-μs period, the resulting duty cycle is 0.1 (10%), and the desensitization factor is $10_{\log}10(0.1) = -10$ dB. In the Power Meter Transmission Coefficients dialog, select "Two Port at Power Meter" and then click on "Two Port Config…" In this next dialog, click "Insert" and for the first frequency (which is the lowest frequency supported by the VNA), click in the "Transm. Coefficients" column and enter "–10.0" Then click "Insert" again and set the frequency to the highest frequency of the VNA (here, 8 GHz). Use the same transmission coefficient (–10.0). Then click "Close" twice, remaining in the power calibration menu. This will apply a correction factor of –10 dB across the entire operating range of the instrument.
3. Attach the power sensor at the desired DUT input calibration plane. Then click the "Power" button under "P1" and click "Start Cal Sweep" to begin the source power calibration. You should see a trace that is approximately –10 dBm. (This process also calibrates the port 1 reference receiver.)
4. Remove the power sensor and attach a thru at the DUT calibration plane. Click "Meas. Receiver" under "P2" and click "Start Cal Sweep" to commence a power calibration of the b_2 measurement receiver. At the completion of this process, you should see the b_2 measurement receiver trace overlaying the a_1 reference receiver trace. Click "Apply" to exit the power calibration process. Save the combined S-parameter and power calibration using the calibration manager (see Section 7.3.1.1 for the button sequence).

If this power calibration approach does not give satisfactory results (because of an extremely low duty cycle and hence a high signal desensitization), you can perform the power calibration using a CW signal. You should still use the pulse modulator, but set it into an always-on (low attenuation) mode. Then repeat power calibration step 1. For step 2, we need to account for the pulse desensitization that will impact the receivers when the pulse waveform resumes. So we select "Two Port at DUT" rather than "Two-Port at Power Meter" in the "Transm. Coefficients..." dialog. Also, the desensitization factor is 20 dB for the narrowband example, so we modify the settings accordingly. Continue with steps 3 and 4. At the end of the power calibration process, restore the pulse modulator to pulse operation.

At the completion of this process, you should see the a_1 and b_2 traces overlaying each other with a power value of −10 dBm. S_{21} should show 0 dB across the trace. Now install the DUT in place of the thru adapter and begin making narrowband pulse average measurements.

7.3.2 Point-in-Pulse (Wideband) Measurement Setup

Configuring a wideband point-in-pulse measurement is relatively easy. Depending on your VNA, configure the appropriate setup based on Figure 7.9. Install a thru adapter in place of the DUT. Ensure that the pulse generator's "Trigger Out" is connected to the "External Trigger IN" on the back of the VNA, and that the pulse generator is configured to apply the appropriate voltage levels to the pulse modulator (e.g., TTL "hi" and "low" levels). Keep in mind that for many pulse modulators, a "TTL Hi" turns the modulator off (high attenuation state), while a "TTL Lo" turns the modulator on (very low attenuation).

1. Start with a VNA PRESET. Configure the desired measurement (start/stop frequency, number of points) as you would for a conventional CW sweep. For this preliminary step, set the IF BW based on the full pulse width, using (7.4) as a guide. For our example of a 10-μs-wide pulse, the calculated bandwidth is 158.5 kHz, so we select the next highest standard IF filter value, 200 kHz.

$$\textit{filter bandwidth}\left(kHz\right) = 10^{\left(-1.1\cdot\log_{10}\left(\textit{sample time}(us)\right)+3.3\right)} \tag{7.4}$$

2. On the pulse generator, initially set the pulse period to 100 ms.
3. On the VNA, Change the trigger source to "External" and the trigger mode to "Point trigger:"

   Channel > Trigger > External

   Channel > Trigger > Sequence > Point

   Set the VNA trigger slope in accordance with your pulse generator: Channel > Trigger > Slope > [Rising Slope | Falling Slope | High Level | Low Level]. (Note that rising slope or falling slope are the most common.) At this point, the VNA should be sweeping, even if the screen shows only noise.

4. On the VNA, create traces for a_1 (source port 1) and b_2 (source port 2), both in dB mag. Apply the same scaling factor by using the Trace Manager (see Chapter 3 for button sequences). In a second display area, create a S_{21} trace (dB mag). If your system is set up correctly and the pulse generator, pulse modulator, and VNA are properly synchronized, you should see flat, noise-free traces. a_1 and b_2 should have roughly the same amplitudes, and S_{21} should be close to 0 dB.
5. The next step is to determine the VNA's internal processing time, which will impact the minimum duty cycle that can be supported. Temporarily install a BNC Tee on the pulse generator "Trigger Out" and connect one side of the Tee to the VNA "Ext Trig In" and the other side of the tee to the oscilloscope's "Ext Trig In." Connect pin 6 of the 24-pin "User Control" connector on the back of the ZVA (or ZNB) to the Channel 2 input of the oscilloscope (high-impedance setting). You should see the VNA busy signal on the oscilloscope as shown in Figure 7.10.

   In this example, the oscilloscope shows that the minimum pulse period that can be supported is 850 μs. Shorter pulse periods can be used, but the VNA will then ignore a number of pulses until it has

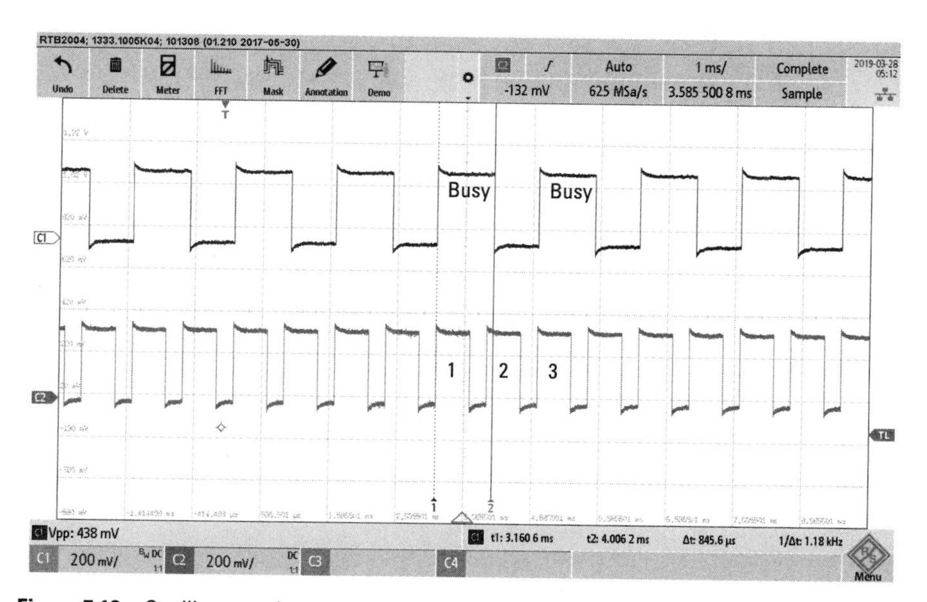

Figure 7.10 Oscilloscope-based timing measurements. Top trace: the VNA busy signal is 845 μs long. Bottom trace: pulse generator output.

completed internal processing of its last measurement point. In Figure 7.10, pulse 2 begins just before the busy line is de-asserted, so the VNA waits until the next pulse to resume measurements.

6. The next step is to determine the appropriate IF filter bandwidth to achieve the narrow timing aperture desired for the point-in-pulse measurement. Again using (7.4), if I choose a pulse aperture of 1 μs, the equation yields a filter BW of 1,995.3 kHz. I select the ZNB's next higher standard IF filter bandwidth (2 MHz). To test this value, save the current a1 trace to memory, and increase the scaling per division to 0.5 dB. Then reduce the pulse width of the pulse generator from 5 μs down to 1 μs, in 0.1-μs steps. You should see the trace amplitude remain constant until you reduce the pulse width below 1 μs, at which point the short pulse duration is no longer sufficient to fill the available receiver measurement aperture. If necessary, apply trace averaging (Channel > Power BW AVG > "Average" tab > set averaging factor to 10 and turn averaging on). Verify that you still see peak power with this pulse aperture. You are now ready to begin calibrating the wideband point-in-pulse measurement.

7.3.2.1 S-Parameter Calibration

For S-parameter calibration, choose Channel > Cal > "Start Cal" tab > "Start... (Manual)" button. Select the one-path, two-port calibration type, and "Next." Choose the DUT connector type and gender, along with your calibration kit model, and commence the calibration as you would for a conventional S-parameter calibration. At the completion of the calibration, click "Apply." It would be a good idea to save this calibration using the calibration manager: Cal > "Use Cal" tab > Cal Manager … (Click the "Add" button to add the current calibration that you just performed to the calibration pool.) Rename the calibration in the pool as desired.

7.3.2.2 Power Calibration

Power calibration is the trickiest part of the entire measurement process. This is because the average-power sensor always operates in CW mode and thus sees pulse desensitization, whereas the wideband filters in the VNA see the peak pulse power with no desensitization. However, this is easy to compensate, as long as you know the pulse duty cycle. We will start with the easy calibration first. Here, easy means that you are able to use the same pulse settings you chose in Section 7.3.1. In this case, the power sensor is going to display the average pulse power, which will always be less than the peak power. The amount of desensitization may be calculated from (7.2). Another way of looking at this is to consider that this desensitization represents attenuation, and it reduces the power sensor reading from the true peak pulse power. So we compensate the power sensor for this attenuation in order to obtain accurate peak power readings.

1. Ensure that an appropriate power sensor is connected to the VNA, either a diode sensor (e.g., NRP18S) or a thermal sensor (NRP40T). Channel > Cal > "Start Cal" tab > "Power Cal…" and set the maximum iterations to 10 and the tolerance to 0.1 dB.
2. Click on the "Transm. Coefficients." Assuming a 10-μs pulse width and 100-μs period, the duty cycle is 0.1 (10%), and the desensitization factor from (7.2) is −10 dB. In the Power Meter Transmission Coefficients dialog, select "Two Port at Power Meter" and then click on "Two Port Config…" In this next dialog, click "Insert" and for the first frequency (which is the lowest frequency supported by the VNA), click on the "Transm. Coefficients" column and enter "−10.0." Then click "Insert" again and set the frequency to the highest frequency of the VNA (here, 8 GHz). Use the same transmission coefficient (−10.0). Then click "Close" twice to remain in the power calibration menu.
3. Attach the power sensor at the desired DUT input calibration plane. Then click the "Power" button under "P1" and click "Start Cal Sweep" to begin the source power calibration. You should see a trace that is approximately −10 dBm. (This process also calibrates the port 1 reference receiver.)
4. Remove the power sensor and attach a thru at the DUT calibration plane. Click "Meas. Receiver" under "P2" and click "Start Cal Sweep" to commence a power calibration of the b_2 measurement receiver. At the completion of this process, you should see the b_2 measurement receiver trace overlaying the a_1 reference receiver trace. Click "Apply" to exit the power calibration process. Save this combined S-parameter and power calibration using the calibration manager (see Section 7.3.1.1 for the button sequence).

If the easy power calibration approach does not give satisfactory results (because of extremely low duty cycle and hence high signal desensitization), you can perform the power calibration using a CW signal. You should still use the pulse modulator, but set it to an always-on (no signal attenuation) mode. Then change the ZNB trigger temporarily to "Free Run." Repeat power calibration steps 1, 3, and 4. (You will not have any pulse desensitization, because a CW signal is being applied to the power sensor.) At the end of the power calibration process, be sure to set the trigger back to external trigger and set sequence to point trigger. Also, restore the pulse modulator to the pulse operation.

At the completion of this calibration process, you should see the a_1 and b_2 traces overlaying each other, with a power value of −10 dBm. The S_{21} trace should display 0 dB across the frequency span used for the wideband point-in-pulse measurement. Now install the DUT in place of the thru adapter and begin making point-in-pulse measurements.

With the current settings, point-in-pulse measurements will be made at the beginning of the pulse. You may position the narrow measurement aperture anywhere within the pulse by adjusting the pulse delay (Channel > Trigger > Delay).

7.3.3 Pulse Profile Measurement Setup

Pulse profile measurements with a ZVA or ZNA may use any of the configurations shown in Figure 7.9. Install a thru adapter in place of the DUT. If employing an external pulse generator, connect its "Trigger Out" to the "External Trigger In" on the back of the VNA, and apply appropriate voltage levels to the pulse modulator (e.g., TTL "Hi" and "Low" levels). Keep in mind that for many pulse modulators, a "TTL Hi" turns the modulator off (high attenuation state), while a "TTL Low" turns the modulator on (very low attenuation). An analog signal generator with integrated pulse modulator or a vector signal generator with IQ arbitrary waveform generator (AWG) playback may also be used, as shown in Figure 7.9(c). In this scenario, the ZVA and external generator should share a common 10-MHz reference. For maximum flexibility, the ZVA should be configured to directly control the signal generator's frequency and amplitude. To do this, go to System > System Config and scroll along the tabs at the top until you see "External Generators." (This tab is off the right side of the screen and must be scrolled to reach it.)

Select the appropriate interface (GPIB or Ethernet/VXI-11) and add the appropriate instrument address. Then click "Add Other..." and select your signal generator from the Driver list. If your instrument is not shown, or does not behave as expected, refer to the appropriate section of the ZVA user's manual for details about modifying a driver file. For a missing generator, it is a good idea to select a driver of a similar model from the same vendor and then modify SCPI command(s) associated with, for example, power level or frequency, and then save the driver under a new name in the same directory (see Section 6.4.2.1).

Finally, the ZVA requires a TTL trigger signal from the external signal generator. This may be supplied from a pulse generator (in the case of an analog signal generator) or by a marker in the baseband section of a vector signal generator.

Although pulse profile measurements are, by definition, performed at a single frequency, it is often desirable to investigate pulse behavior at different frequency points across the DUT's operating range. For this test approach, it makes sense to perform a single S-parameter and power calibration that covers all measurement frequencies. Then pulse profile measurements can be made at any frequency within this range.

1. Start with a VNA PRESET. Configure the desired frequency points of interest via either a linear sweep (start/stop frequency, number of points) or a segmented sweep. Set the IF BW to 1 MHz.
2. Use a CW signal for calibration. (If this is not possible due to DUT/setup constraints, the wideband pulsed calibration technique of Section 7.3.2 may be used.) Apply a DC voltage to the pulse modulator that places it in the always-on state. Add traces for a_1 source port 1, b_2 source port 1, and S_{21}. You are now ready for calibration.

7.3.3.1 S-Parameter Calibration

For S-parameter calibration, choose the Channel > Cal > "Start Cal" tab > "Start...(Manual)" button. Select the one-path, two-port calibration type and "Next." Choose the DUT connector type and gender, along with your calibration kit model, and commence the S-parameter calibration. At the completion of the calibration, click "Apply."

7.3.3.2 Power Calibration

Since we are using a CW signal for calibration, there is no difference between the average power measured by the power sensor and the peak power level observed by the VNA receivers (no desensitization).

1. Connect an appropriate power sensor to the VNA. Use either a diode sensor (e.g., NRP18S) or a thermal sensor (NRP40T). Channel > Cal > "Start Power Cal" tab > "Source Power Cal..." and in "Modify Settings..." set the "Maximum Number of Readings" to 10, and the "Tolerance" to 0.1 dB. Click "OK" to close this dialog.
2. Attach the power sensor at the desired DUT input calibration plane. Then click "Take Cal Sweep" to begin the source power calibration. If you are using the default (0 dBm) power level setting on the ZVA, you should see a trace that is approximately 0 dBm. (This process also calibrates the port 1 reference receiver.) Click "Close" to close the "Source Power Cal" dialog.

3. Remove the power sensor and attach a thru at the DUT calibration plane. Click the "Receiver Power Cal" softkey and ensure that the dialog shows b_2 as the wave quantity to calibrate and the calibration source is set to port 1. The "Reference Power Value" should also be set to "Reference Receiver." Click "Take Cal Sweep" to commence a power calibration of the b_2 measurement receiver. Then click "Close." You should see the b_2 measurement receiver trace overlaying the a_1 reference receiver trace, and S_{21} should be zero. Save this combined S-parameter and power calibration using the Cal Manager by selecting "Channel" > "Calibration" > "Cal Manager." Click on the highlighted calibration in the upper left window (Calibration State) and then click "Add ->" to transfer this calibration to the calibration pool. (To later recall the calibration, highlight the desired calibration in the upper right window (calibration pool), and click "Apply" to apply it to the currently highlighted channel in the "Calibration State" window, and then click "Close.") This will guard against accidental loss (e.g., accidentally pressing "Preset").

7.3.3.3 Making Pulse Profile Measurements

With calibration complete, you are ready to make pulse profile measurements. It is best to start with a thru in place of the DUT. Once you get the pulse profile measurement configured properly, you can substitute your DUT.

1. Change from a CW RF source to pulsed RF source. (This will require setting the pulse generator/pulse modulator to the appropriate pulse width and period, or changing an external signal generator into pulse mode.)
2. Change the trigger settings on the ZVA to synchronize measurements with the pulse generator's trigger signal:

   Channel > Sweep > Trigger > External (Trigger on Rising or Falling edge, as required)

   At this point, the VNA should be sweeping.
3. Change the ZVA to the Pulse Profile mode:

   Channel > Sweep > Sweep Type > [More] > Pulse Profile
4. Adapt the Pulse Profile settings for your pulse. In this example, the pulse width is 2 μs, so we set the instrument as follows:

   - Channel > Sweep > Sweep Type > [More] > Define Pulse Profile...

- Under Time Parameters, set the sweep start time to $-1\ \mu$s, and the stop time to $9\ \mu$s for a total display time of $10\ \mu$s (and to take advantage of the horizontal grid scaling).
- Under "Stimulus," set the power level and center frequency as desired. Set the "No. of Points" according to the "Optimum No. of Points" (in this case, 801). See Figure 7.11.

This completes the basic steps for setting up a pulse profile measurement.

7.3.3.4 Advanced Pulse Profile Measurements

Figure 7.12(a) shows a more challenging scenario, where there is appreciable time delay within the DUT. Here, the 2-μs-wide pulse is delayed by $1\ \mu$s.Trc1 (dotted line) measures wave quantity a1 source port 1, which is the power entering the DUT. Trc2 (solid line) shows wave quantity b2 source port 1, which measures the power exiting the DUT. In this case, S21 measurements only yield meaningful results where the pulses overlap (e.g., the highlighted region). Outside this region, the S21 ratio measurement is meaningless.

While we can at least make a measurement in the overlap region, it is not hard to envision a situation where you have a DUT with sufficient group delay such that there is no overlap at all. There are two ways to handle this situation. First, you can use "Shift Stimulus" in the Receiver Settings dialog to remove the delay and allow the pulses to overlap (Figure 7.12(b)). Here, we shift the b2 receiver response by $1\ \mu$s to eliminate the delay.

Alternatively, you can select a portion of the input pulse as a reference for the entire measurement. In Figure 7.13(a), the section limit lines are turned on and the "Section Start" is positioned $1.3\ \mu$s into the pulse seen by the a1 receiver. The section stop is positioned at $1.5\ \mu$s.

When "Evaluation Mode" is set to "Mean Value of Pulse Section," the average amplitude (and phase) will be calculated from this 0.2-μs aperture and

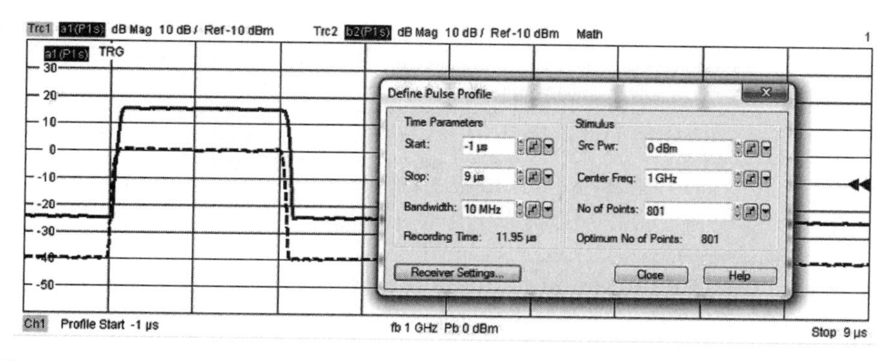

Figure 7.11 Pulse profile settings.

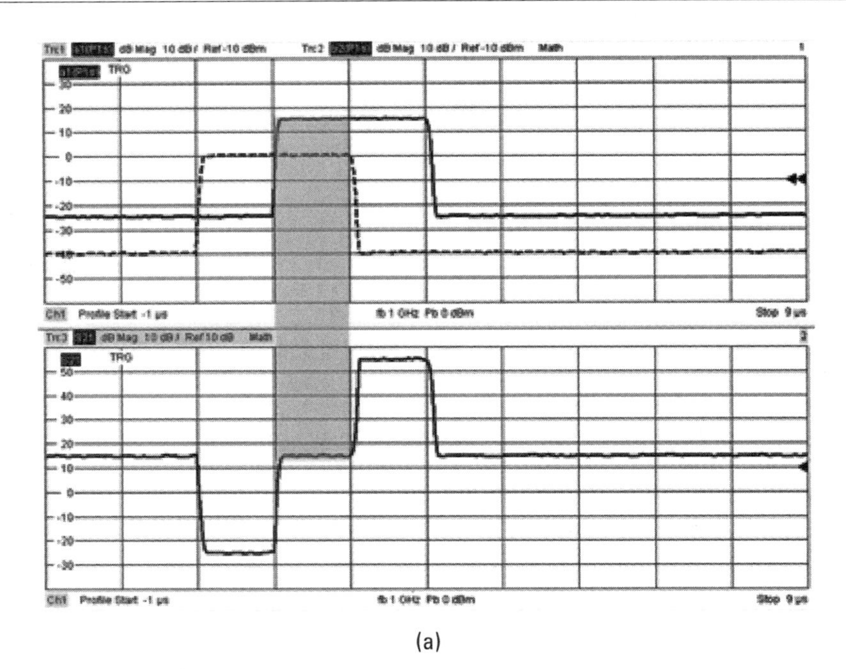

(a)

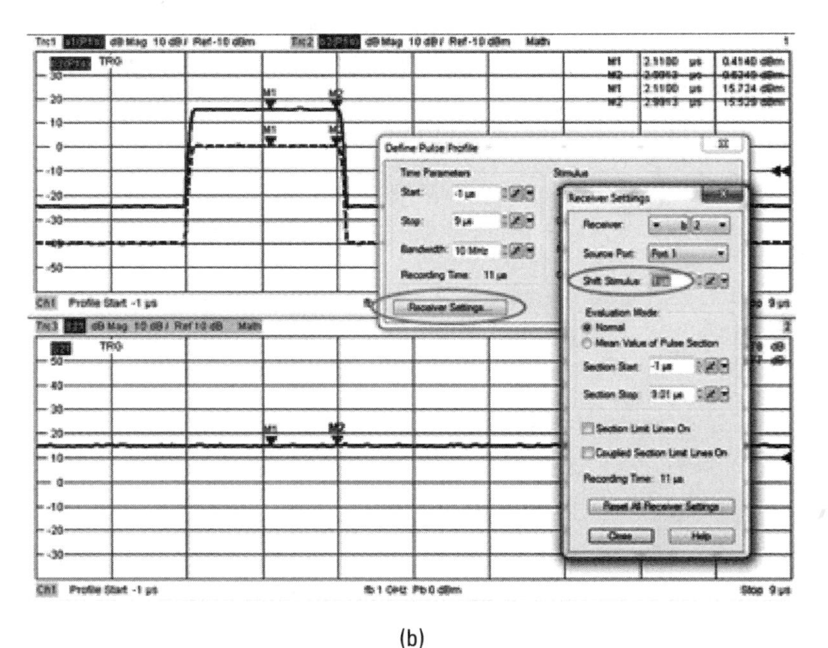

(b)

Figure 7.12 (a) DUT delay limits accurate S21 measurements to the gray overlap region. (b) Time delay is removed by using "Shift Stimulus" in the Receiver Settings dialog.

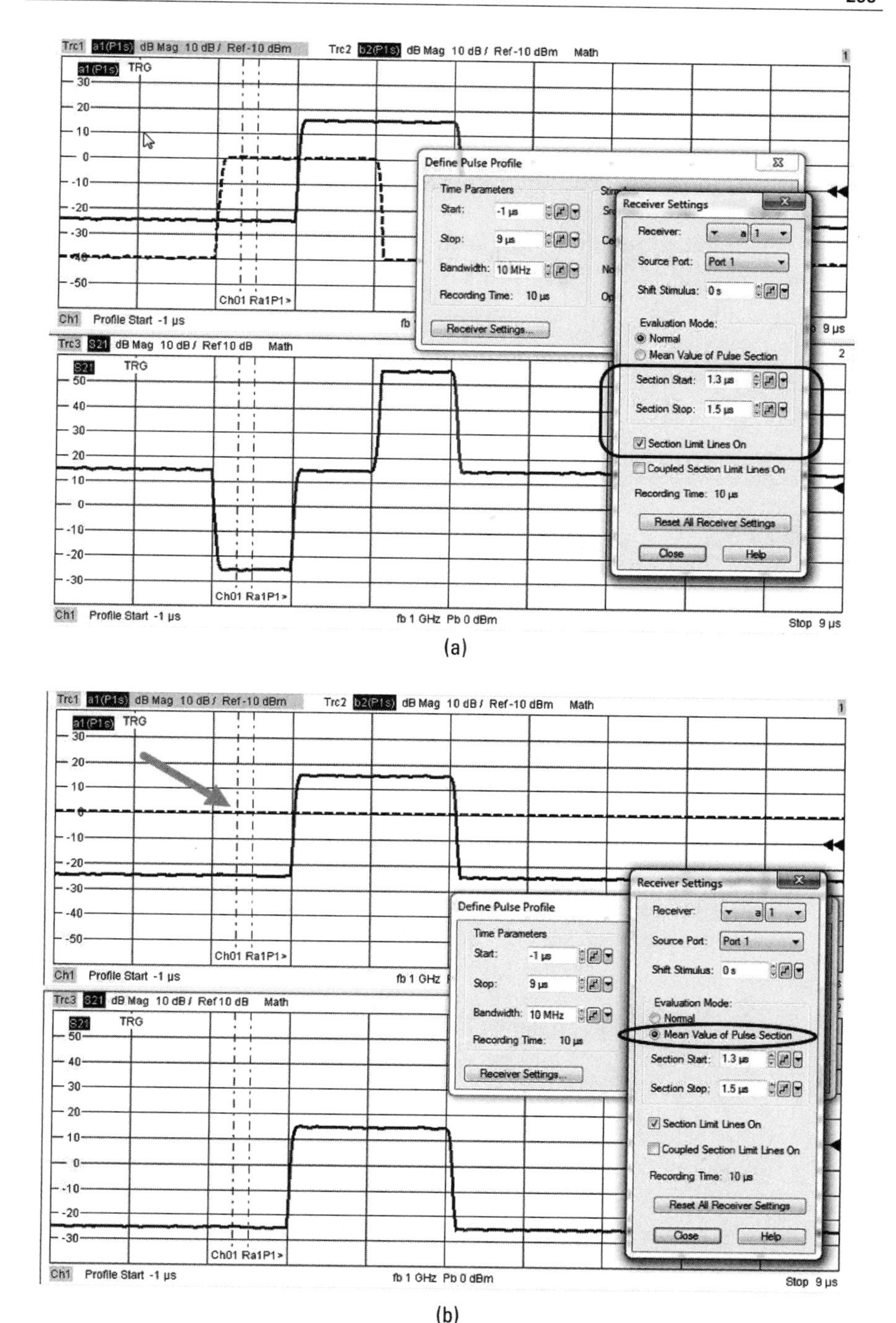

Figure 7.13 (a) Setting aperture for the mean value evaluation. (b) Results after activating the mean value function.

applied over the entire 10-μs measurement sweep (Figure 7.13(b)). Note that S21 now shows proper gain across the entire 2-μs pulse.

Appendix 7A: Fourier Analysis Review

If $f(t)$ is a periodic function of time with period T, then it may be decomposed into a series of sine and cosine functions (and one DC value), using Fourier series representation:

$$f(t) = \frac{a_0}{T} + \frac{2}{T}\sum_{n=1}^{\infty}\left(a_n \cos\omega_n t + b_n \sin\omega_n t\right) \tag{7A.1}$$

where

$$\omega_n = \frac{2\pi n}{T} \tag{7A.2}$$

To determine the a_n values, multiply both sides of (7A.1) by $\cos(w_n t)$ and integrate over one period. All right side terms will vanish except the a_n term, since

$$\int_{-T/2}^{T/2} \sin\omega_j t \cos\omega_n t \, dt = 0 \; all \; j \tag{7A.3}$$

$$\int_{-T/2}^{T/2} \cos\omega_j t \cos\omega_n t \, dt = 0 \; j \neq n \tag{7A.4}$$

So after multiplying, we can rewrite (7A.1) as:

$$\int_{-T/2}^{T/2} f(t)\cos\omega_n t \, dt = \frac{2a_n}{T}\int_{-T/2}^{T/2} \cos^2 \omega_n t \, dt \tag{7A.5}$$

The right side of (7A.5) can be rewritten as:

$$\frac{2a_n}{T}\int_{-T/2}^{T/2} \frac{1+\cos 2\omega_n t}{2} dt = \frac{a_n}{T}\left[t + \frac{\sin 2\omega_n t}{2\omega_n}\right]_{-T/2}^{T/2} = a_n \tag{7A.6}$$

Yielding, from (7A.5) and (7A.6):

$$a_n = \int_{-T/2}^{T/2} f(t)\cos\omega_n t \, dt, n = 0,1,2,3,\ldots \tag{7A.7}$$

Similarly, multiplying both sides of (7A.1) by $\sin(w_n t)$ and integrating over one period yields:

$$b_n = \int_{-T/2}^{T/2} f(t)\sin\omega_n t\, dt, n = 1,2,3,\ldots \tag{7A.8}$$

The Fourier series of (7A.1) with Fourier coefficients determined by (7A.7) and (7A.8) may also be represented in complex exponential form.

$$f(t) = \frac{1}{T}\sum_{n=-\infty}^{\infty} c_n e^{j\omega_n t} \tag{7A.9}$$

where

$$c_n = \int_{-T/2}^{T/2} f(t)e^{-j\omega_n t}\, dt \tag{7A.10}$$

This format is particularly useful for pulse analysis. Consider Figure 7A.1(a) where a periodic pulse waveform is presented in the time domain. The pulse width is τ and the pulse period is T.

Here, the time-domain waveform may be defined as:

$$f(t) = A_m \text{ for}\left(-\frac{\tau}{2} < t < \frac{\tau}{2}\right)$$
$$f(t) = 0 \text{ Otherwise}$$

The Fourier coefficients become:

$$\begin{aligned} c_n &= \int_{-\tau/2}^{\tau/2} A_m e^{-j\omega_n t}\, dt = \left[-\frac{A_m}{j\omega_n} e^{-j\omega_n t}\right]_{-\tau/2}^{\tau/2} \\ &= -\frac{A_m}{j\omega_n}\left(e^{-\frac{j\omega_n \tau}{2}} - e^{\frac{j\omega_n \tau}{2}}\right) \end{aligned} \tag{7A.11}$$

$$c_n = A_m\left(\frac{e^{\frac{j\omega_n\tau}{2}} - e^{-\frac{j\omega_n\tau}{2}}}{j\omega_n}\right) = \frac{2A_m}{\omega_n}\left(\frac{e^{\frac{j\omega_n\tau}{2}} - e^{-\frac{j\omega_n\tau}{2}}}{2j}\right) = \frac{2A_m}{\omega_n}\sin\left(\frac{\omega_n \tau}{2}\right) \tag{7A.12}$$

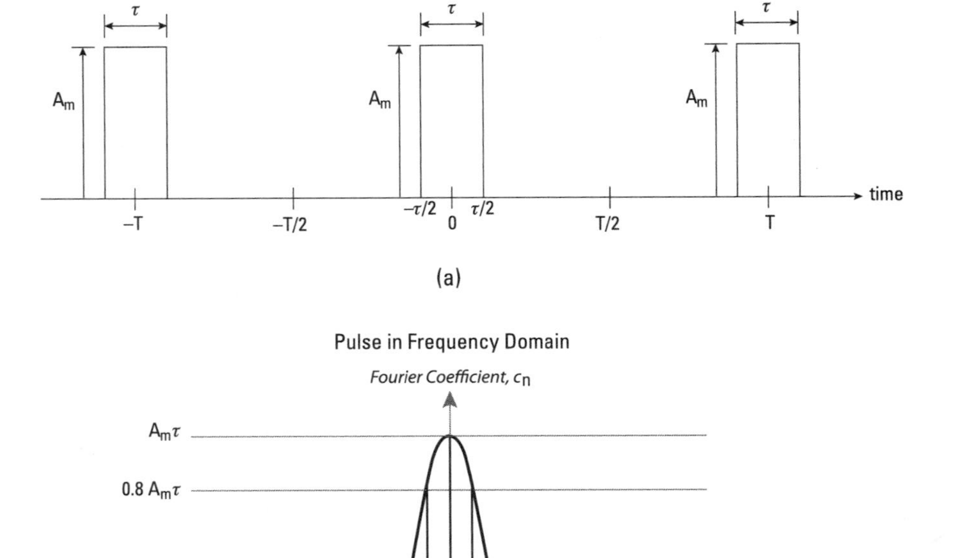

Figure 7A.1 (a) Continuous pulse stream in time domain. (b) Pulse in frequency domain, highlighting Fourier coefficients and sinc function envelope.

Finally, multiplying numerator and denominator by τ, we end up with:

$$c_n = \tau A_m \frac{\sin\left(\dfrac{\omega_n \tau}{2}\right)}{\left(\dfrac{\omega_n \tau}{2}\right)} = \tau A_m sinc\left(\frac{\omega_n \tau}{2}\right) \tag{7A.13}$$

The envelope of the continuous sinc() function (=sin(x)/x) is plotted in Figure 7A.1(b). Superimposed on it are the spectral components of $f(t)$ (7A.9), occurring at radian frequencies that are multiples of $\omega_n = 2\pi\omega n/T$ each with an amplitude c_n (from (7A.10)). This frequency response is symmetrical about the y-axis with maximum value of tA_m occuring at $f(0)$. The spectral lines are spaced $(2\pi(n+1)/T - 2\pi n/T) = 2\pi/T$ apart. The sinc$(\omega_n t/2)$ function has zeros at integer multiples of $\omega_n = 2\pi n/\tau$.

Appendix 7B: Odd and Even Functions

Cosine is an even function and displays even symmetry about the origin (vertical line $x = 0$). Sine is an odd function and displays odd symmetry with respect to the origin (vertical line $x = 0$). Note that both of these functions have equal areas above and below the line $y = 0$ (Figure 7B.1, top and middle).

If we add these respective positive (dark) and negative (light) areas together, they will cancel for both the sine and cosine. Hence, performing integration

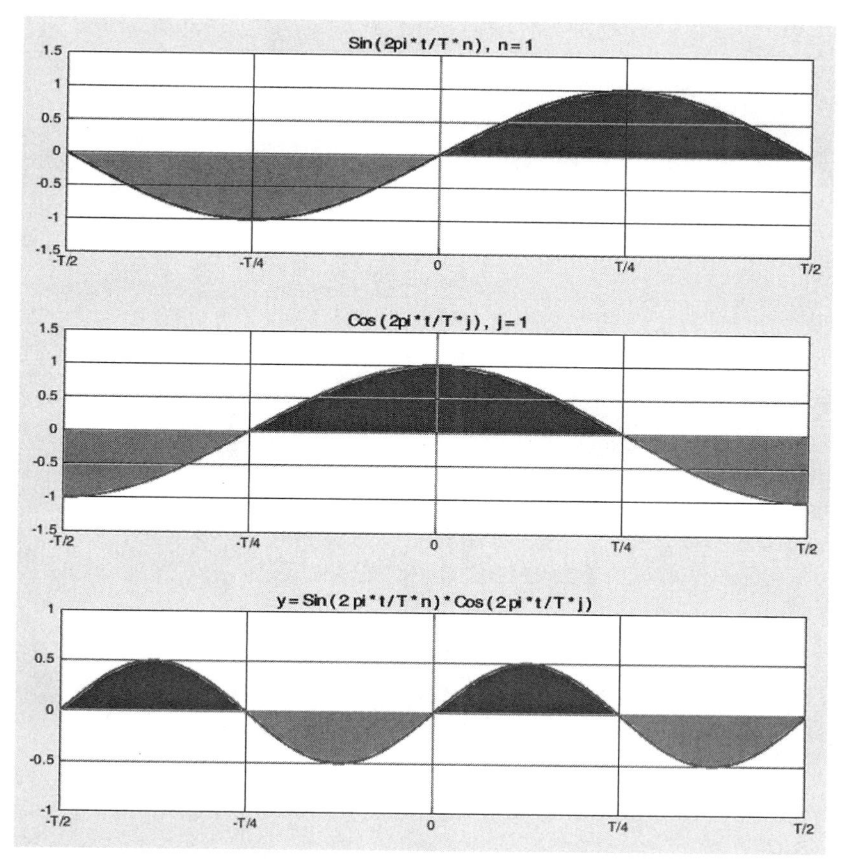

Figure 7B.1 (a) Sine function with odd symmetry highlighted. (b) Cosine function with even symmetry highlighted. (c) Product of an odd function (Sine) and even function (Cosine) is another odd function. In all three cases, the integrated area under the curve (from -T/2 to T/2) is zero.

over the period $-T/2$ to $T/2$ is going to yield a value of zero for both functions. Similarly, multiplying an odd function (sine) by an even function (cosine) as in (7A.3) will yield an odd function (Figure 7A.1, bottom) with odd symmetry. This resulting waveform also has equal positive and negative areas, and hence will always integrate to zero for any integer multiples of the fundamental frequencies $\omega_n = 2\pi n/T$.

Similar symmetry occurs in (7A.4) for all $j \neq n$. Figure 7B.2 shows an example for $n = 3$, $j = 2$. The bottom graph shows the result of multiplying the two cosines (with different frequencies) together, where the odd symmetry of the resulting waveform results in complete cancellation over the fundamental period ($-T/2$ to $T/2$).

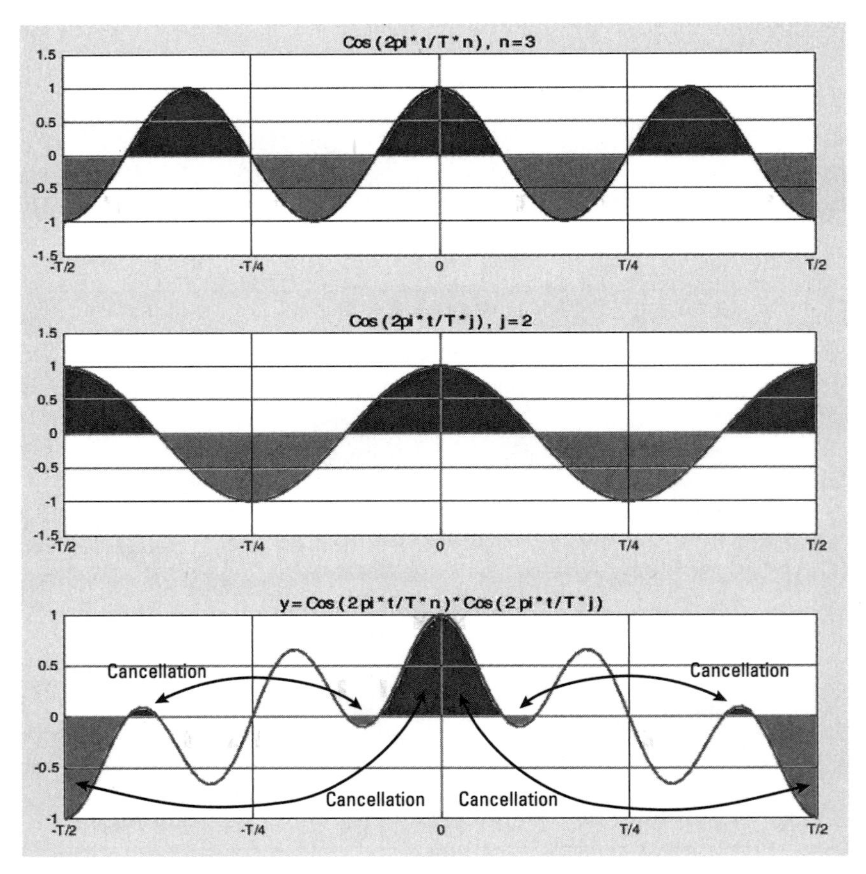

Figure 7B.2 Example indicating the result of multiplying even functions having different frequencies. (a) Cosine of 3 times the fundamental frequency. (b) Cosine of 2 times fundamental frequency. (c) Product of these two functions integrate to zero (equal areas above and below y=0).

Figure 7B.3 shows that something interesting happens for the unique case $j = n$. Here, we see that the result is always a raised cosine of twice the frequency of the multiplier (= $2j$ or $2n$). There is still symmetry about the y-axis, but the function never dips below the x-axis. The area under the waveform is hence always positive and so the integral is always greater than zero.

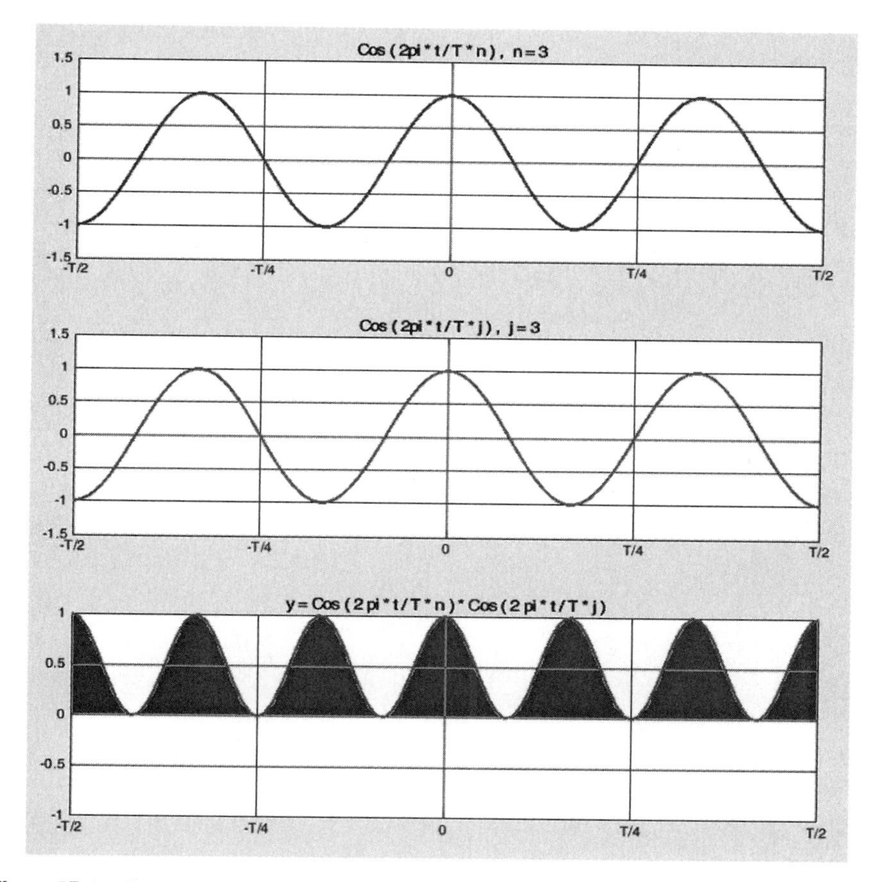

Figure 7B.3 Example indicating the result of multiplying even functions having the same frequency. Cosines at (a) and (b) have frequencies that are twice the fundamental. (c) Shows the product of these two functions which have areas that are entirely above y=0, yielding a positive (nonzero) integration.

8

Time-Domain Reflectometry and Distance-to-Fault Measurements

8.1 Introduction

One-port devices commonly used in digital electronics such as cables, traces, connectors, and various combinations of these have traditionally been characterized using an instrument called a time-domain reflectometer (TDR). A TDR generates a pulse and transmits it into the DUT (usually a cable or trace) and listens for the reflections. It operates similarly to a radar and measures any and all reflections back to the source that were generated by discontinuities in the transmission medium. The TDR measures the time for the signals to leave the source and return, resulting in electrical length or distance. These reflections will be generated from any mismatches or physical discontinuities in the transmission path.

As digital devices have gone higher and higher in frequency, the traditional approach to TDR measurements has become impractical. Modern VNA techniques allow measurement of PCB and system impedance characteristics in addition to S-parameters and other standard VNA measurements. In this chapter, we begin by measuring simple devices such as a series-connected cable in series and move towards more complex measurements such as PCB traces, connectors, and vias.

One-port devices measured with a time-domain reflectometer typically have a reduced set of measurement requirements compared with the one-port device measurements in Chapter 3. While all the measurements previously

Table 8.1
Example Devices and Unique Measurement Requirements for TDR and Distance-to-Fault Measurements

Device/ Application	Section	Unique Measurement Requirements (Beyond IL and RL
Cable	8.1.4	Electrical length, impedance versus distance, connector cable junction impedance
PCB traces	8.1.5	Impedance versus distance and input and output impedance
Connectors and launches	8.1.6	Impedance, junction impedance
Vias	8.1.7	Via input and output impedance

discussed can most certainly still be measured with a VNA (e.g., S-parameters), other measurements are more commonly made with a TDR. For example, a device may have a requirement to measure the characteristic impedance, Z_0, over the length of the cable or trace (usually 50Ω) as well as at the transitions between the cable or trace vias. Z_0 of different types of interconnection devices such as connectors may also be necessary and impedance discontinuity measurements are also possible. Most modern VNAs have a TDR mode that provides higher performance and more capabilities than available with a legacy-dedicated TDR instrument. While all these measurements are fairly straightforward, the setup of the VNA may require harmonic frequency spacing as well as a defined impedance at the cable end. Further discussion of more complex TDR topics, such as differential measurements, skew, and crosstalk, are deferred to Chapter 10.

VNAs are favored for one-port, time-domain measurements for a few reasons. First, unlike a stand-alone TDR instrument, VNAs can make measurement on devices at much higher frequencies as high as 1 THz. Second, effective TDR measurements typically require greater bandwidths than the maximum frequency of the DUT. This is because the TDR needs to generate a square wave with an appropriate rise time for the required measurements, typically 3 or 5 times the desired frequency. These larger bandwidths limit the noise floor of the system as the frequency increases. As noise floor example, a 10-GHz clock typically may require 50 GHz (50e9 Hz) of bandwidth (to include the third and fifth harmonics). This increases the minimum noise floor of the system by 107 dB ($10_{\log}$(50e9)). So for an ideal system with a noise floor of kTB = –174 dBm/Hz, the noise floor required for a traditional 50-GHz TDR measurement would be –174 + 107 = –67 dBm, dramatically higher. This will make it impossible to measure signal levels lower than –67 dBm. A VNA does not share this limitation, because its TDR response is calculated from a frequency-domain measurement. In this measurement, each point in the frequency sweep is individually measured using a bandwidth defined by the user, leading to a much lower noise floor (see Figure 8.1). In addition, the energy of a TDR's pulse is spread across all the discrete spectral lines of the pulse signal, whereas in a VNA, the entire source power is concentrated at each frequency point, further enhancing the SNR.

8.1.1 Steps for Establishing a Single-Port TDR Measurement Baseline

While Chapter 3 gave examples of single-port measurements using the Rohde & Schwarz (R&S) ZVA network analyzer, this chapter will use an alternative VNA, the R&S ZNB. The ZNB is better suited for TDR measurements, which require lower-frequency capabilities. These lower frequencies might go down to less than 1 MHz. This is measured more accurately with the ZNB as its architecture uses a bridge instead of a coupler for signal separation. A bridge gives it better low frequency performance as it has lower loss at frequencies below several hundred megahertz, unlike a coupler.

- *Step 1:* Determine the appropriate interface between the VNA and the DUT. If a cable is required, use only a high-quality, phase-stable test cable with appropriate connectors (and sexes) to interface with the VNA and the DUT. Avoid adapters and try to use metrology-grade coaxial connecters when possible. See Table 8.2 for help in selecting the correct connector.

  Unfortunately, RF and microwave connectors used on VNAs are rarely used for high speed digital devices. Signal integrity (SI) measurements tend to use high-speed, multipin digital connectors instead. This means that interfaces may be required between the calibration plane and the input to the DUT. Other applications (such as discrete component

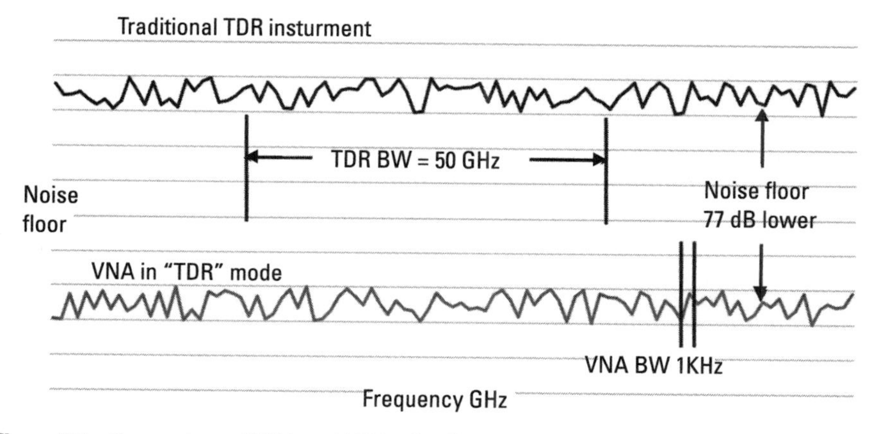

Figure 8.1 Comparison of VNA and TDR noise floor.

Table 8.2
Metrology-Grade Coaxial Connectors

Connector	Frequency Range	Compatibility
SMA	DC to 24 GHz	Not considered a metrology grade connector try to avoid using; use 3.5 mm
3.5 mm	DC to 34 GHz	Mates with 2.92 mm
2.92 mm	DC to 40 GHz	Mates with 3.5 mm
2.4 mm	DC to 50 GHz	Mates with 1.85 mm
1.85 mm	DC to 70 GHz	Mates with 2.4 mm
1.0 mm	DC to 110 mm	N/A

tests) generally require a test fixture and appropriate calibration and de-embedding techniques to move the calibration plane to DUT terminal. Many of these structures will require software tools for fixture removal such as Atitec ISD or Packet Microwave SFD to measure a test coupon on a PCB and create the de-embedding file for the device measurement. These techniques are discussed in more detail in Chapter 10.

- *Step 2:* Connect the device with required adapters and interfaces.
- *Step 3:* Start from an instrument preset. Configure the start frequency, stop frequency, and number of points (or equivalently, frequency step size) appropriate for the measurement requirements. Understand that when making TDR measurements, the mathematics of Fourier analysis may require that the measurement uses a harmonic grid, which is formed by a set of equidistant frequency points f_i $(i = 1 \ldots n)$ with spacing Δf and the additional condition that $f_i = f$. In other words, all frequencies f_i are set to harmonics of the start frequency f_1. If a harmonic grid, including the DC value (zero frequency), is mirrored to the negative frequency range, the result is again an equidistant grid. The point symmetry with respect to the DC value makes harmonic grids suitable for lowpass time-domain transformations (step or impulse), but are not required for bandpass response. This is shown in Figure 8.2. The VNA will tell you if you violate these criteria.

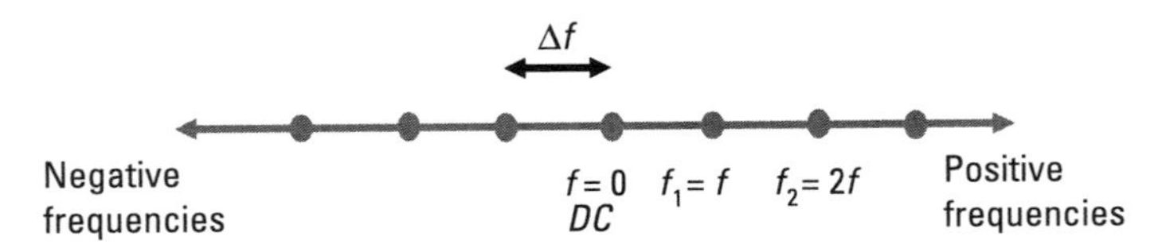

Figure 8.2 Harmonic grid point spacing.

In the lowpass response mode, it is easier to set the start frequency, stop frequency, and frequency step size (under the sweep menu) rather than number of points to ensure a harmonic grid. To use the full range of the 20-GHz ZNB, the settings in Table 8.3 may be used:

You can use a smaller step size if required by your measurement. For lowpass response, the first frequency point (start value) divided by the frequency step size must be an integer to ensure a harmonic grid. You can verify that the grid is harmonic by pressing the sweep button and checking the frequency step size. Again, if you are in bandpass mode, a harmonic grid is not required. Alternatively, you can have the instrument adjust the start, stop, or gap frequency automatically.

The minimum resolution (spacing) is determined by the highest measurement frequency. In this case, 20 GHz yields a resolution (in lowpass mode) of 1/(20 GHz) = 50 ps or 590 mils in air. (In the bandpass mode, the formula is $2/(F_{stop} - F_{start})$.)

The maximum range or distance that you can measure is determined by the number of points. In this case, 2,000 points gives a range of 100 ns or 98 feet in air. To measure longer devices, simply increase the number of points or otherwise decrease the frequency step size. For bandpass or lowpass impulse responses, the unambiguous time or distance is given by the reciprocal of the frequency step size. For a lowpass step response, it is half this value (e.g., 50 ns or 49 feet).

- *Step 4:* If applicable, verify that causality is not violated. Set the MEAS mode to $Z \leftarrow S_{11}$ to display impedance as a function of distance. Verify no impedance discontinuities before $t = 0$ as shown in Figure 8.3.
- *Step 5:* Set the source level and receiver bandwidth. Since most TDR-based SI applications are used to measure passive devices (cables, connecters, PCB traces), the instrument default settings for PWR (−10~0 dBm) and BW (1~10 kHz) are sufficient. Wider bandwidths are used for low-loss DUTs and production measurements requiring speed. Narrower bandwidths are used for lossy DUTs requiring higher dynamic range.

Table 8.3
VNA Default Settings for TDR Measurement

Stimulus Functions	
Start frequency (button)	10 MHz
Stop frequency	20 GHz
Sweep frequency step size	10 MHz
Resulting number of points	2,000

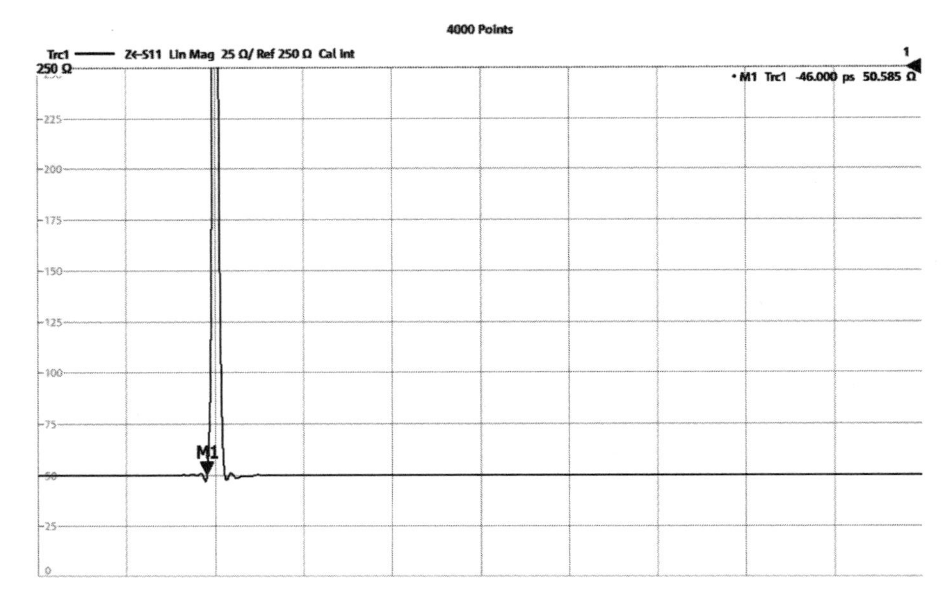

Figure 8.3 TDR causality.

- *Step 6:* Select the time-domain transform by pressing the trace button followed by the time domain tab. Configure the time-domain transform settings as described in Table 8.4 (note typical default values in bold).
- *Step 7:* Perform a single-port calibration using an appropriate calibration kit or autocal unit (e.g., 3.5-mm calibration kit for SMA or 3.5-mm connectors). Save the calibration and/or the test setup after completion.

Table 8.4
Default TDR Trace Settings for ZNB

TRACE CONFIG (button)	
Time domain (tab)	
Type	**Bandpass impulse** Lowpass impulse Lowpass step
Lowpass settings	Verify grid is harmonic Set the DC value: this is the end of the cable or DUT, short open, or load (1 open, –1 short, 0 load) or select continuous extrapolation
Impulse response	No profile (rectangle) Low first sidelobe (Hamming) Normal profile (Hann) Steep falloff (Bohmann) Arbitrary sidelobes (Dolph-Chebyshev)

- *Step 8:* After completing the calibration, it is important to independently verify the calibration by installing an open or a short standard from a different calibration kit or from a verification kit. Configure the instrument to measure S_{11} in trace format Smith (Smith chart). Does the trace look as expected for a short or open standard? If so, move the cable around and see if the response remains stable. If either the magnitude (the distance from the center of the chart to any part of the trace) or phase (angle from the horizontal plane to any part of the trace) changes, you must reexamine your setup or you will make poor or invalid measurements. Is the cable properly torqued at both ends? Is the cable worn or damaged, or is the center conductor of the connector (pin or receptacle) dirty, damaged, or recessed? Check the latter with a precision connector gauge. (Refer to Section 2.2.1.2 for a discussion of connector care.)
- *Step 9:* If the calibration is valid, connect the DUT and configure the traces for the desired measurement. The starting point will be an S_{11} measurement. The appropriate formats are shown in Table 8.5.

Now that we have completed the setup of the VNA for time-domain measurements, we will discuss the many parameters that can be varied when making VNA TDR measurements. The flexibility of the VNA can be quite daunting with many different settings that can be changed in the TDR mode. Fortunately, default values are usually sufficient for most TDR measurements, but it is beneficial to know what these settings do and when to use them.

8.1.2 Time-Domain Basics

When setting up the time-domain measurements with the VNA, there are two different measurement modes: bandpass and lowpass. The key distinction between them is in the results they provide. The bandpass mode is the most

Table 8.5
Common TDR Measurement Settings

Desired Measurement	MEAS	VNA Format
Return loss versus distance	S_{11}	dB Mag
Impedance versus distance	Z<–S_{11}	Real
Electrical length	Any	To show meters on the *x*-axis instead of time, press any stimulus button start, time-domain *x*-axis, distance (verify that the offset embed has the correct permittivity (1 for air))

general-purpose operating mode. It allows arbitrary frequency ranges to be defined and does not require a DC value. It is well-suited for band-limited devices, such as bandpass or highpass filters. However, bandpass mode has distinct limitations:

1. The step response is not supported (because it lacks a DC value).
2. It provides only half the resolution of the lowpass mode.
3. Complex impedance information is unavailable.

These limitations are avoided by using lowpass mode, which incidentally is the mode employed for most SI measurements. The key requirements for lowpass mode include: harmonic frequency grid; extrapolated DC value, expressed as a reflection coefficient (1 for open, –1 for short, 0 for 50 Ω); and windowing function.

When operating in the lowpass mode, a harmonic grid is required. This means that the stop frequency is an integer multiple of the start frequency, and the equal frequency steps of the harmonic grid will allow it to have a measurement point at DC (zero frequency). Such a frequency grid can be defined by the user or automatically calculated by the instrument. Note that if you let the instrument do it automatically, it typically will change some of the frequency sweep parameters.

8.1.1.1 Lowpass Versus Bandpass Mode

Bandpass mode restricts the measurement to impulse response only (no step response), but gives the user the ability to define arbitrary start and stop frequencies and noninteger frequency steps. It is most commonly used for devices that have a band-limited response. Figures 8.4 and 8.5 contrast the lowpass and bandpass impulse responses. In these examples, the same start and stop were used for the measurements, yielding nearly the same responses, but with the constraint that only the lowpass response provides complex impedances.

8.1.1.2 Step Response Versus Impulse Response

The time-domain response is generated by taking the inverse Fourier transform of the frequency-domain response. The VNA will do this for you using the time-domain transform function by merely checking the time-domain box under the Trace Configuration menu. The most common time-domain measurements made with the VNA are transmission-line impedances or cable fault location detection. For transmission-line characteristic impedances, it is recommended to use the step response. This mode shows the impedance of the line, trace, or cable as a function of distance. It also presents the data similar to how a signal integrity engineer might see it with a square-wave input. Otherwise, it is recommended to use the impulse response as it is more general-purpose and

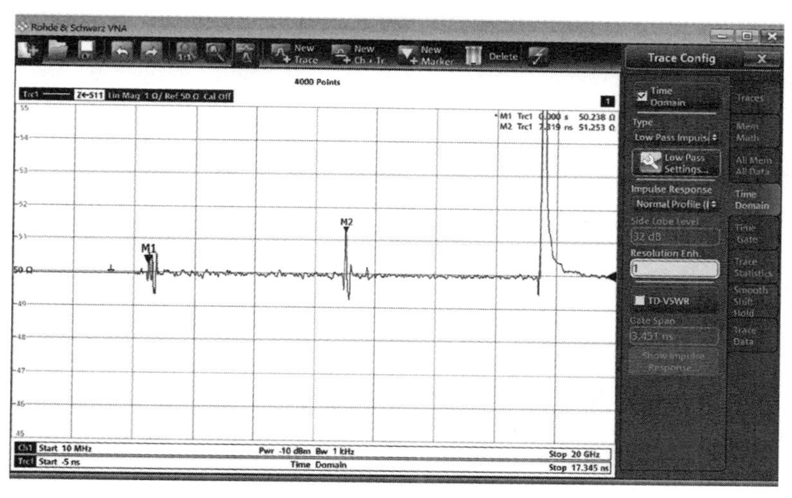

Figure 8.4 Lowpass impulse response.

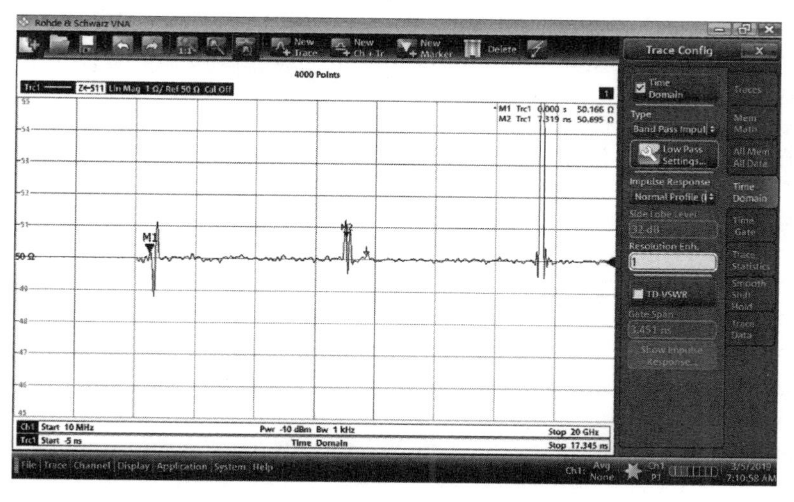

Figure 8.5 Bandpass impulse response.

can be used for TDR in the bandpass mode. Figure 8.6 shows the lowpass step response. Here, you can see how the impedance gradually increases from the launch point/calibration plane marker M1 to the end of the cable, where the open termination is encountered at 14.443 ms (marker M3).

8.1.1.3 Lowpass DC Values or Terminating Impedance

It is crucial to set the DC value of impedance at the end of the cable correctly for lowpass measurements. Since the VNA cannot actually measure down to

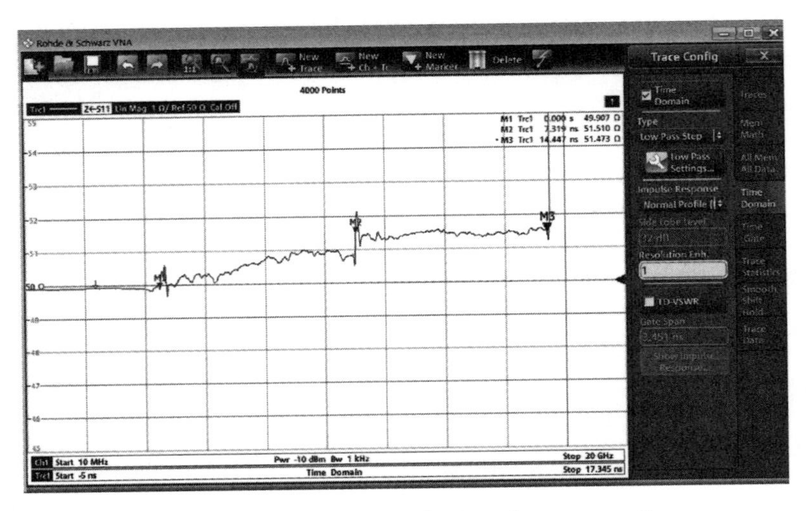

Figure 8.6 TDR lowpass step response measuring impedance versus time.

DC, it allows a user to provide this value manually, which may be obtained by different methods (e.g., measurement with a volt/ohm meter, interpolated from available RF measurements, or simply assigned by the user). Since the instrument is a VNA, this DC value is traditionally described in terms of its reflection coefficient. The formula for reflection coefficient is given by:

$$\Gamma = \frac{Z_L - Z_O}{Z_L + Z_O} \tag{8.1}$$

where Γ = reflection coefficient, Z_L = load impedance (ohms), and Z_o = characteristic impedance (ohms).

An open-circuit (infinite impedance) has a reflection coefficient of +1, while a short-circuit (zero impedance) has a reflection coefficient –1. A 50Ω load should have no reflections in a 50Ω system (reflection coefficient 0). Figure 8.7 shows the dialog box for measurements obtained from two cables connected together with a 3.5-mm barrel and an unterminated (open) end. Here the DC value is correctly set to 1, and the line impedance is displayed as approximately 50Ω, with some mismatch discontinuities observed where the cables are connected together, and there is clearly an impedance step to infinity at the end of the cable.

Incorrect DC values give the results shown in Figure 8.8. Here the DC value is set to 1 representing a short. This causes the real part of Z_0 to be displayed as 0Ω, and the mismatch between the two cable sections is barely noticeable. However, the length of the cable is still correct

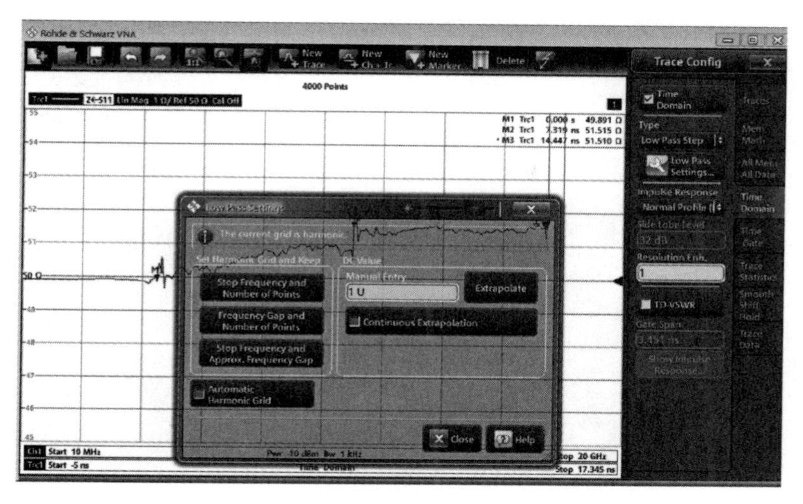

Figure 8.7 Lowpass step response DC value dialog box open (1).

Finally, in Figure 8.9, the DC value is set to zero. This causes the real part of Z_0 to be incorrectly displayed as 16Ω. The expected mismatch between the two cables is more noticeable than in Figure 8.8, but the magnitude of the mismatch is still incorrect and the cable length measurement is unaffected.

8.1.1.4 Windowing

To generate an infinitely narrow pulse in the time domain (an impulse), infinite bandwidth is required. Obviously, no instrument can generate or measure an infinite pulse. In addition, the cost of any instrument is typically a function of its maximum frequency. Because of this, the pulses that can be generated in the time domain will have a finite pulse width. This finite pulse width will be accompanied by ringing or sidelobes. These sidelobes can cause echoes in the response that do not actually exist (i.e., erroneous data). To reduce these sidelobes, windowing functions are used in conjunction with the lowpass time-domain function. The different windowing functions are a trade-off between pulse width and sidelobe suppression. The R&S ZNB includes windowing functions described in Table 8.6.

For most common applications it is recommended to use the normal profile, which gives a good trade-off between pulse width and sidelobe suppression.

8.1.1.5 Resolution Enhancement Factor

One unique capability of R&S VNAs is the resolution enhancement factor. In most VNAs, the resolution of the measurement is solely dependent on the maximum frequency of the network analyzer. In this case, it is 20 GHz and is defined as discussed previously. What the resolution enhancement factor does

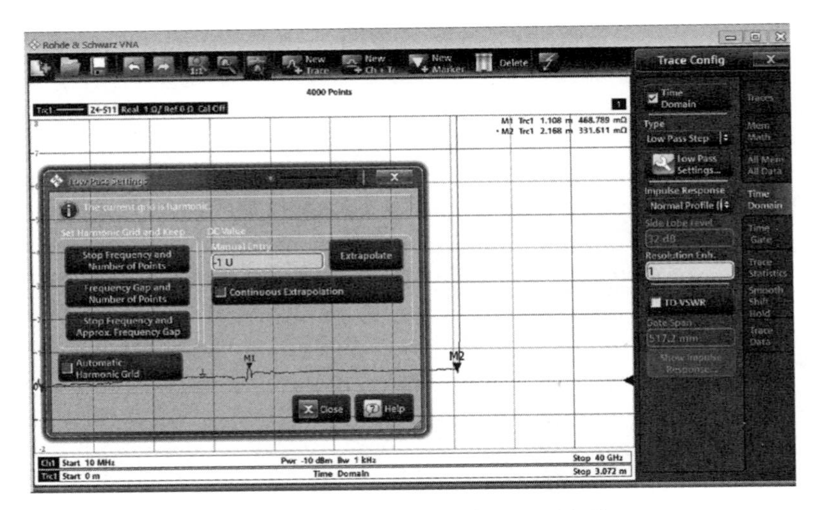

Figure 8.8 Lowpass step response DC value dialog box short (–1).

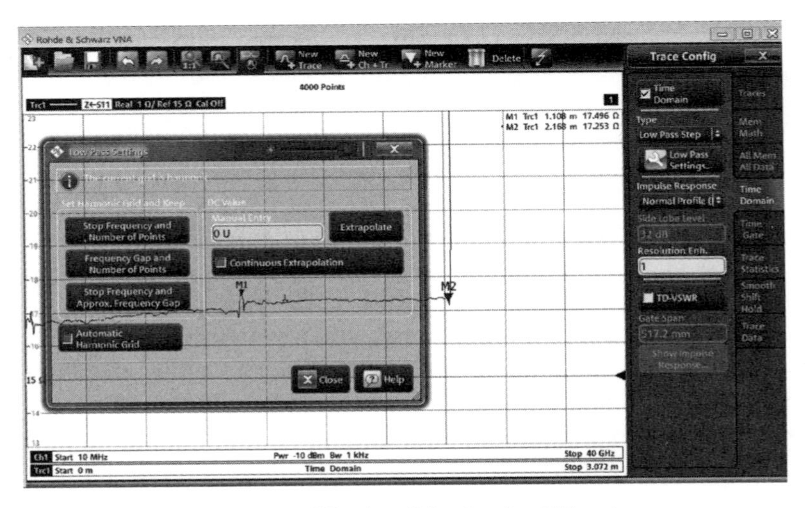

Figure 8.9 Lowpass step response DC value dialog box load (0).

is to scale this maximum frequency by a user-specific integer factor. This allows finer resolution without the expense of a higher-frequency analyzer and the associated higher performance connectors. It uses a linear prediction method which is quite effective for measuring passive structures such as PCB traces and interconnects. Distance to fault measurements typically use a factor of 1 for best accuracy. Shown in Figure 8.11 is a comparison of the bandwidth of a TDR measurement in the TDR mode with different resolution enhancement factors. Shown are factors of 1, 2, 4, and 8.

Table 8.6
TDR Windowing on ZNB Network Analyzer

Windowing Function	Width (Resolution)	Approximate Sidelobe	Approximately 3-dB BW	Figure 8.10
No profile (rectangle)	Best or narrowest	10 dB (worst)	22 ps	Trace 1
Low first sidelobe (Hamming)	Wider	40 dB + First	32 ps	Trace 2
Normal profile (Hann)	Wider	30 dB first	36 ps	Trace 3
Steep falloff (Bohmann)	Lower	40 dB+	42 ps	Trace 4
Arbitrary sidelobes (Dolph-Chebyshev)	Variable depending on sidelobe level	Variable (user defined)	26 ps	Trace 5

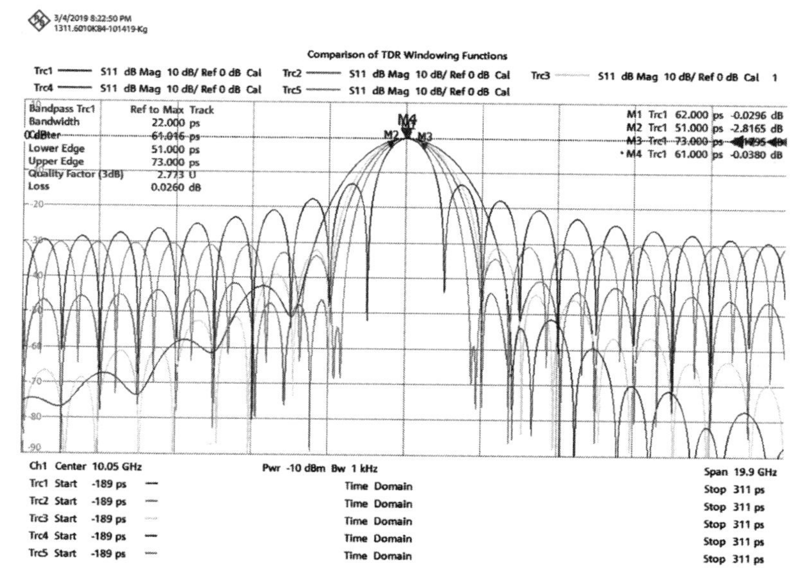

Figure 8.10 Comparison of TDR windowing functions.

8.1.1.6 Concept of Time in TDR Measurements

The interpretation of time and distance in the TDR mode depends on the measurement type. For reflection measurements (S-parameters or ratios with equal port indices (S11, b2/a2, etc.)), the time axis represents the propagation time of a signal from the source to the DUT and back. For transmission measurements, it represents the propagation time from the source through the device to the receiver.

In addition, the calibration plane is time zero and shows 0 phase and unity gain between the incoming and outgoing waves.

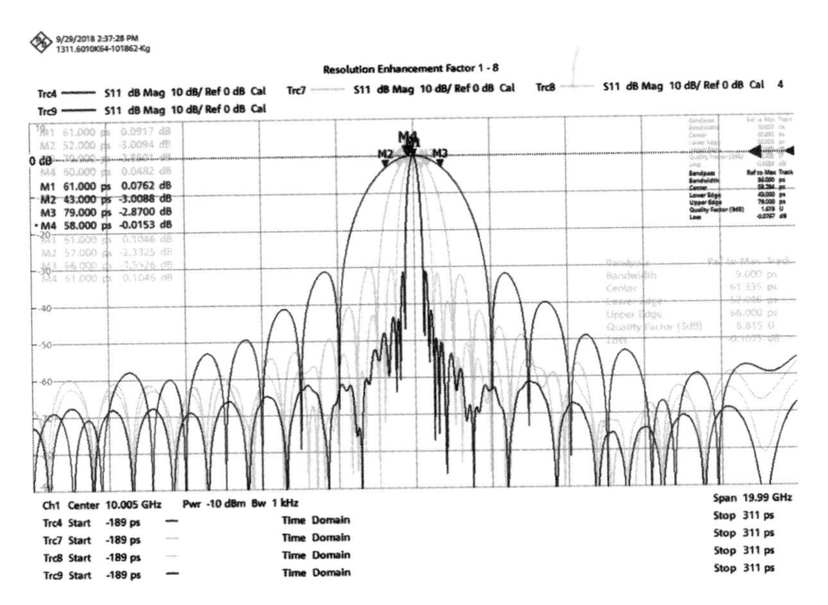

Figure 8.11 Comparison of resolution enhancement factors of 1 to 8.

8.1.1.7 Concept of Distance in TDR Measurements

The distance between the source and the DUT is calculated from the propagation time, the velocity of light in a vacuum, and the velocity factor of the receiving port:

$$Distance = \langle Time\rangle * c_0 * \langle Velocity\ Factor\rangle \text{ for transmission measurements}$$

$$Distance = 1/2 * \langle Time\rangle * c_0 * \langle Velocity\ Factor\rangle \text{ for reflection measurements.}$$

The factor 1/2 accounts for the return trip from the DUT to the receiver. When measuring distance with the VNA by putting the *x*-axis units into distance, the VNA will adjust the response for this round-trip propagation by dividing the distance by 2 to give the correct physical geometry and distance of the device.

8.1.1.8 Concept of Distance Resolution in TDR Measurements

For an instrument with an infinite bandwidth, an infinitely narrow pulse can be derived, obviously infinite bandwidth is not possible, so over the useable frequency range of the instrument, a weighting function is applied that takes a value of 1 over the operating frequency range of the instrument. For the ZNB20, this would be 100 kHz to 20 GHz and a value of 0 otherwise. The limited f_{max} of the VNA in the case of the ZNB20, 20 GHz, widens the pulse in the time domain. This weighting or multiplication in the frequency domain corresponds to convolution in the time domain with Dirac function or:

$$\frac{\sin(x)}{x} \tag{8.2}$$

the width of these pulses is inversely proportional to the span by the equation.

For the lowpass response

$$\Delta T = \frac{1}{\Delta F} \tag{8.3}$$

and for bandpass response

$$\Delta T = \frac{2}{\Delta F} \tag{8.4}$$

8.1.1.9 Concept of Range in TDR Measurement

When operating in the lowpass mode, a harmonic grid is required with the aliasing distance (or ambiguous range) determined by the total number of measurement points divided by the frequency span (in hertz). The aliasing distance is the maximum distance that can be measured for any DUT. To increase the

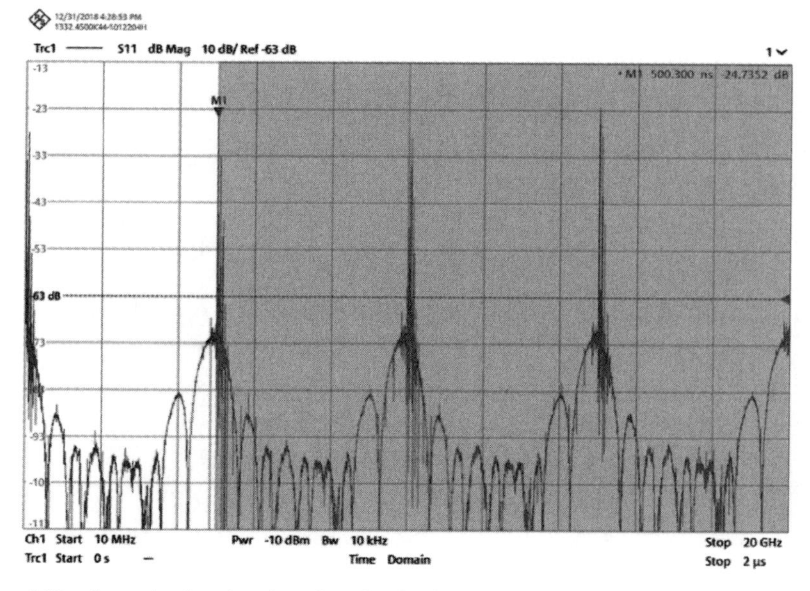

Figure 8.12 Example showing time-domain aliasing in a low-pass impulse response.

alias free range, increase the number of points and the start frequency is an integer multiple of the frequency step size.

This frequency grid can be defined by the user or automatically calculated by the instrument. If you choose to let the instrument do it automatically, understand that it will typically change at least one of the frequency sweep parameters (start, stop, step, or number of points). The frequency grid must be a harmonic grid for the lowpass mode, which also requires a measurement (extrapolated or user-provided) at DC (zero frequency).

As an example, Figure 8.13 shows the lowpass impulse response for a TDR obtained with a 10-MHz to 20-GHz frequency sweep, using 2-MHz frequency steps. The first unambiguous range extends to 1/(frequency step), or 500 ns (Marker 1).

Likewise, Figure 8.13, shows a similar unambiguous range for the bandpass impulse response (500 ns).

However, in Figure 8.14, the unambiguous range for the lowpass step response is only 250 ns, which is half the value obtained for either the lowpass or bandpass impulse responses.

8.1.3 Time-Domain Gating

One of the most powerful and useful tools in the VNA is the time gate. It allows you to remove unwanted reflections or limit the physical distance over which measurements are made, essentially moving the calibration reference plane

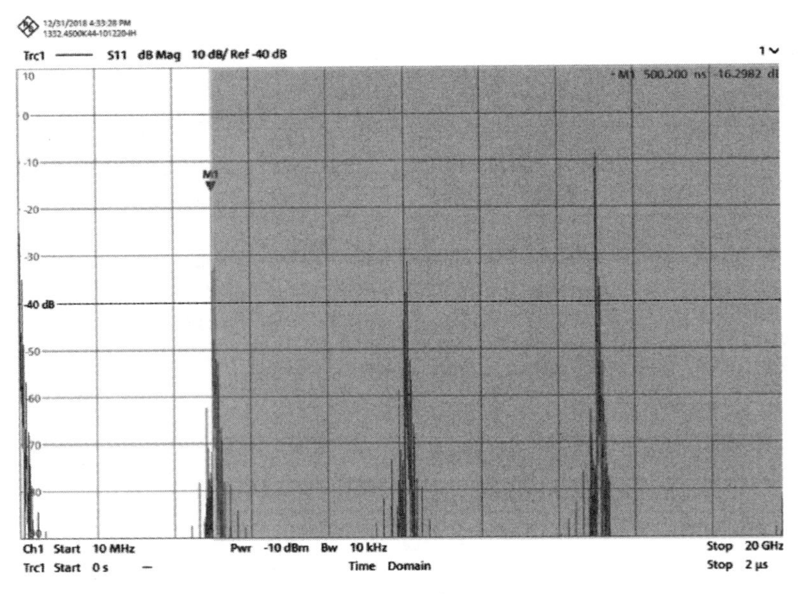

Figure 8.13 Ambiguity range for bandpass impulse response.

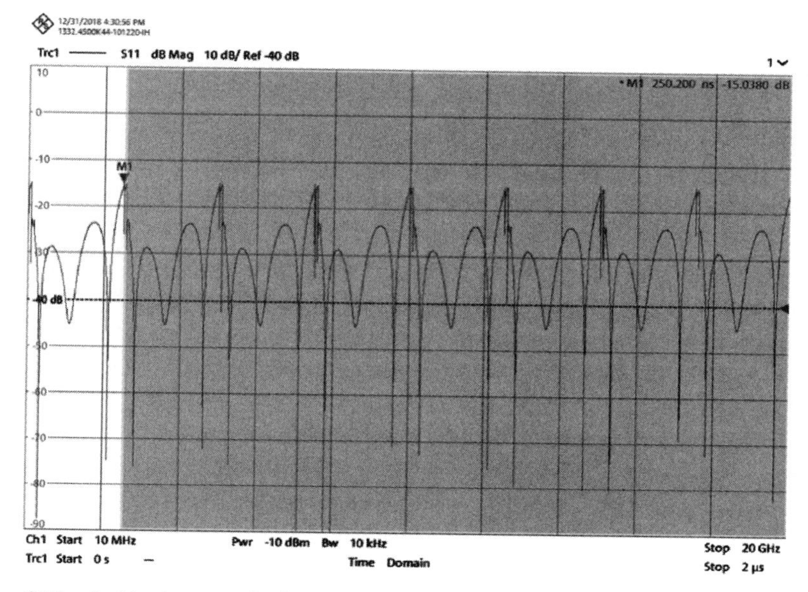

Figure 8.14 Ambiguity range for lowpass impulse response.

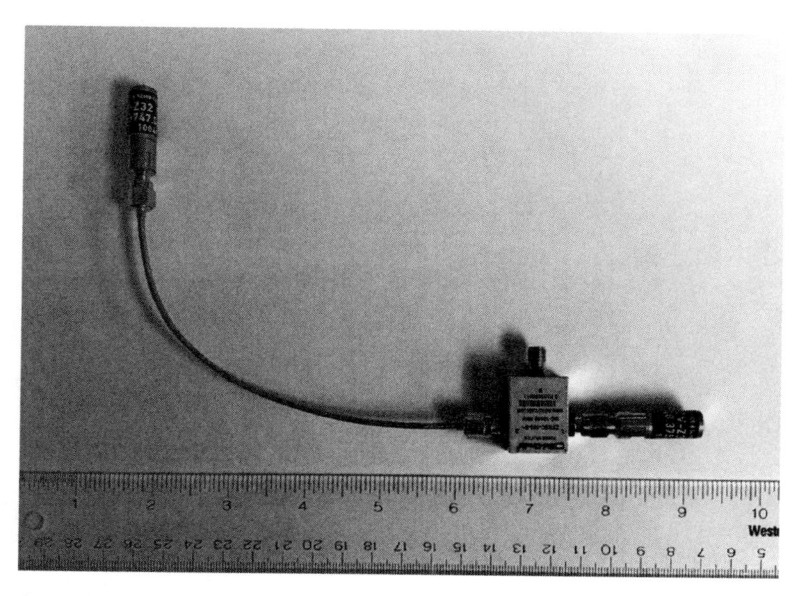

Figure 8.15 Ambiguity range for lowpass step response.

from wherever the calibration was performed, to any arbitrary distance within the range of the current TDR settings of the instrument. You simply turn on time-domain gating and move the gates to where you want the reference plane to be. In simple cases, you may know the geometry of your PCB and the exact distances to where you want to establish the gate. In other instances, you may want to move the gate to just past a step or discontinuity and look at the input or output impedance just beyond that discontinuity. This is done in the time domain, but you can switch back to the frequency domain with gating on and look at the S-parameters at any arbitrary point within your DUT.

An example to illustrate the capability of TDR gating, Figure 8.16 shows a power splitter with a terminated length of cables attached to split port leg and a termination on directly on the other leg. The resulting S_{11} versus frequency measurements is given in the top of Figure 8.16 while the bottom of Figure 8.16 shows Z_{11} versus distance showing a constant 50Ω at all distances.

To show the utility of the gating function, we will remove the termination on the port at the end. The measurements shown in Figure 8.17 are repeated and shown in Figure 8.18 but now with the cable open. For the remaining frequency response plots, the upper portion will remain the same and the lower TDR portion will be modified step by step.

8.1.4 Cable Measurements

For accurate and repeatable measurements, the use of high-quality, phase-stable cables cannot be understated. In this case, we are talking about the measurement cables that are used for calibration, not the cable under test. Poor mea-

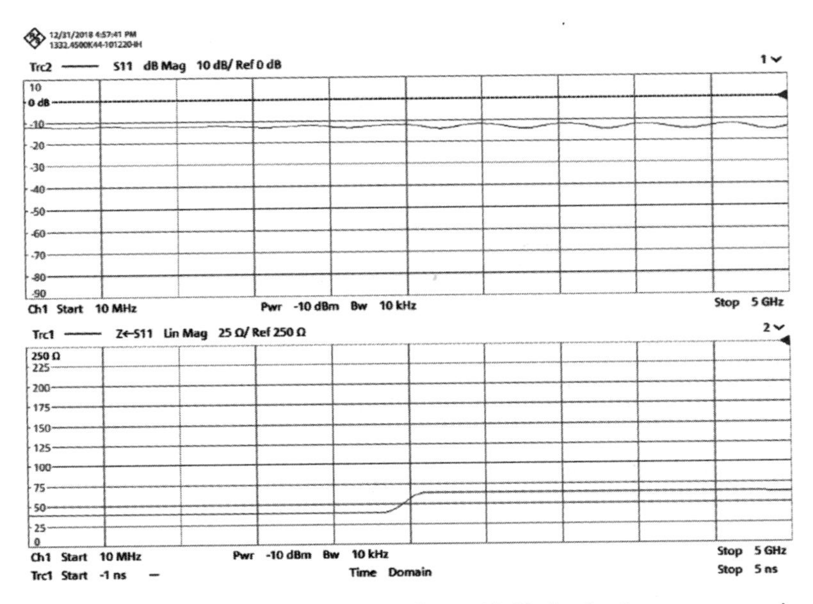

Figure 8.16 Frequency response of power splitter with 50 ohm load on one arm and an unterminated coaxial cable on the other.

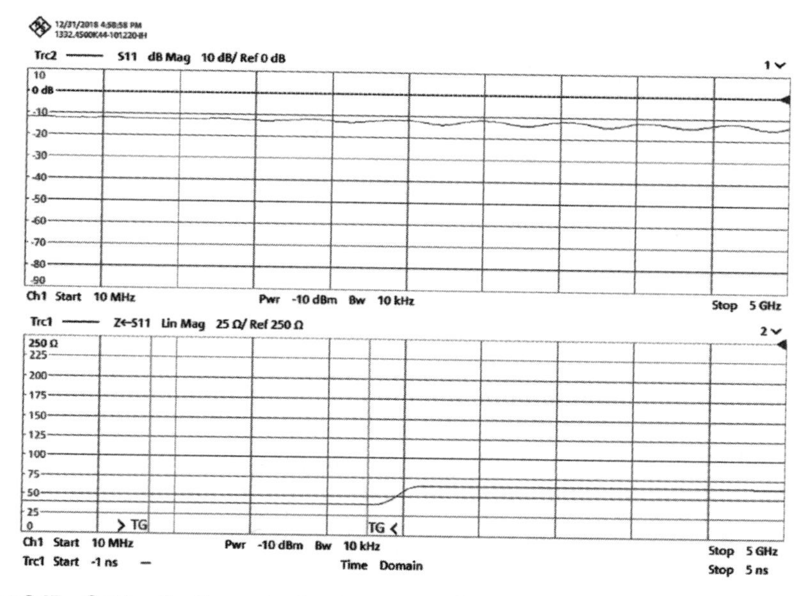

Figure 8.17 Setting the time gates for the power splitter network.

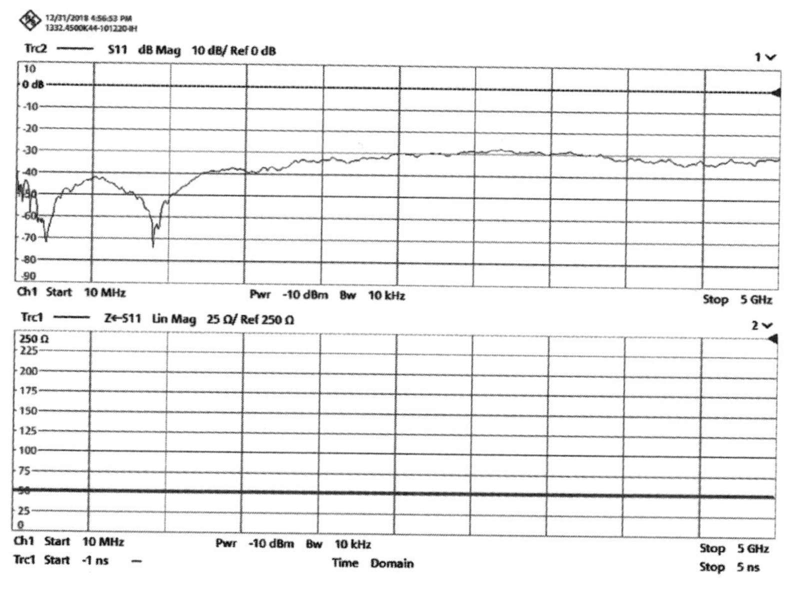

Figure 8.18 Power splitter with cable on one leg.

surement cables may degrade the overall return loss by 10 or 15 dB even after calibration. To make matters worse, low-quality test cables are typically not phase-stable; bending the cable will dramatically change its S-parameters.

As discussed in Chapter 3, passive, one-port measurements of cables having low insertion loss and high return loss are quite challenging. A typical measurement setup is shown in Figure 8.19. TDR functionality gives some unique capabilities to make measurements as a function of distance and remove unwanted reflections from the measurement using gating. The most commonly measured cable parameters are impedance, impedance discontinuities, and accurate length measurements for phase matching.

Calibration should be used to establish a zero phase and unity amplitude plane as in standard VNA measurements. Users often display real or imaginary impedance values on the *y*-axis and electrical length or distance on the *x*-axis. In VNA time-domain analysis, the default VNA display is time on the *x*-axis and the magnitude of the reflection coefficient on the *y*-axis. The time shown on the *x*-axis is the round-trip travel time of a wave along the conductor, revealing all impedance discontinuities along its length. Both axes can display alternate units. The *x*-axis can show the distance along a DUT, as long as the dielectric (permittivity or velocity factor) of the DUT is known. Some permittivities of common materials are described in Table 8.7.

To change the value displayed on the *x*-axis, enter the value for relative dielectric constant and switch from time to distance.

When measuring cables, there are a number of significant considerations. First, is the cable connectorized and, if so, on both ends? If it is, you can put a known load on the output such as a short or a load to give a slightly more accurate measurement of the time-domain response. Figure 8.20 shows a simple plot of S_{11} versus time.

8.1.5 PCB Measurements

PCB trace measurements are generally similar to cable measurements and present many of the same challenges. In addition, PCB traces of interest may not be conveniently terminated in standard coaxial connectors. To circumvent these problems, RF probing, de-embedding, and time-domain methods may be ap-

Table 8.7
Permittivity of Common Dielectric Materials

Material	Permittivity
FR-4	4.8
Teflon	2.1
Silicon Dioxide	3.9
Silicon	11.7
Duroid	2.3
Rogers RO3003	3
Alumina	9.8

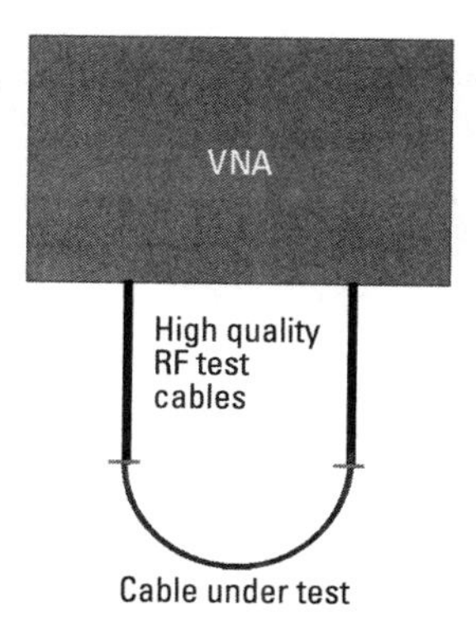

Figure 8.19 Typical cable measurement setup.

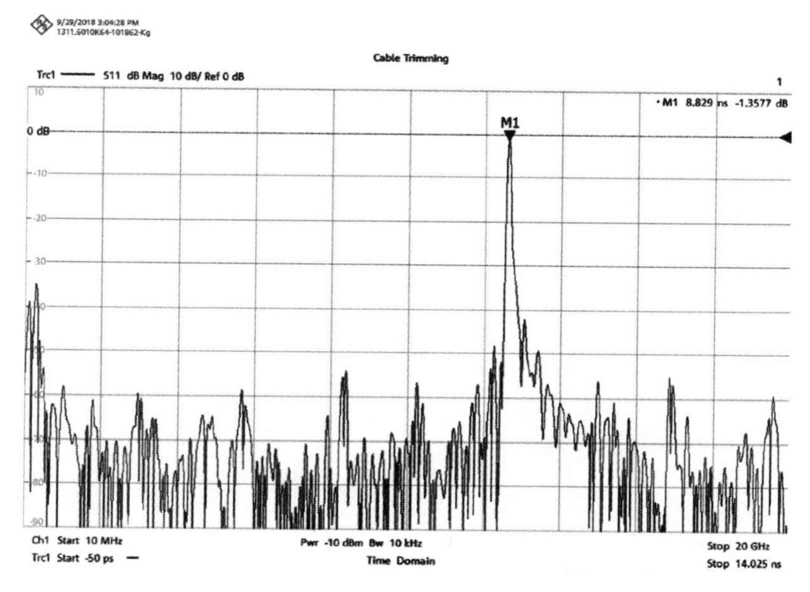

Figure 8.20 Typical cable measurement setup.

plied. For example, TDR offers the easiest method of de-embedding, since the desired reference plane can often be identified by inspection of discontinuities in the TDR response, and TDR gating can then be applied to move the reference plane. For example, in the case of a soldered pigtail, you would want to remove the effects of the pigtail.

In the case of stripline structures, it becomes even more challenging as the trace that you are measuring is buried within the inner layers of the PCB and completely inaccessible other than by the vias that feed them.

8.1.5.1 PCB Transmission Lines and Vias

Traces on PCBs often need to traverse the top and bottom of a PCB and may even be on inner layers of the PCB. To change layers on a PCB, a via is generally used. There are different types of vias with different dimensions and paths. A basic via is a hole that goes through the entire board to the bottom with a specified dimension and is plated all the way through. Blind vias are much more difficult to fabricate because they do not go all the way through the board. If these structures are not designed or fabricated correctly, they will look inductive and can cause signal distortion.

When signal integrity engineers or PCB designers examine trace impedance in the time domain, they usually display it as a function of real impedance versus distance. For a single-ended trace, the impedance should be 50 with no reactance. Differential traces commonly exhibit a 100Ω real impedance, although certain applications such as computer backplanes sometimes use 90Ω. This impedance was originally defined by microprocessor manufacturers trying to optimize PCB fabrication processes. In this case, 90Ω defined the minimum trace with for standard etching processes, thereby lowering costs and improving yields.

Figure 8.21 shows an evaluation board equipped with RF connectors and probing pads for measuring a variety of different trace structures. Figure 8.22 shows a side view of the connectors and the PCB stack.

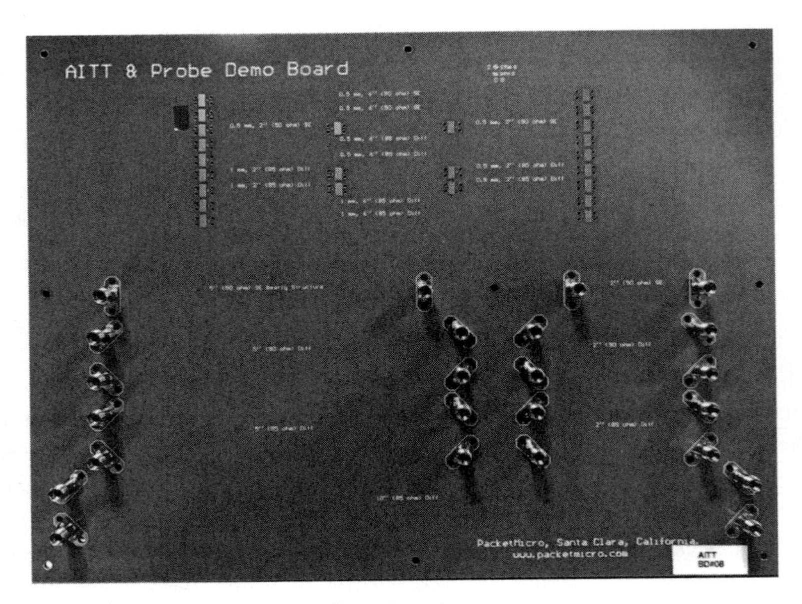

Figure 8.21 PCB trace measurement demo board.

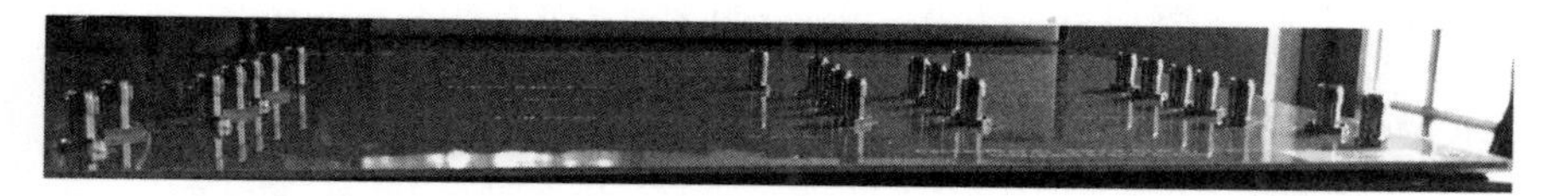

Figure 8.22 PCB trace measurement demo board.

Measurement results for the 2-inch, 50Ω, single-ended and 2-inch, 100Ω traces are shown in Figures 8.23 and 8.24. Marker 1 is the lowest impedance point midway within the input connector. Marker 2 is the beginning of the 2-inch trace with marker 3 the end of the trace. Finally, M4 shows the lowest impedance point midway within the output connector.

Time-domain gating can be used to isolate portions of the traces. The start gate will be moved to where M2 is and the stop gate will be moved to where M3 is. The resulting traces are shown in Figure 8.25. In Figure 8.26, we return to the frequency domain but with the gating turned off, while Figure 8.27 shows the frequency-domain measurement with the gating turned back on.

The previous exercise is repeated with a differential pair. The start gate will be moved to where M2 is and the stop gate will be moved to where M3 is. The resulting traces are shown in Figure 8.28. In Figure 8.29, we return to the frequency domain but with the gating turned off, while Figure 8.30 shows the frequency-domain measurement with the gating turned back on.

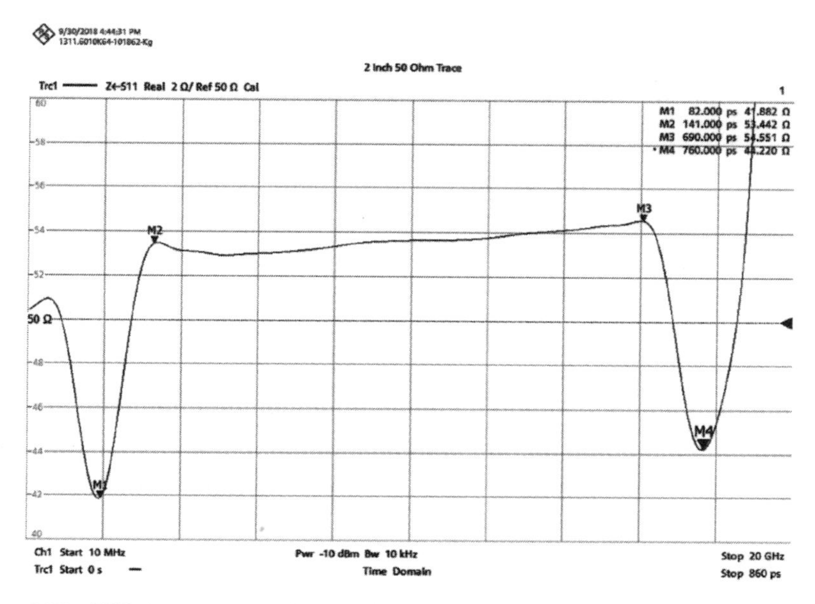

Figure 8.23 TDR measurement of 2-inch 50Ω test coupon.

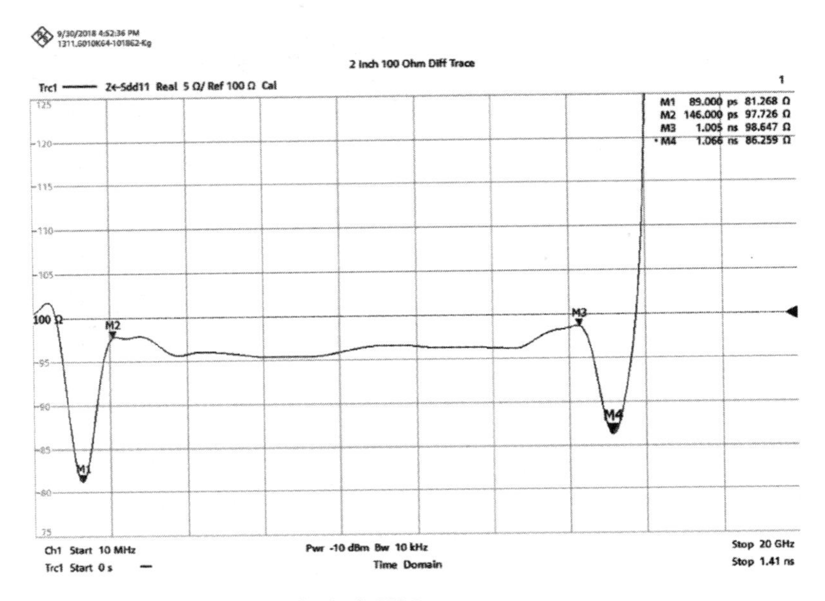

Figure 8.24 TDR measurement of 2-inch 100Ω test coupon.

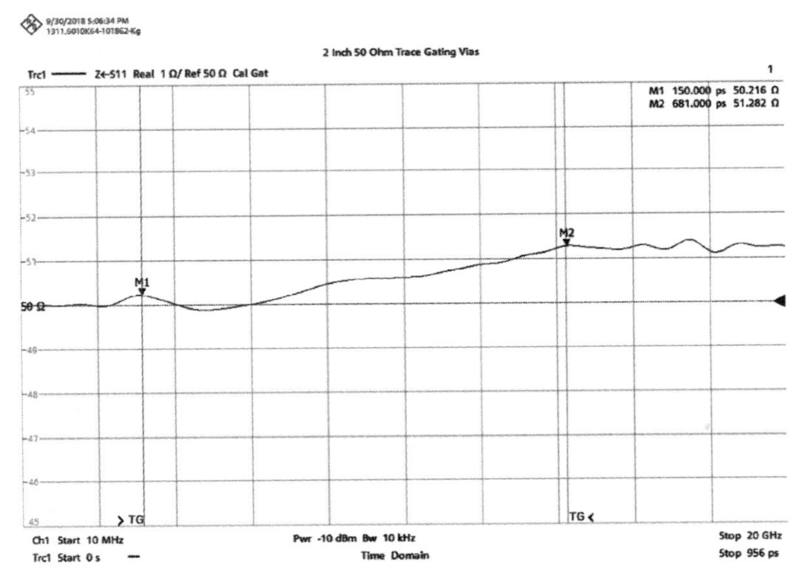

Figure 8.25 TDR measurement of 2-inch 50Ω test coupon with gating removing connectors..

8.1.6 Advanced De-Embedding Tools

Some customers have requirements of very precise de-embedding of input and output feed networks for devices on PCBs including the input and output con-

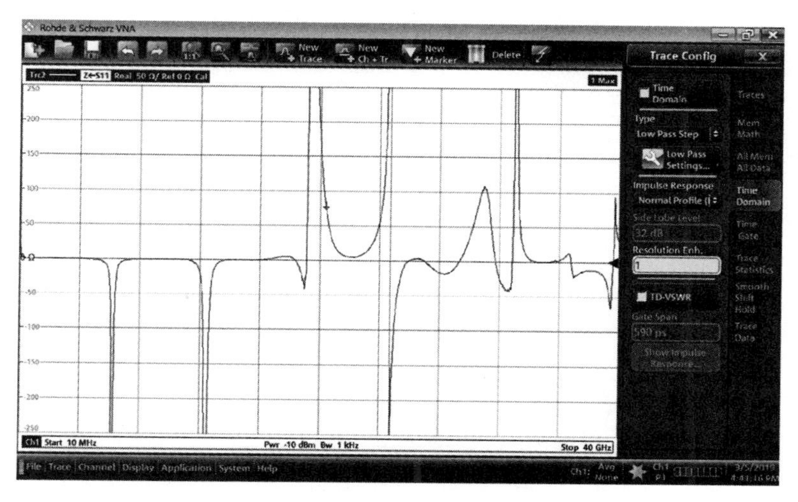

Figure 8.26 Frequency-domain measurement of 2-inch, 50Ω test coupon with gating off.

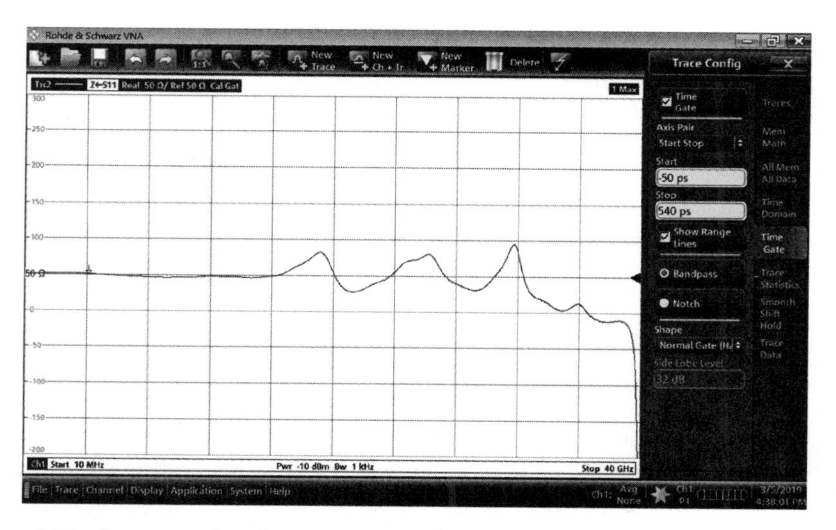

Figure 8.27 Frequency-domain measurement of 2-inch, 50Ω test coupon with gating applied to remove the effect of connectors.

nector as well as input and output traces. There are a number of software tools optimized for these applications. We have used the following:

1. *Ataitec ISD or in situ de-embedding:* This tool enable the user to fabricate a 2-X thru on the PCB to include the input connector and trace and output connector and trace as a test coupon. The tool measures

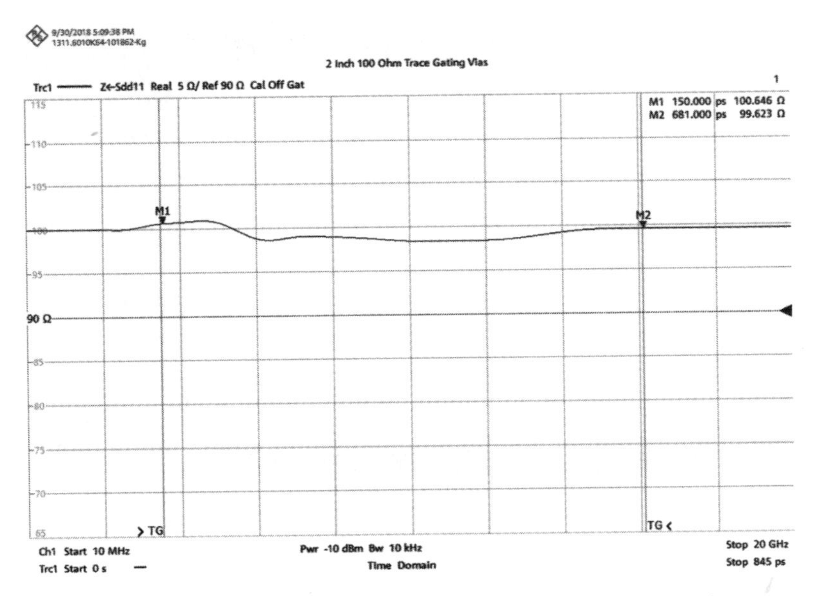

Figure 8.28 TDR measurement of 2-inch, 100Ω test coupon with gating removing connectors.

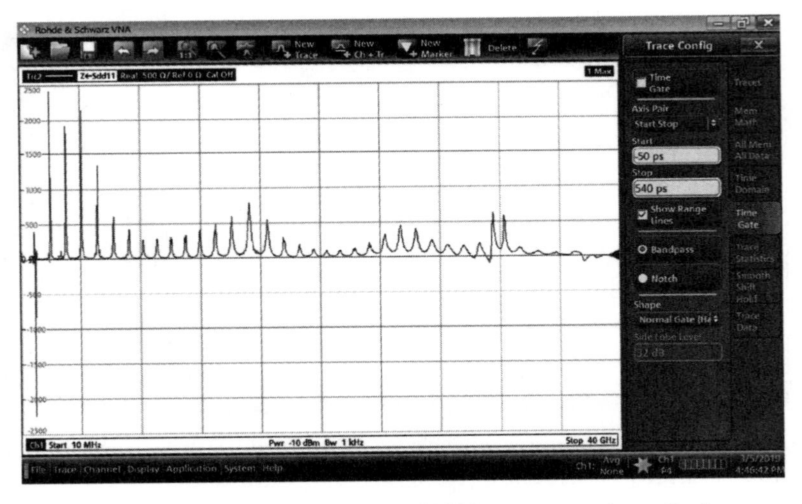

Figure 8.29 Frequency-domain measurement of 100Ω test coupon (no gating).

these traces and creates an S-parameter file to use for de-embedding. It is highly regarded within the SI community.

2. *Packet micro smart fixture de-embedding (SFD) tool:* This tool is similar to the Ataitec tool with a somewhat slightly more modern interface. It has a number of software tools available for common interconnect

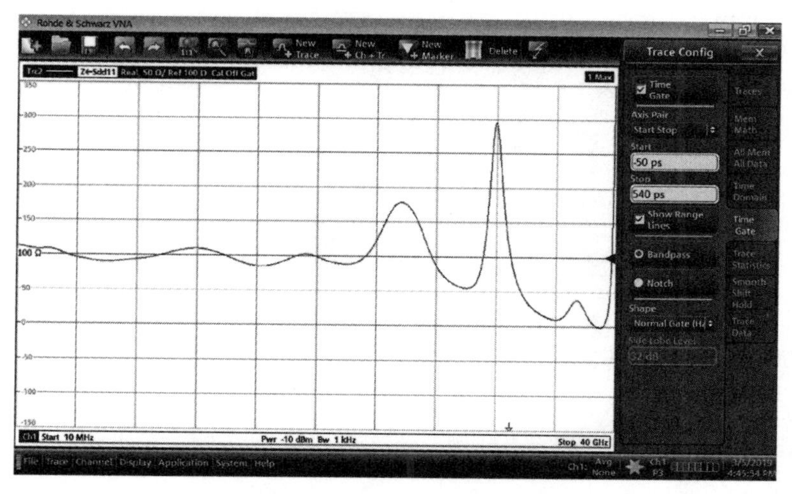

Figure 8.30 Frequency-domain measurement of 2-inch 100Ω test coupon with gating applied.

interfaces such as High-Speed USB 3.1. It is a newer tool with very good results.

3. *Keysight automatic fixture removal:* This is a similar tool specifically designed for Keysight VNAs and has a similar functionality to Ataitec.

Ataitec ISD and Packet Micro SFD are featured in Chapter 10.

8.1.7 DUTs with Pigtails

When PCBs are first fabricated there are often components on the board that need to be characterized on the board to verify the footprints for the device used on the PCB. In our experience, devices such as LNAs, PAs, surface acoustic wave (SAW) filters, and antennas impose tight placement and alignment tolerances for vias, silkscreening, and plating. Often, if these requirements are not met exactly, the device will not perform within specifications and may sometimes even operate in an unsafe manner, leading to device damage.

To characterize these devices, it is commonplace to solder a "pigtail" or short length of cable onto the associated PCB trace to feed the device of interest as shown in Figure 8.31. These pigtails may involve one or multiple ports.

This is a perfect application for time domain gating. The gate can be moved past the pigtail and even beyond the PCB trace to the device input, and then the TDR mode will be turned off returning to the frequency domain with the gate remaining on. This results in a first-order approximation of the input and output response of the device.

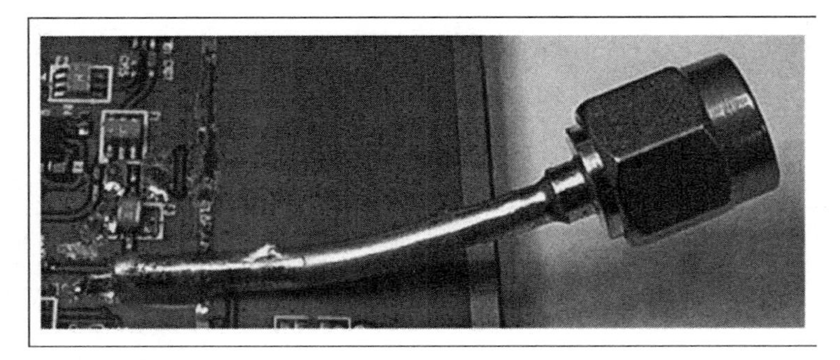

Figure 8.31 Test board with pigtails.

8.2 Best Practices

The following sections are intended to be used prior to starting a VNA time-domain measurement. The key aspect of Sections 8.2.1 and 8.2.2 is to know what to expect prior to measurement. This will save you countless hours of recalibrating, presetting, redoing setups, and perhaps damaging your DUTs.

8.2.1 Device Characteristics

This section is really intended as a stand-alone reference to be kept handy near the VNA when setting up new measurements. Remember to always save measurement states as well as calibrations, and be sure to save different versions as you proceed through your measurement setup. As with all measurements discussed throughout this book, for best measurement results, it is important to know the basic characteristics of your DUT before making any measurements. This is for a number of reasons:

1. Many devices, particularly active devices, have maximum input levels before their characteristic changes, sometimes quite dramatically such as harmonics or nonlinear measurements.
2. Key parameters of your device geometry: expected mismatches or impedance changes should be known at least roughly a priori. This information should be readily available from computer-aided design (CAD) layout files or can be measured at least roughly with calipers.
3. What are the key characteristics of your board material, connectors, or adapters? For loss tangent? For ε_r or dielectric constant?
4. When making TDR measurements, the setting of the instrument defines the minimum resolution as well as the maximum range as was

discussed, changing these parameters after calibration can reduce the accuracy or invalidate your calibration.

5. What are the things that are going to be measured and do they need to be measured simultaneously?

8.2.2 VNA Settings

1. When making TDR measurements, having a base instrument setup that is saved on the instrument with known setting can significantly reduce setup time. For TDR on a ZNB 20, typically use settings described in Table 8.8.
2. Dynamic range is a key benefit of using a VNA. So optimizing IFBW settings, power levels, and even windowing functions can significantly improve measurement accuracy and repeatability.
3. What are the specifications of your cables? We have done measurements whereby using a phase stable cable versus a low-cost SMA cable improved device return loss by nearly 20 dB.
4. I recommend that you look at your device response before you calibrate. This is not as crucial as in days past before auto calibration, but it frequently eliminates the need for multiple recalibrations.
5. Finally, have some sense of the required accuracy or uncertainty of the measurement desired. People often speak of this but rarely do any consequential analysis.
6. Ideally have a means for verification. Verification standard validate your measurement after calibration. A simple device such as a 15-dB

Table 8.8
ZNB Suggested Settings for TDR

ZNB Setting		
Start Frequency	10 MHz	
Stop Frequency	20 GHz	
If Bandwidth	1 kHz	
Number of Points	2,000	
Time Domain	Off	For calibration, on for measurement
Time-Domain Trace Config	Lowpass step	
Impulse Response	Normal profile	
Resolution Enhancement Factor	1	
Time Gate	Off	For calibration, on when required

attenuator is a good device for validation. Keep its response stored as a trace in memory. Often engineers have validation structures on their PCBs with known values.

7. Given the device measurement requirements, do we need multiple channels?
8. How should the data be presented? Simple S-parameters? Smith chart? Z_0 versus distance? Step response, impulse response?
9. DC impedance value?
10. Is the data going to be postprocessed or imported into a simulator or other instrument? If so, what format is required?

8.2.3 Correlations Between Frequency Domain and Time Domain

One of the most powerful features of using a VNA with a time-domain feature is the ability to go back and forth between the time and frequency domains. Both can give very useful pieces of information. The time domain lets you move the reference plane without a new calibration or de-embedding files, it also lets you remove or hide unwanted responses with time-domain gating. The frequency domain gives a better representation of circuit responses to input sinusoids and easier to accurately measure phase characteristics and design-matching networks. Unlike a TDR, the VNA uses Fourier transforms to go from the frequency to the time domain and back.

As has been discussed, all of the networks discussed thus far are linear and time invariant. Linear means a network where the voltages, dV/dt, current, and dI/dt are all linearly related. Time-invariant means the response of the network does not change with respect to time (i.e., the response will be the same regardless of when the signal is applied). Each linear time-invariant network can be represented by its impulse response $h(t)$ or in the frequency domain by its transfer function $H(f)$ and are related by the Fourier transform:

$$H(f)=\int_{-\infty}^{\infty} h(t)e^{-j2\pi ft}\,\mathrm{dt} \tag{8.5}$$

The converse is also true that the spectral response $H(f)$ measured in the frequency domain can be represented in the time domain by the inverse Fourier transform:

$$h(f)=\int_{-\infty}^{\infty} H(f)e^{j2\pi ft}\,\mathrm{dt} \tag{8.6}$$

9

Millimeter-Wave and Waveguide-Based Measurements

9.1 Introduction

The pioneering spirit is alive and well and expanding rapidly at millimeter-wave frequencies. This may be a bit surprising to users of spectrum below 6 GHz. From a "flatlander's" perspective, this lofty, untamed and sparsely inhabited millimeter-wave terrain has, until recently, been suitable only for maverick researchers and astrophysicists facing enormous and costly technical challenges. Consider that the free-space wavelength at 6 GHz is 50 mm; this shrinks to only 5 mm at 60 GHz and becomes considerably shorter in other dielectric media. This means that slight dimensional changes have a big impact on phase. Losses also increase with frequency, with certain bands accelerating absorption due to molecular and atomic resonances [1]. The combination of these factors place stringent requirements on things like cabling for millimeter-wave applications, which require high phase stability (and preferably low loss), and therefore tend to be as short as possible.

Higher frequencies also impose tight dimensional tolerance and conductor smoothness requirements as components shrink to microscopic proportions with infinitesimal skin depth. Against this technical milieu, the financial impact escalates exponentially with frequency.

Yet, in spite of these considerable technical and economic obstacles, the bands from 30 to 300 GHz are experiencing a recent surge of interest, fueled in part by the promise of fresh, wide swaths of spectrum for high-speed wireless communications (5G NR), commercial and military interest in materials measurements, and active/passive detection technology (automotive radar and body-scanning technologies).

Modern VNAs have evolved to meet new millimeter-wave applications with two-port and four-port models having measurement capability up to at least 67 GHz. Furthermore, measurements up to 1.4 THz (or higher) are possible by pairing VNAs with commercially available millimeter-wave frequency extenders. Even these powerful instruments are not immune to the consequences of millimeter-wave operation. Output power tends to taper off at higher frequencies. Receiver noise figure and mixer conversion loss also increase with frequency, leading to a net decrease in dynamic range. Phase noise of the VNA's RF and LO sources influences the trace noise of measured quantities, like waves (measured by a single receiver) and S-parameters (the quotient of two waves) to some degree. However, this influence is rather small in the normal operating frequency range of the VNA and even in the case of moderate frequency multiplication (for example, VNA millimeter-wave converters up to about 300 GHz.) The phase noise of the VNA's IF signal, which causes some trace noise, is generally much smaller than the phase noise of the VNA source signal. (Here, we benefit from the fact that the phase noise of VNA RF sources and LOs are pretty well correlated, at least at low offset frequencies from the carrier.) For both RF and LO synthesizers, phase noise close to the carrier is dominated by the frequency reference, which is shared by both sources and tends to cancel out to a high degree. But noise cancellation decreases with offset distance from the carrier (and hence with increasing measurement bandwidth) and also increases with frequency multiplication factor.

Below about 300 GHz, trace noise is mainly due to broadband receiver noise. This noise is mostly uncorrelated, so almost no cancellation occurs and trace noise increases with decreasing receiver signal-to-noise ratio (SNR). In higher mm-wave bands, however, phase noise becomes a major contributor to trace noise. For example, if a reasonably clean external RF generator with phase noise 20 dB better than the VNA source is used as the LO for a 750-GHz mm-wave converter, the resulting trace noise will nevertheless be larger than when using a LO signal derived from the VNA itself. This is because the external generator's close-in phase noise is not that well correlated to the VNA source. But this will vary depending on the way the references of the VNA and external generator are coupled. Also, newer 1-GHz references offer the potential for lower trace noise than with a traditional 10-MHz reference.

In spite of these challenges, modern millimeter-wave measurement systems make extensive use of leading-edge device technology to achieve remarkable performance. The VNA enables millimeter-wave measurements to be performed in a similar manner to lower-frequency measurements.

Still, the VNA itself represents only one of three critical factors required for successful millimeter-wave measurements. The second factor involves the other system components: cables, connectors, adapters, and millimeter-wave converters, where required. Sacrificing quality here can have a big impact on measurement performance, repeatability, and long-term stability. The third and final factor, operator experience and technique, is important for all VNA measurements, but is essential at millimeter-wave frequencies to achieve high-quality, repeatable results.

In this chapter, we focus on VNA measurement and calibration techniques for millimeter-wave frequencies, with an emphasis on wafer probing. We also de-mystify millimeter-wave converters by studying their architecture and showing how they are used to solve several common measurement applications. For completeness, we review relevant waveguide theory, because it remains a very popular, low-loss transmission line at millimeter-wave frequencies. We conclude the chapter with a useful, but less familiar application: spectrum analysis (and image-rejection) using VNAs and millimeter-wave converters.

9.2 On-Wafer Measurements

Similar to probing at lower frequencies, millimeter-wave probing is useful for device characterization, verification of prototypes and preproduction samples, and determining process yield and batch-to-batch consistency. At millimeter-wave frequencies, conventional signal launch mechanisms become problematic because of shrinking device dimensions, increasing loss (with distance), and difficulty providing an effective low-reflection launch between the substrate and (expensive) millimeter-wave coaxial connectors. In addition, higher-order modes may be excited at the microstrip to coax interface [2]. As a result, a majority of applications use on-wafer measurements at these frequencies. Coaxial or coplanar probes provide connection between the VNA test cables and the DUT. These probes are calibrated as part of the VNA measurement system (including cables) by using calibration substrates right at the probe tips.

Fortunately, many of the same S-parameter error-correction techniques found at lower RF may be used at millimeter-wave frequencies. The most popular among these are TRL and UOSM. But instead of coaxial calibration standards, these millimeter-wave standards take the form of metal pads and metal or resistive elements patterned on substrate or laminate materials (Figure 9.1).

There are two primary differences between calibration substrates and conventional coaxial calibration standards. The first is that the model coefficients for the substrate-based standards (open, short, match, thru, reflect, and line) are rarely included as part of the VNA firmware, and thus must be imported

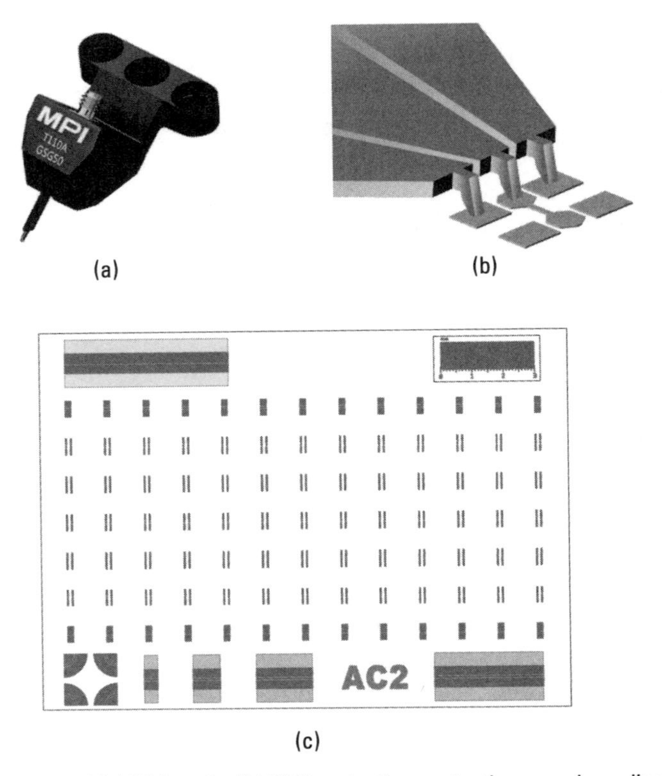

Figure 9.1 Example of (a) GSG probe (b) GSG probe tips contacting a coplanar line structure and (c) calibration standards on a substrate. (Images courtesy MPI Corporation.)

(or created manually). Fortunately, this is a very easy process (described below for the R&S ZVA).

1. Access the "Calibration Kits" menu in the VNA, and define a "New Calibration Kit": Channel > Cal > Cal Kits. Select "Available Connector Types" and click on "Add Type." Give it a unique name, choose "Sexless" and Offset Model "TEM" and then OK (Figure 9.2). This establishes a connector template.
2. Add a calibration kit for your newly defined connector. Click on "Add Kit" and under "Connector Type," select your new connector type (here, it is "MySubstrate").
3. Under "name," give it a meaningful description. For example, "GSG_125" for a 125-μm pitch ground-signal-ground probe
4. Next click "Add Standard" and select a "Reflect" standard for a TRL calibration kit. Enter the model parameters for the standard based on

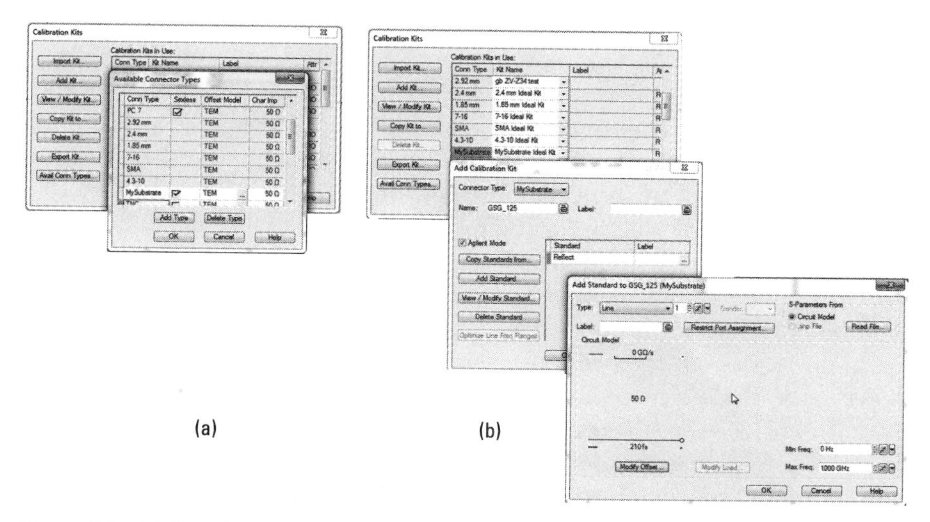

Figure 9.2 (a) Creating a new connector type. (b) Adding standards to a user-defined calibration kit.

manufacturer's data. Click "OK" and repeat for all required calibration kit standards (Figure 9.2).

The second difference between substrate-based calibration and traditional coaxial calibration is that considerable practice is required to make reliable and repeatable on-wafer connections (touchdowns) on the substrate. Too much contact pressure can damage probes or substrates. Too little pressure will result in poor contact/high contact resistance and consequently invalid calibration. Probe positioning is also very important to obtain correct phase during thru and line measurements.

After completing a calibration, it is a good idea to use an independent verification standard (often included on the calibration substrate) to validate the resulting calibration. Poor results are indicated by sudden jumps in amplitude or phase versus frequency, excessive trace noise, or excessive/unrealistic loss.

One other challenge posed by on-wafer measurements is that sometimes devices have test pads oriented 90° with respect to each other. Such arrangements require orthogonal coplanar probing, with calibration structures that support this geometry. The bends in coplanar line structures tend to generate undesired propagation modes. Conventional calibration methods (SOLT, TRL, LRM) do not account for these modes. By using the SOLR (short, open, load, reciprocal thru) technique (also known as UOSM), the characteristics of the unknown thru are determined during calculation of the error correction matrix [3].

9.2.1 Power Calibration

Power calibration poses an interesting challenge for on-wafer measurements. This is because power measurements are usually established at a coaxial plane using a coaxial power sensor. How does one properly account for the additional loss (as a function of frequency) of the input and output wafer probes?

In reality, only the loss of the port-1 probe is needed to calibrate the port-1 reference receiver and the port-2 measurement receiver. (The port-2 receiver power calibration automatically accounts for the losses through the port-2 probe.) Obviously, if accurate power measurements are required in the reverse direction (e.g., port-2 generator to port-1 measurement receiver), the port-2 wafer probe S-parameters will also have to be characterized.

To characterize a single wafer probe, two one-port calibrations are required. Both of these calibrations should be performed with identical sweep settings (start frequency, stop frequency, and number of points) using the same VNA test port (for optimum consistency). The calibration plane for the first ("Outer Cal") is established at the end of the port-1 coaxial test cable, using the OSM standards from a coaxial calibration kit. The chosen kit should match the connector type (e.g., 3.5 mm, 2.4 mm, 1.85 mm). The second ("Inner Cal") is performed on-wafer at the tip of the port-1 RF probe. Both calibrations should be saved (with descriptive names) in the Calibration Manager. A free external software program[1] called S2P Extractor uses these one-port calibrations to extract two-port S-parameters for the wafer probe. Thereafter, a conventional coaxial power calibration can be performed at the end of the port-1 test cable, and the S-parameter file can be used to correct the power calibration for the probe loss. After completing the power calibration, the test cable is reconnected to the wafer probe and a two-port on-wafer S-parameter SEC is performed. Finally, an on-wafer thru connection is established and the port-2 receiver power calibration is completed. The system is now ready to make accurate two-port S-parameter and power measurements.

A step-by-step procedure for extracting the wafer probe S-parameters and de-embedding the probe attenuation from the power calibration is presented in Sections 9.2.1.1 through 9.2.1.3.

9.2.1.1 Using the S2P Extractor to Obtain RF Probe S-Parameters

The following is the procedure to obtain RF probe S-parameters:

- *Step 1:* After installing the S2P Extractor program on the instrument, launch it. The ZVA path is: System > System Config > External Tools…

1. https://vna.rs-us.net/applications/s2p_extractor.html. This program may be installed directly on an R&S VNA or on a PC or laptop.

> "RSA S2P Extractor.lnk" softkey. (The ZNB path is System > Applic > "External Tools" tab > "RSA S2P Extractor.lnk" softkey.)

- *Step 2:* Click on "Outer Cal" and select your previously performed coaxial calibration (one-port OSM).
- *Step 3:* Click on "Inner Cal" and select your previously performed on-wafer calibration.
- *Step 4:* Click on "Ports" and select "Port 1".
- *Step 5:* Click "Generate." You will see a dialog "Save Touchstone files..." Give the file a descriptive name, for example, "GSG_125_P1" and select the directory path where you wish to store the resulting two-port Touchstone file. This file will be required for source power calibration (discussed next).

9.2.1.2 Power Calibration for RF Probes with ZVA

The following is the procedure for power calibration of RF probes with ZVA:

- *ZVA Step 1:* Connect a suitable power sensor to one of the front-panel USB connectors. Select "Channel > Cal > Start Power Cal > Source Power Cal..." Click "Modify Settings..." and set the "Maximum Number of Readings" to 10 and the "Tolerance" to 0.1 dB, then "OK."
- *ZVA Step 2:* Click on "Power Meter Corr..." and at the bottom of the dialog, click on "Import Data..." and import the wafer probe S-parameter file previously saved in Section 9.2.1.1. Click "Open" and then click "S21" as the trace to use for power correction. The transmission coefficient table will be automatically populated with the loss data (versus frequency) for the wafer probe.
- *ZVA Step 3:* In the same "Power Meter Correction" dialog, select "Take Transmission Coefficient into Account" and click "In front of" "Device Under Test (During Measurement)" (see Figure 9.3(a)). Click "OK" to close the dialog.
- *ZVA Step 4:* Ensure the Power sensor is connected to the end of the coax cable for the port-1 wafer probe. Click "Take Cal Sweep" and perform a port-1 source power calibration. At the completion of the sweep, click "close" and use the Channel > Cal > "Cal Manager" to save the power calibration to the calibration pool.
- *ZVA Step 5:* Reconnect the VNA port-1 test cable to the port-1 wafer probe. Perform an on-wafer two-port S-parameter calibration. Add this

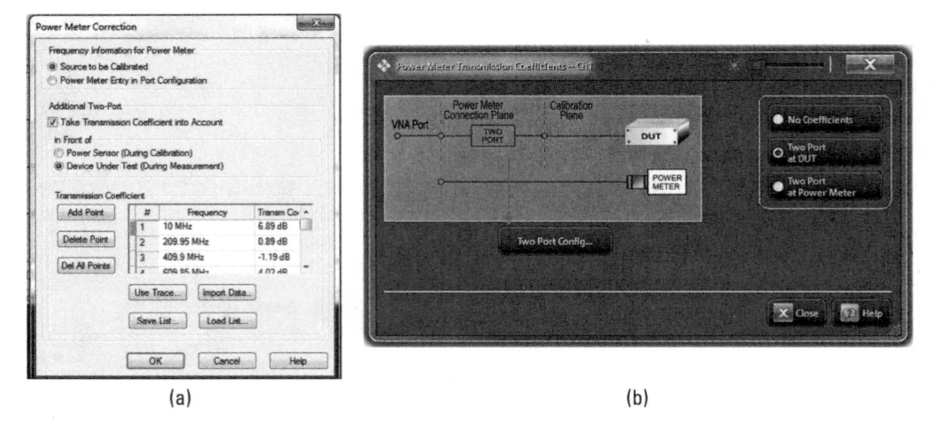

(a) (b)

Figure 9.3 Power meter correction dialog for source power correction of wafer probe loss. (a) ZVA dialog. (b) ZNB dialog.

S-parameter calibration to the calibration pool by saving the calibration to the same filename as step 4.

- *ZVA Step 6:* Configure the probes to make a thru measurement on the calibration substrate. Then select "Channel > Cal > Start Power Cal > Receiver Power Cal..." and ensure the receiver power calibration dialog is set to "b2" for the wave quantity to calibrate and "Port 1" for the calibration source. Add this receiver power calibration to the calibration pool by saving the calibration to the same file as steps 4 and 5. This concludes the ZVA power (and S-parameter) calibration process.

9.2.1.3 Power Calibration for RF Probes with ZNB

The following are steps for power calibration for RF probes with ZNB:

- *ZNB Step 1:* Ensure a suitable power sensor is connected to one of the front-panel USB connectors. Select "Channel > Cal > "Start Cal" tab > Power Cal..." Set the "Maximum Number of Readings" to 10 and the "Tolerance" to 0.1 dB.
- *ZNB Step 2:* Click on "Transm. Coefficients..." and click on "Two Port at DUT" (Figure 9.3(b)). Then click on "Two Port Config..." and click "delete all" if there are any values in the table. Then click "Import File" and navigate to the wafer probe S-parameter file previously saved in Section 9.2.1.1. Click "Open" and then click "S21" as the trace to use for power correction. The transmission coefficient table will be automatically populated with the loss data for the RF probe. Click "Close" two times to exit both dialogs.

- *ZNB Step 3:* Ensure that the power sensor is connected to the end of the coax cable for the port-1 wafer probe. Click on "Power" below "P1" to perform a port-1 source power calibration. At the completion of the sweep, click the "Use Cal" tab and select "Cal Manager…" to save the power calibration to the calibration pool.
- *ZNB Step 4:* Reconnect the VNA port-1 test cable to the port-1 RF probe and click "Start Cal" tab and the "Start… (manual)" button to perform an on-wafer two-port S-parameter calibration. Add this S-parameter calibration to the calibration pool by saving the calibration to the same filename as step 3.
- *ZNB Step 5:* Configure the probes to make a thru measurement on the calibration substrate. Then select "Channel > Cal > Start Cal > Power Cal…" and click on the "Meas. Receiver" below P2 to commence the receiver calibration. At the conclusion of the sweep, add this receiver power calibration to the calibration pool by saving the calibration to the same file as steps 3 and 4. This concludes the ZNB power (and S-parameter) calibration process.

9.3 Over-the-Air Measurements

Some material measurements are performed in three-dimensional space (radiated measurement) rather than in a coaxial or waveguide chamber/resonator (conducted measurement). For these applications, narrow RF beams are used to focus the millimeter-wave energy onto the DUT material and avoid unwanted reflections from other surfaces. Additionally, the test setup is often constructed is such a way as to ensure that a plane-wave quiet zone is established across the entire surface of the material to be measured. The test setup may be installed on rails with precision positioners to constrain the source and/or receiving antennas to motion along a single dimension, moving them closer together or further apart. Additionally, another positioner associated with the material holder allows precision adjustment of its angle with respect to the wavefront. Mechanical stability and repeatability are of utmost importance. Additionally, the entire setup should be mechanically isolated from vibrations. For calibration, over-the-air (OTA) TRL is used, where the calibration kit consists of a zero-offset reflect standard, a zero-length thru, and a 90° line standard (defined at the geometric center frequency of interest). The calibration process is described below.

- *Step 1:* Position the empty material holder frame at the measurement plane and ensure that the antennas are placed at equal distances (*d*) from it, as shown in Figure 9.4(a). The distance, *d*, must be sufficient

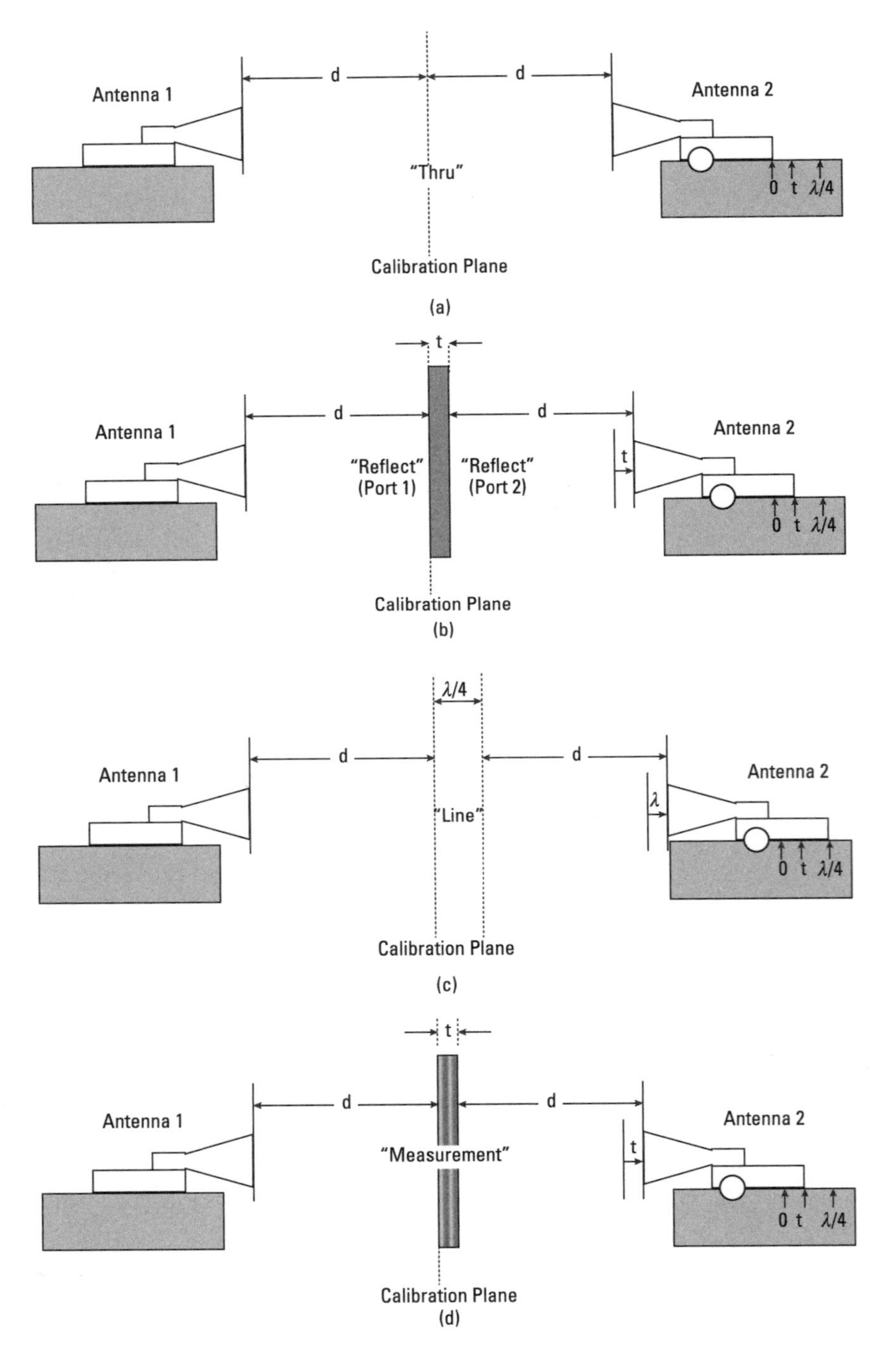

Figure 9.4 Over-the-air (OTA) TRL calibration, with horizontal positioner settings for (a) thru standard, (b) reflect standard, and (c) line standard. Setup (d) shows position of material under test.

to ensure that measurements are obtained in the far-field (plane waves impinging on the material holder). Make sure that the antennas and material holder are orthogonal to the wavefront. Then perform a TRL zero-length thru calibration sweep.

- *Step 2:* Insert a metal plate (with appropriate surface smoothness for the frequency of interest) in the material holder frame. Ensure that the surface of the plate facing port 1 is aligned with the measurement plane for port 1 (Figure 9.4(b)). Perform the port-1 reflect calibration sweep.
- *Step 3:* Adjust the position of the port-2 antenna to compensate for the plate thickness and maintain the proper distance (*d*) to the metal plate (Figure 9.4(b)). Perform the port-2 reflect calibration sweep.
- *Step 4:* Remove the metal reflect plate from the material holder frame. Adjust the position of the port-2 antenna away from the measurement plane by a quarter-wavelength: 90 electrical length in free space for the geometric center frequency of the measurement band (Figure 9.4(c)). Then perform the line calibration sweep.

This completes calibration. If the port-2 antenna is repositioned to the thru position, the VNA should show zero loss and zero phase. If the port-2 antenna is moved to position "t," then the VNA will show the loss and length of a volume of air represented by the material holder frame.

To establish a zero-amplitude, zero-phase reference condition, export the S-parameters of this air sample as a two-port Touchstone file. Then de-embed this S-parameter file. When the material under test is inserted into the frame, its loss and phase characteristics (relative to air) will be measured.

9.4 Millimeter-Wave Converters

External converters are popular for extending the working range of a VNA into microwave or millimeter-wave frequencies. These banded solutions take advantage of the fact that many millimeter-wave applications simply do not require DC-to-daylight sweep capability. Instead, an application will focus on only a small portion of the electromagnetic spectrum. Fortunately, much of this spectrum was classified long ago in terms of rectangular waveguide fundamental operating modes. Consequently, millimeter-wave frequency ranges today are still identified in terms of waveguide dimensions/designators and/or military radar-band designators. For example, applications requiring spectrum between 50 and 70 GHz use V-band converters equipped with WR15 waveguide flanges, while applications in the 60 to 90-GHz range use E-band converters equipped with WR12 waveguide flanges. Clearly, there is considerable frequency overlap between these two converters, so the choice as to which converter to use for a particular application may come down to a specific flange type called out in the requirements document, or careful comparison of the converter specifications to determine which one offers the best performance for the overlapping frequencies of interest.

Another important benefit of frequency converters is that they can be positioned close to the DUT, avoiding costly attenuation and phase instability/thermal drift associated with even moderate-length test cables at millimeter-wave frequencies.

The frequency converter approach can also be considerably less expensive than a VNA that provides DC-to-daylight sweep coverage. Consider that a 67 GHz VNA can be three times more expensive than a similarly-equipped instrument in the sub-10-GHz range.

Finally, a growing number of millimeter-wave applications require frequencies that commercial VNAs simply cannot support. Commercial VNAs are generally limited to an upper frequency of around 70 GHz. Commercial frequency converters extend this reach up to 1.4 THz (or higher) through the use of waveguide-banded converters. An example of the frequency bands and waveguide dimensions is presented in Table 9.1.

9.4.1 Millimeter-Wave Converter Architecture

A typical millimeter-wave converter for W-band (75–110 GHz) is shown in Figure 9.5. It provides the millimeter-wave version of a conventional VNA test port, and contains the same functional blocks: a millimeter-wave directional coupler, two receivers (REF and MEAS), and a millimeter-wave source. The solution leverages two microwave generators from the VNA to provide LO and RF signals at subharmonics of the desired operating frequency. For this W-band example, one generator provides the RF signal in the subharmonic frequency range (12.5 to 18.333333 GHz), which is multiplied by a factor of 6 to produce millimeter-wave stimulus between 75 and 110 GHz. The millimeter-wave directional coupler separates the forward-going and reflected millimeter-wave signals, and harmonic mixers downconvert these signals to a fixed IF (approximately 279 MHz in this example). A second VNA microwave generator provides the LO signal for the ×8 harmonic mixer in the subharmonic frequency range (9.340125 to 13.715130 GHz), which is 1/8 × (RF – IF).

Figure 9.6 shows two converters physically connected to a VNA equipped with direct source and receiver access. During a sweep, the RF and LO generators advance in lockstep such that the MEAS and REF mixers produce a signal at the fixed IF of 279 MHz. The MEAS and REF receivers in the VNA must be coherent in order to measure the relative phase between signals at different

Table 9.1
Millimeter-Wave Frequency Bands, Waveguide Dimensions, and Converter Models

Operating Frequency (GHz)	Letter Designator	Cutoff Frequency (GHz)	Designation		Internal Waveguide Dimensions				R&S Model Number
			EIA	RCSR	a	b	a	b	
			(US)	(UK)	(mm)		(mil)		
50–75	V	39.8616	WR-15	WG 24	3.7592	1.8796	148.0	74.0	ZVA-Z75
60–90	E	48.3567	WR-12	WG 25	3.0988	1.5494	122.0	61.0	ZVA-Z90E
75–110	W	58.9951	WR-10	WG 27	2.5400	1.2700	100.0	50.0	ZC110
90–140	F	73.7439	WR-8	WG 28	2.0320	1.0160	80.0	40.0	ZC140
110–170	D	90.7617	WR-7	WG 29	1.6510	0.8255	65.0	32.5	ZC170
140–220	G	115.6767	WR-5	WG 30	1.2954	0.6477	51.0	25.5	ZC220
170–260	(V)	137.198	WR-4	WG 31	1.0922	0.5461	43.0	21.5	
220–325	J(H)	173.5151	WR-3	WG 32	0.8636	0.4318	34.0	17.0	ZC330
325–500	Y	268.1596	WR-2.2	—	0.5588	0.2794	22.0	11.0	ZC500
500–750		393.3008	WR-1.5	—	0.3810	0.1910	15.0	7.5	
750–1,000		589.9512	WR-1	—	0.2540	0.1270	10.0	5.0	

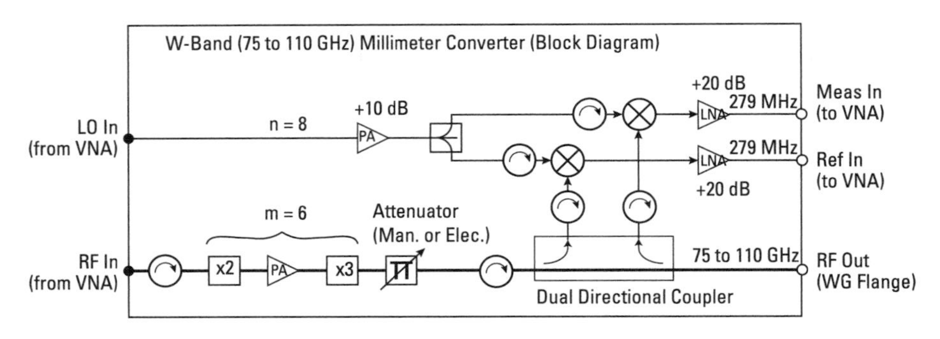

Figure 9.5 Block diagram of a W-band frequency converter.

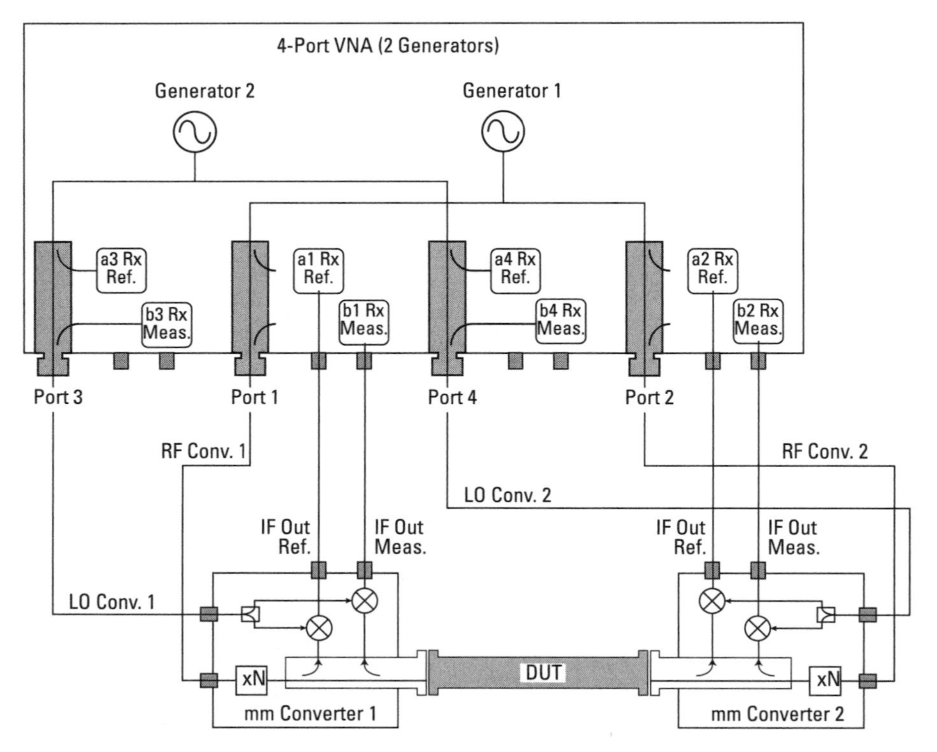

Figure 9.6 Millimeter-wave converters connected to a VNA.

receivers. (This coherence requirement applies to conventional lower-frequency VNAs as well.)

Theoretically, it is true that a VNA's LO generator does not need to be phase-locked to an RF generator. This is because S-parameters are ratio measurements, for example, S_{21} = (Measurement receiver at Port 2)/(Reference receiver at Port 1), so any phase change imposed by a slightly drifting LO (relative to the generator) will be applied to all receivers in exact proportion since all mixers share the same LO, but this is not a sustainable practice at millimeter-wave frequencies: For example, if the DUT is a cavity resonator, a slight drift in the generator, multiplied 6 times to the millimeter-wave frequency, can have a big impact on measurement results (phase and amplitude). Even for a less-demanding application (e.g., cable attenuation measurement), it is possible for an unlocked generator to easily drift out of the millimeter-wave receiver's bandwidth. For this reason, when any external generators are required for a millimeter-wave application, it is important to share a common 10-MHz reference with the VNA to ensure optimal measurement stability and accuracy.

9.4.2 Millimeter-Wave Converter Applications

Configuring a VNA for a coaxial- or waveguide-based millimeter-wave measurement is straightforward. In most high-end VNAs, a setup menu is available for selecting a particular converter (Figure 9.7). Once selected, the VNA automatically establishes the required subharmonic LO and RF, and also establishes the fixed IF, so all the user has to do is choose millimeter-wave start and stop frequencies and the number of frequency sweep points. Often, a special table is built into the VNA firmware for a given manufacturers' converters, which dithers the LO and IF at certain sweep points to avoid known mixer spurs. But this dithering takes place invisibly to the user.

When connecting the millimeter-wave converters to the VNA, the following suggestions will help to ensure optimum measurement performance.

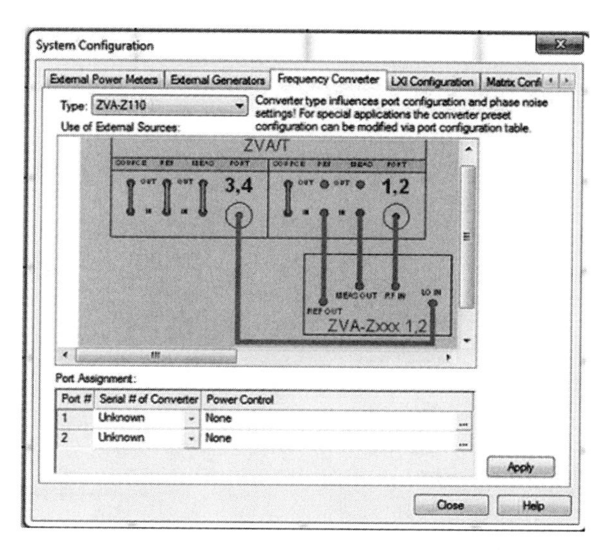

Figure 9.7 ZVA setup menu for configuring millimeter-wave operation.

1 Pay attention to millimeter-wave converter power level requirements. Since these converters use combinations of multipliers and amplifiers to reach desired frequencies, the input RF and LO power ranges must be maintained for optimum performance. Take special care not to exceed the maximum power levels, and make all VNA setting adjustments prior to connecting the converters.

2. Use only high-quality, phase stable cables for the LO and RF connections. Remember that any phase change due to cable flexure will be multiplied 6 times for the 75–110-GHz converter (up to 36 times for a 325–500-GHz converter) [4].

3. Align the waveguide flanges using the levelling feet on the converters to ensure proper mating of waveguide flanges.

4. Arrange the converters to minimize RF and LO cable movement. Cables should ideally be constrained so that their motion patterns are repeatable between calibration and measurement.[2] For example, some laboratories mount their converters on a rail system (or directly to a wafer probing system) to achieve this goal.

5. Maintain the measurement environment at constant temperature. This should include the VNA, cables, converters, and calibration kits.

6. Ensure that all equipment has had sufficient time to warm up and thermally stabilize prior to calibration and operation.

7. Take the time necessary to perform a proper calibration. Rushing procedures at millimeter-wave frequencies will ironically result in lost time as calibration and/or measurements will have to be repeated.

9.4.2.1 Small-Signal (Two-Port) Measurements

Small-signal two-port S-parameter measurements use the setup as shown in Figure 9.6. Perform calibration using an application-appropriate calibration kit and calibration method. For example, millimeter converters typically default to waveguide kits that match their operating frequency range. However, probe vendors may supply their own software[3] that supports calibration methods unique to the wafer-probing environment. The software communicates with the VNA during the calibration process and, upon completion, transfers all error-correction matrix terms into the VNA. From that point on, the VNA can make stand-alone measurements.

There is no substitute for operator experience at millimeter-wave frequencies, especially when repeatable measurements are required. This experience

2. The IF cables, operating at less than 300 MHz are much less prone to phase instability issues.
3. Several prominent examples include WinCal XE from FormFactor (formerly Cascade Microtech) and QAlibria from MPICorporation.

may start with something as basic as instinctively using proper torque levels and tightening technique for all coaxial connections in the millimeter-wave setup, as well as avoiding RF adapters and adhering to the best practices listed above. As previously mentioned, dynamic range is somewhat reduced at millimeter-wave frequencies, which encourages the use of smaller IF bandwidths and consequently longer sweep times. Any measurement instability (e.g., thermal drift or dielectric relaxation due to cable bending) will be apparent as a discontinuity in amplitude and/or phase from sweep to sweep.

9.4.2.2 Amplifier Measurements Using Two Tones for IMD

Two-tone measurements at millimeter frequencies require additional hardware and test equipment. A setup[4] for two-tone measurements is shown in Figure 9.8.

Here, an additional millimeter-wave converter provides the second millimeter-wave stimulus tone. This converter, in turn, requires a microwave generator to provide a unique subharmonic source for this second tone. A VNA with four internal generators uses the third or fourth generator for this task. Otherwise, an external generator is required. Other waveguide components are also needed (e.g., a waveguide "magic tee" for combining the two tones together).

Since this is a nonlinear application, accurate power calibration (of both generators and receivers) is required. While commercial power sensors are available up to 110 GHz[5] and provide easy integration with commercial VNAs, power meters above this frequency range[6] take on a bit more of a science project feel, but nevertheless enable accurate power measurements up to at least 3 THz.

Once setup and calibration are complete, measurements become relatively straightforward using the same procedures and dialogs as for lower-frequency IMD measurements. The only caveat involves millimeter-wave spurs and images for swept-frequency measurements. The former can be checked by performing the measurement sweep with each tone individually, looking for the presence of unwanted signals at the IMD product frequencies. The latter can be checked by using both tones without a DUT to check for unwanted images. In both cases, it is wise to include a baseline noise-floor sweep with all sources off [5, 6].

---

4. Alternatively, a millimeter-wave signal generator could be used instead.
5. The Rohde & Schwarz NRP110T is a thermal sensor with 1.0-mm coaxial connector for wideband measurements (DC to 110 GHz). NRP75TWG, 90TWG and 110TWG are thermal power sensors with WR-15, WR-12, and WR-10 flanges for measurements in the respective waveguide bands. All use USB connectors for instant recognition and seamless operation with the ZVA, ZNB, and ZNA families.
6. A good example of a higher-frequency measurement solution is the Erickson PM5 Power Meter from VDI. This calibrated calorimeter-style power meter is good for applications ranging from 75 GHz to > 3 THz.

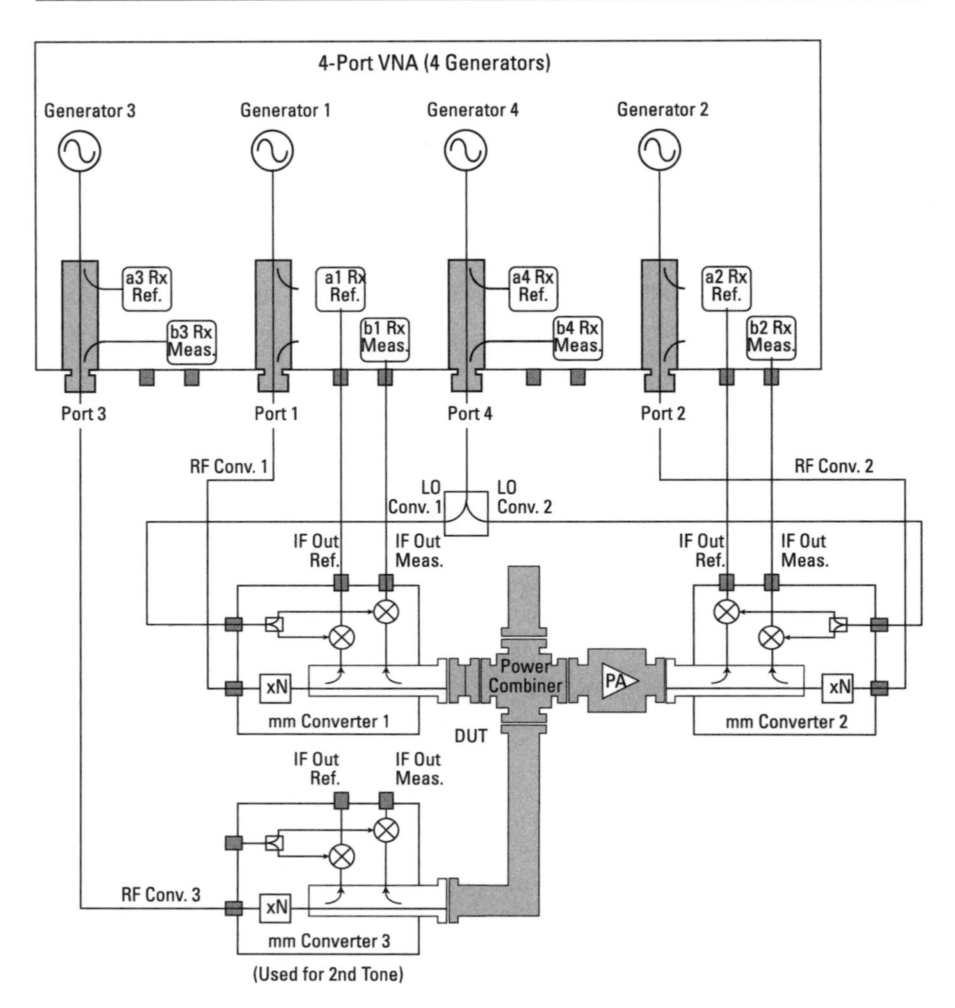

Figure 9.8 A two-tone setup for IMD measurements at millimeter-wave frequencies.

9.4.2.3 Millimeter-Wave Mixer Measurements

A difficult application at millimeter-wave frequencies involves measurement of a millimeter-wave mixer. Two possible application setups are shown in Figure 9.9, depending on the LO power requirements of the DUT. Here, the port 1 millimeter-wave converter provides the RF signal, and an external signal generator (setup a) or internal generator (setup b) provides the LO drive. The biggest challenge is when the MUT IF (400 MHz) is different from millimeter-wave converter IF (279 MHz). Success means configuring correct frequency relationships for each measurement parameter: LO and RF port impedance match, RF-to-LO feedthrough, and RF-to-IF conversion loss. Table 9.2 illustrates the

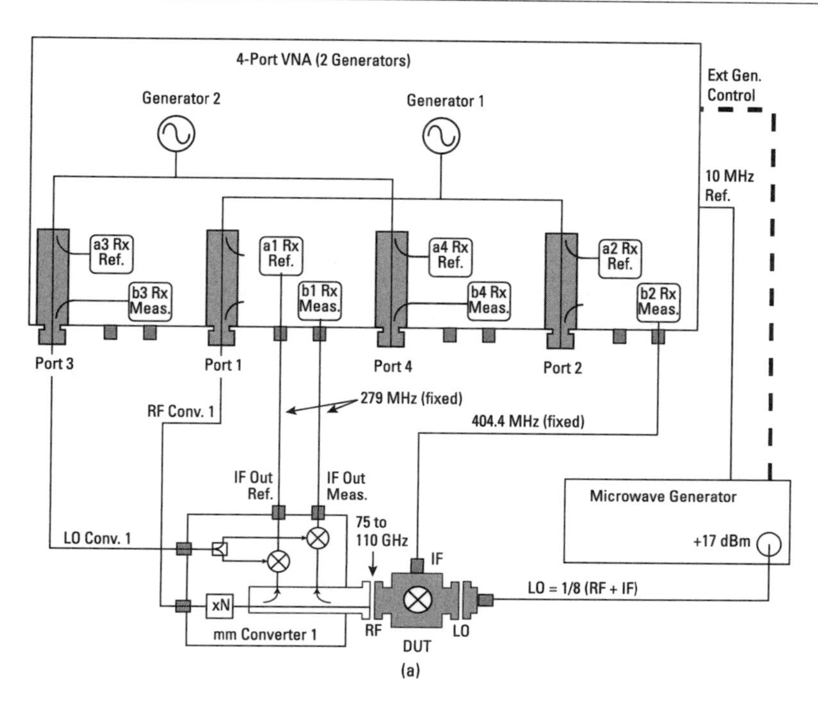

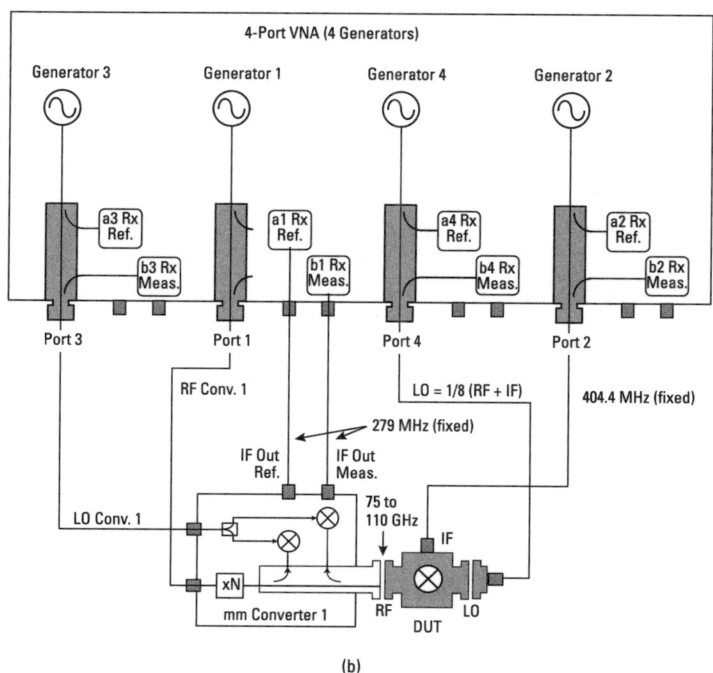

Figure 9.9 Mixer setup. (a) Configuration for high LO power using external microwave generator. (b) Configuration for low LO power using VNA with four internal generators.

Table 9.2
Millimeter-Wave Mixer Measurements

	Setup a	**Setup b**
Measurement	(+13 dBm mixer)	(0 to +7 dBm mixer)
RF port return loss	S11	S11
LO port return loss	—	S44
IF port return loss	—	S33
RF to LO feedthrough	—	S41
LO to RF feedthrough	—	S14
RF to IF conversion loss	b2/a1 Src Port 1	b2/a1 Src Port 1
1-dB compression Pt (RF)	b2/a1 Src Port 1	b2/a1 Src Port 1
Conversion loss as function of	b2/a1 Src Port 1	b2/a1 Src Port 1
LO drive level	Sweep Ext. Gen. pwr	Sweep Gen 4 pwr

different measurements that can be performed with the two test setups. Full measurement details are provided in [7].

9.4.3 Waveguide Calibration

Waveguide-based converters use many of the same calibration methods as lower-frequency coaxial systems. The two most popular methods are TOSM and TRL. Additionally, after selecting a particular waveguide-based millimeter-wave converter, the instrument will also default to an appropriate waveguide calibration kit. For example, when selecting the W-band ZVA-Z110 millimeter-wave converter, the R&S ZVA automatically selects a W-band waveguide calibration kit, the ZV-WR10.

Waveguide kits are conceptually much simpler to define and build than coaxial kits, in part because waveguide flanges are sexless, and the reference plane is established at the flange face. A typical kit has only three standards: a short, a match, and a shim, which are sufficient for both one-port and multiport calibrations. There is no thru standard,[7] because the two flanges can be directly mated together to establish a 0-dB loss, 0° phase reference plane. The short is a flat reflecting plate, connected directly at the reference plane to provide a 0° offset, 0-dB loss reflection. The match standard consists of RF-absorptive material with minimal reflectivity. Since good match standards are difficult to build at millimeter-wave frequencies, sliding matches are sometimes employed. With these, it is assumed that there will be some reflection at each frequency point due to the nonideal absorber, and the resulting reflection coefficient will have both amplitude and phase displacements away from the center of the Smith

7. At higher millimeter-wave frequencies, an additional thru shim is required because the thickness of a quarter-wave shim, on its own, has insufficient mechanical integrity.

chart (50Ω). By turning the linear adjustment knob in the sliding match to physically increase the distance between the absorber material and the reference plane, the amplitude and phase angle of these small reflections will change. By averaging measurements obtained at five or more positions along the sliding match, the resulting vector-averaged reflection coefficient will be much closer to 50Ω than any individual measurement.

The shim standard is unique to waveguide. It has a very carefully machined thickness of $\lambda/4$ at the geometric center frequency of the waveguide band. Why $\lambda/4$? Consider that we already have a short standard for waveguide (a purely reflective sheet at the reference plane) that reflects all the energy impinging on it. If we were to simply remove this plate in an effort to create an open-circuit condition, it would instead act like a truncated horn antenna and launch the energy into space. So instead we apply a short-circuit through a quarter-wave shim. This shim transforms the waveguide short into an open by performing a quarter-wave impedance transformation to the opposite side of the Smith chart, with no radiated energy. As a bonus, the shim is also used as a $\lambda/4$ line standard for two-port TRL calibrations.

9.4.3.1 Determining Length of a $\lambda/4$ Line Standard in Waveguide

The most critical part of defining a waveguide calibration kit is determining the length of the $\lambda/4$ line standard. In this section, we present relevant waveguide theory to derive the appropriate equations, and use drawings to help explain the important concepts of group velocity and phase velocity. We conclude with an example based on calculating the $\lambda/4$ line dimensions for the WR10 waveguide. This process can easily be applied to any size of rectangular waveguide:

- *Step 1:* Determine the waveguide cutoff frequency. Waveguide cutoff frequency information is available from many online sources.[8] The cutoff frequency is associated with the most common rectangular waveguide propagation mode, TE_{10}. If you know the waveguide dimensions, you can also calculate the cutoff frequency from the following formula [8]:

$$(f_c)_{mn} = \frac{1}{2\pi\sqrt{\mu\varepsilon}}\sqrt{\left(\frac{m\pi}{a}\right)^2 + \left(\frac{n\pi}{b}\right)^2} \tag{9.1}$$

where m and n represent the mode number of the waveguide field configurations for the TE^z mode (transverse electric field, propagation in the z direction). For WR-10, $m = 1$ and $n = 0$, and $\frac{1}{\sqrt{\mu\varepsilon}} = v$ for speed

8. See, for example, www.everythingrf.com/tech-resources/waveguides-sizes.

of electromagnetic waves in a particular medium. When the medium is vacuum, we have $\frac{1}{\sqrt{\mu_o \varepsilon_o}} = c$, which is 299,792,458 m/s [9]. Propagation through other media will reduce the speed. In our rectangular waveguide application, we assume an air-filled waveguide, so we could add a speed correction factor: the refractive index of air. Determining this value involves understanding the complex relationship between wavelength and atmospheric conditions (air temperature, pressure, and humidity) [10]. Generally, using $\nu = 2.997e8\ m/s$ is sufficient for many waveguide calculations. Using this value, we compute the TE_{10} mode cutoff frequency:

$$(f_c)_{10} = \frac{\nu}{2\pi}\sqrt{\left(\frac{\pi}{a}\right)^2} = \frac{\nu}{2a} = \frac{2.997e11\ mm/\sec}{2(2.54mm)} = 58.99\ GHz \tag{9.2}$$

- *Step 2:* Determine the geometric center frequency. Waveguide is normally operated in the frequency range where only the lowest propagating mode exists, with all higher modes being beyond cutoff. Hence, the lower corner of the band f_l is determined by the cutoff frequency of the desired propagating mode (with some conservation factored in), while the upper corner f_u is determined by the cutoff frequency of the next higher propagating mode. Fortunately, this information is already available for common waveguide, as seen in Table 9.1. Armed with these values, the geometric center frequency[9] is easily calculated from

$$f_0 = \sqrt{f_l f_u} \tag{9.3}$$

For WR-10, we get:

$$f_0 == \sqrt{(75\ GHz)(110\ GHz)} = 90.83\ GHz \tag{9.4}$$

- *Step 3:* Determine the waveguide phase velocity. Separate from electromagnetic propagation in air, a guided wave has some interesting properties. In particular, electromagnetic energy can only propagate down the guide at frequencies above the cutoff for a particular mode. In addition, the wavelength of an electromagnetic signal in waveguide is different

9. A 90° length is selected for the band center, because the TRL method usually works over a range where the line represents 20 (at the low end) and 160° (at the high end).

from the same signal in free-space, and the waveguide wavelength is dispersive, meaning it is frequency-dependent. This arises because the guided wave travels down the waveguide at an angle, bounded by the guide walls. The situation is depicted in Figure 9.10(a).

The velocity v is assumed to be the propagation velocity of an electromagnetic wave in air. In waveguide of a particular frequency and mode, the guided plane wave travels in the direction determined by angle Ψ, which is calculated from:

$$\psi = \cos^{-1}\left[\sqrt{1-\left(\frac{f_c}{f}\right)^2}\right] \tag{9.5}$$

Notice that the angle Ψ approaches 0° as f approaches ∞ (i.e., a free-space plane wave). When it reaches 90° (which occurs at), the wave ceases to propagate down the line.

Since the guided wave in the drawing propagates at an angle, we need to consider two related velocities. The group velocity v_g is the speed at which electromagnetic energy is transported along the length of the waveguide (z-direction). It is less than the speed of light, and is given by:

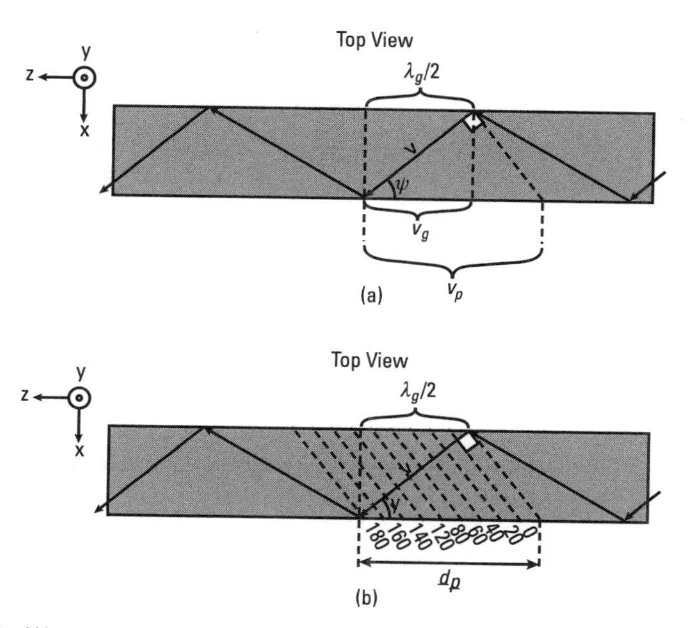

Figure 9.10 Waveguide propagation factors and (a) angular relationships and (b) phase velocity.

$$v_g = v\cos\psi \tag{9.6}$$

Conversely, the phase velocity v_p represents the velocity that must be maintained (geometrically) to account for the moving phase front of the guided wave. In the depiction above (Figure 9.10(b)), the phase is observed at various points along the phase front. Clearly, the phase front has covered 180° over the distance d_p (*z*-direction) in the same amount of time that the guided wave has traveled the shorter distance of $\frac{\lambda_g}{2}$. The velocity of the phase front, called the phase velocity, is given by:

$$v_p = \frac{v}{\cos(\psi)} \tag{9.7}$$

For our particular example, with f_c = 58.99 GHz and the geometric frequency f = 90.83 GHz (geometric center of W-band), then Ψ = 40.5°, and v_p = 1.315).

- *Step 4:* Determine the wavelength in rectangular waveguide. With the result of step 3, we can calculate the wavelength in waveguide by considering the velocity of the phase front:

$$v_p = 1.315v = f\lambda \tag{9.8}$$

Since the phase velocity v_p of a guided wave is greater than plane wave propagation in space, its associated wavelength will be longer than the corresponding free-space wavelength. With f = 90.83 GHz and v = 2.997*e*11 mm/s, we obtain

$$\lambda = \frac{(1.315)\left(2.997e11\frac{mm}{s}\right)}{90.83\ GHz} = 4.339 \text{ mm} \tag{9.9}$$

- *Step 5:* Determine the shim width. Recall that the objective was to calculate a physical length in waveguide that exhibits 90° of electrical length (delay) at the band center; 90° represents ¼ of one wavelength (360°), so we simply divide the waveguide wavelength (from step 4) by 4, yielding a quarter-wave shim thickness of 1.085 mm.

9.4.3.2 Defining a Waveguide Calibration Kit Using R&S VNAs

When working with waveguide, you may need to define your own waveguide calibration kit. Fortunately, this is an extremely simple process (once the line dimensions have been calculated per Section 9.4.3.1). The process is described here for the R&S ZNA network analyzer:

- *Step 1:* Define a "New Connector Type". Channel > Cal > "Cal Devices" tab > "Cal Connector Types…" In the dialog, click "Add" and rename the new connector type, for example, WR10. Click on "Line Type" and change from "TEM" to "Waveguide." Set the cutoff frequency to the published value for WR10 (see Table 9.1), for example, 58.9951 GHz. Close the dialog.
- *Step 2:* Add a new "Cal Kit" for your newly defined connector: Channel > Cal > "Cal Devices" tab > "Cal Kits…" Select your newly defined connector type from the first column, and click the "Add" button. The calibration kit receives the default name "New Kit 1." Change it to something more meaningful like "W-band_custom." If you want to enter dimensions in time units (e.g., picoseconds), leave the Agilent Mode checked. If you want to enter dimensions in distance units (e.g., millimeters), uncheck the Agilent Mode box. Here, we uncheck the box because we want to enter our $\lambda/4$ line length in millimeters.
- *Step 3:* Add individual calibration standards to your custom calibration kit. In the "Calibration Kits" dialog, with your new kit, for example, "W-band_custom" highlighted, click "Standards…" In the split-screen dialog, click "Add" in the upper window to add one-port standards. For TRL, you will only need a reflect standard (assumed to be a reflecting plate installed right at the waveguide flange). Under "Type," select "Reflect" from the pull-down menu. Set the "Min Freq" to 75 GHz and "Max Freq" to 110 GHz. If an offset length is required, click "View/Modify…" and enter a delay value. (For this example, there is no offset required because the reflecting plate is located right at the waveguide flange.) In the split-screen dialog, click "Add" in the lower window to add the required two-port standards. For TRL, you will need a zero-length thru (created when the flanges are connected waveguide face to waveguide face) and a $\lambda/4$ line standard.
    - a. Click "Add" in the lower window. The default standard is thru. Simply change the "Min Freq" and "Max Freq" as above. Click on "View/Modify" to verify the default value is zero delay and then click "OK" to close the "View/Modify" dialog.

- b. Click "Add" in the lower window. The default value for this second standard is "Line 1." Highlight Line 1, click "View/Modify," and change the delay value to 1.085 mm (calculated in Section 9.4.3.1). Click "Ok."

This completes the calibration kit creation process. Click on "Ok" at the bottom of the "Kit Standards" split dialog and then click "Close" to dismiss the final "Calibration Kits" dialog. You can now perform a calibration using your newly defined waveguide calibration kit.

If you have a match standard available, then it is easy to add the additional standards necessary for a conventional TOSM calibration, where the open standard in waveguide is actually an offset short constructed by placing the $\lambda/4$ line standard between the waveguide flange and the short standard. (Remember that a short reflected through a quarter-wave matching section looks like an open.) To add a short and an offset short standard to your waveguide calibration kit for TOSM, follow the procedure above for the reflect standard, but select a short and then add an "OFFSET SHORT 1." Finally, add a match standard (or sliding match) standard, depending on the contents of your waveguide kit.

9.4.4 VNA-Based Spectrum Analysis

A spectrum analyzer is ideally suited for displaying RF energy across a frequency span. This is because its quasi-analog sweep architecture enables it to detect and display the RF amplitude of CW signals as well as modulated[10] and noise-like emissions anywhere within the span. A VNA's architecture sweeps much differently. It makes measurements only at discrete frequency points determined by its frequency span and number of points. This mode has been perfected for making S-parameter measurements, where the objective is to capture and compare amplitude and phase information only at frequencies where energy is expected (e.g., the generator frequency). This minimizes sweep time, since a matrix calculation must be performed at each measurement point to deliver error-corrected S-parameters. Another drawback of the VNA is that the sweeping receiver will display both the desired signal and its image. Unlike spectrum analyzers, VNAs do not employ hardware-based image-rejection mechanisms. In reality, a VNA generally does not need this additional hardware because it knows exactly where to expect the desired generator to be at each frequency step. The image only becomes noticeable if you set the generator to a single frequency and sweep the receiver across the entire frequency span, choosing a frequency step that allows the generator to land on the image frequency. This

10. A spectrum analyzer could conceivably miss a pulsed or other transient signal that might be off during the moment that the receiver was sweeping that frequency.

image always appears at a frequency offset of twice the IF, with the receiver LO located halfway between the true generator frequency and the image frequency. This is described in Section 3.5.

The final issue is receiver compression due to signal overload. This condition is common to all receivers (heterodyne and direct RF sampling alike). In spite of these limitations, a VNA can still be used as a spectrum analyzer if a few guidelines are followed.

First, ensure that your receiver bandwidth is wide enough to cover the span between adjacent measurement points, with sufficient overlap. This is shown in Figure 9.11. If the point spacing is too great, and/or the IF filter bandwidth is too narrow, the signals will be missed (Figure 9.11(a)). Using many points and/or a wider IF bandwidth will ensure all signals are captured (Figure 9.11(b)). However, there is a trade-off. More points means more calculations. (This is part of the VNA's architectural DNA and cannot be deactivated.) This sweep-time overhead can be mitigated by applying a single-port calibration or selecting the "cal off" state (essentially the same thing from a speed perspective). The other trade-off involves selecting a wider IF bandwidth instead of using more measurement points. This has the additional benefit of speeding up a sweep, since wider filters require fewer samples (equivalent to a wideband

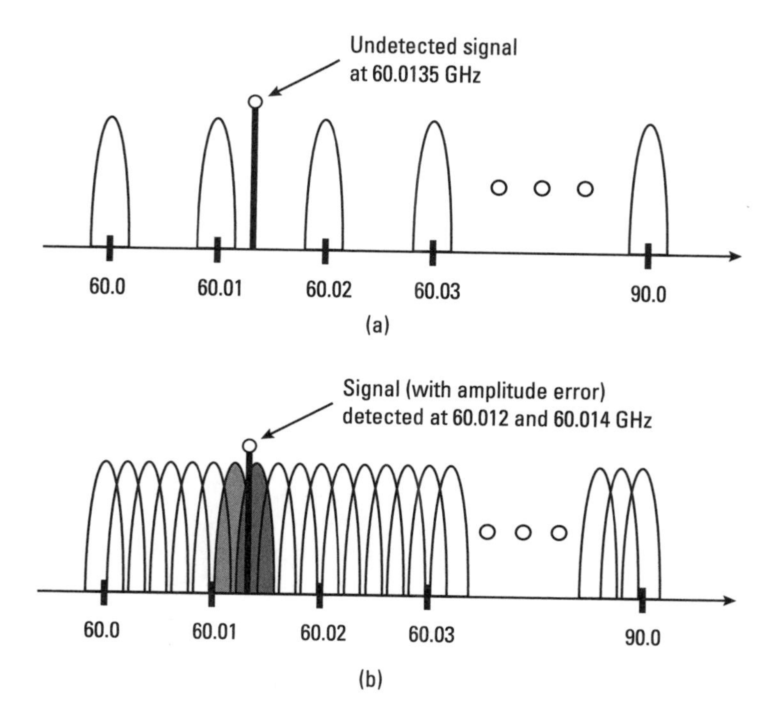

Figure 9.11 Frequency sweep with (a) insufficient and (b) sufficient overlap.

analog filter's shorter settling time). The downside is that a wider bandwidth is going to allow more noise in and hence increase the noise floor. This reduces the VNA receiver's ability to detect low-level signals. Additionally, sufficient overlap is needed to ensure that the amplitude of the signal is reported faithfully. If not, a signal may still be detected off to one side of the filter (in the passband-to-stopband transition), but with reduced amplitude. The degree of amplitude error will depend on exactly where the signal falls within this region.

9.4.4.1 Trace Math for Image Cancellation (Conventional VNA)

The VNA receiver image problem can be eliminated by using two sweeps, one with the receiver LO positioned above the generator (RF) frequency, and the other with the receiver LO positioned below it. Then, by applying trace math, the true signal frequency will remain after cancelling the receiver images. The software preselection setup steps are described here for the ZNB:

- *Step 1:* Start from a preset and set up a sweep using a sufficient number of points and receiver bandwidth to meet the overlap guidelines of Section 9.4.4.
- *Step 2*: Temporarily place the generator in the middle of the band (CW, not sweeping) so you can observe the receiver images (and image cancellation) in the subsequent steps. Select Channel > Channel Config > "Port Config" tab > "Port Settings..." In the dialog, click the box corresponding to "Port 1" in the generator column. Click on "Freq Conversion" for port 1, and under "Mode fb" click "0 Hz" (to disable frequency sweeping) and in the "Frequency Offset" box, type in the center frequency of the current sweep range. Click "OK" to close "Modify frequency conversion" dialog. Under the "RF Off" column, click the boxes next to "Port 2" (and all other ports in the case of a multiport instrument). Click "OK" to close this dialog. Click the "Mode" tab and under "Image Suppr." Click "LO < RF". Click TRACE > Meas > "Wave" tab, and select "b2 source P1." Connect a cable from port 1 to port 2. You should see the true signal and the image.
- *Step 3:* Create a second channel using the opposite LO relationship. Click on "New Ch + Tr" and drag the trace down to a new display area (below trace 1). Select any S-parameter. Click TRACE > Meas > "Wave" tab, and select "b2 source P1" for this new trace (Trc2). It should look the same as Trc1. Click Channel > Channel Config > "Mode" tab and under "Image Suppr." click LO > RF. You should see signals in the middle of the screen for Trc1 and Trc2, and a corresponding image on opposite sides of these traces.

- *Step 4:* Create a software preselection trace math function that will suppress the images. Click on "New Trace" and drag the new trace (Trc3) down to a new display area (below Trc2). Select any S-parameter. Click TRACE > Meas > "Wave" tab, and select "b2 source P1" for Trc3. With Trc3 highlighted (selected), click TRACE > Trace Config > "Mem Math" tab. Select "User defined." Click "Define Math" and in the top window, click the "Min" button, then click "Trc1" from the "operands" box. Then, position the flashing cursor after the comma, and click "Trc2" in the "operands" box. Click "Result is Wave Quantity" in bottom left corner. Then click "Ok" to close the dialog.

The images should now be suppressed from Trc3. To reduce the number of traces cluttering the screen, drag Trc2 and Trc3 to the Trc1 diagram area. Then, under TRACE > Trace Config > "Traces" tab, click on "Trace Manager..." and unclick the boxes in the "On" column corresponding to Trc1 and Trc2. You will be left with only the image-suppressed trace (Trc3) appearing on the screen (the other two are labeled "invisible").

While this function does a nice job of suppressing the image response, it does so with the overhead of requiring an additional sweep. This is yet another trade-off incurred when using a conventional VNA as a spectrum analyzer (without a dedicated spectrum analyzer option).

9.4.4.2 Image Cancellation for Millimeter-Wave Converters

Image cancellation is a bit more challenging when using a VNA with millimeter-wave converters. This is because there are two sets of image frequencies generated: One from the millimeter converters (which use harmonic mixers for downconversion), and a second from the VNA front end. Figure 9.12 shows the challenge: The upper display (Figure 9.12(a)) shows the sweep results for a tone applied at 84 GHz, with a large number of image products. If the software preselection approach of Section 9.4.4.1 is applied, the spectrum is cleaned up significantly (Figure 9.12(b)), but there are still two strong tones present (and several weaker tones), requiring a second image-elimination step. The two signals are separated by 558 MHz, which is twice the IF of the converters (279 MHz). Figure 9.13 shows the frequency relationships. To remove these other images, we need to create additional channels, as shown in Table 9.3.

The process is as follows for the ZVA:

- *Step 1:* Configure the ZVA setup for use with millimeter-wave converters. Select System > System Config > "Frequency Converter" tab. In the pull-down menu, select the appropriate converters and click "Apply" and then "Close."

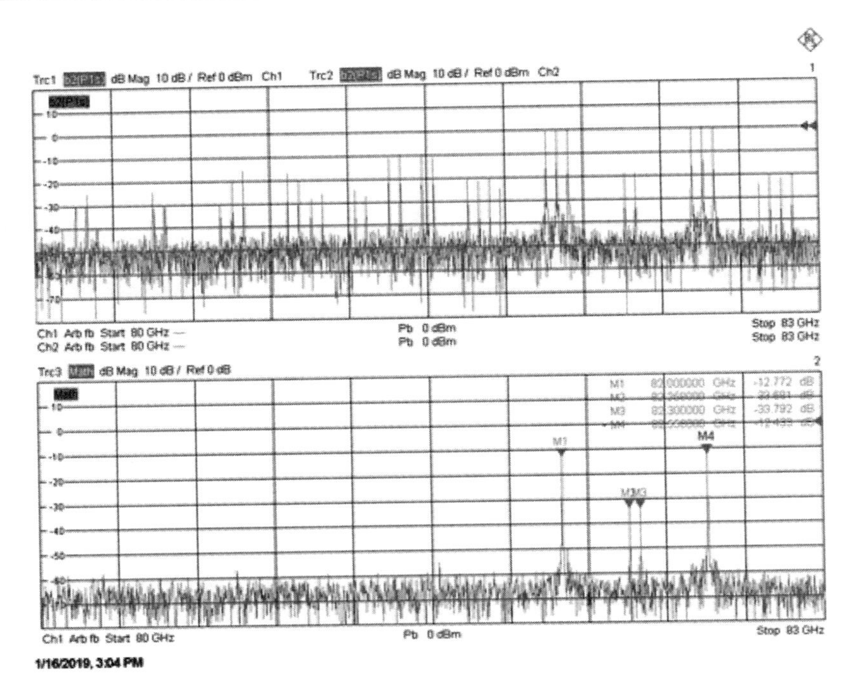

Figure 9.12 At top, trace results for ZVA with millimeter-wave converters sweeping across a span with only a single CW tone present. Below, trace results using software preselection from Section 9.4.4.1.

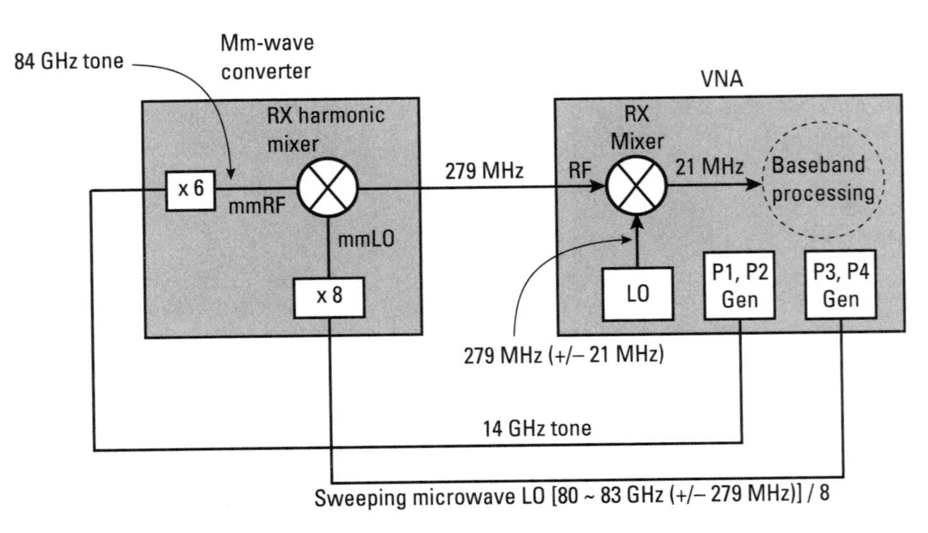

Figure 9.13 Block diagram of millimeter-wave converter and VNA showing frequency relationships.

Table 9.3
Sweep Relationships Required for Image Elimination

	Millimeter-Wave Converter	VNA
Channel 1	mmLO < mmRF	LO > RF
Channel 2	mmLO < mmRF	LO < RF
Channel 3	mmLO > mmRF	LO > RF
Channel 4	mmLO > mmRF	LO < RF

- *Step 2:* Press preset and then configure all required cable connections between the millimeter-wave converters and the ZVA. Configure a millimeter-wave sweep using a sufficient number of points and receiver bandwidth to meet the overlap guidelines of Section 9.4.4.
- *Step 3:* Temporarily place the generator in the middle of the band (CW, not sweeping) so you can observe the receiver images (and image cancellation) in the subsequent steps. Select Channel > Mode > Port Config, and then click the box corresponding to Port 1 in the generator column. Click on "..." under the adjacent entry in the "Frequency" column. Click "0 Hz" on the left side to disable tone sweeping and enter the appropriate subharmonic on the right side. For example, if the denominator on the left side has a "6" in it, divide your desired millimeter-wave tone frequency by 6 and enter it into the space on the right. (For a millimeter-wave tone at 84 GHz, enter "14 GHz" in the dialog.) Click "OK" to close the dialog. Click Mode >"Spurious Avoidance" softkey and under "Image Suppr." click "LO > RF." Click TRACE > Meas > Wave Quantities, and select "b2 src Port1." Connect millimeter-wave converter port 1 to port 2. You should see multiple signals.
- *Step 4:* Create a second channel having a different VNA LO relationship. Click on Channel > Channel Select > "Add Channel + Trace + Diag Area." Click TRACE > Meas > Wave Quantities, and select "b2 src Port1" for this new trace (Trc2). It should look the same as Trc1. Click Channel > Mode >"Spurious Avoidance" softkey and under "Image Suppr." click "LO < RF."
- *Step 5:* Create a third channel using the same VNA LO as Trc1 but a different millimeter-converter LO relationship. Click to highlight Trc1 (which makes it the active trace). Then select Channel > Channel Select > "Add Channel + Trace + Diag Area." This creates Trc3. With Trc3 highlighted, Click TRACE > Meas > Wave Quantities, and select "b2 src Port1." It should look the same as Trc1. Click Channel > Mode >"Port Config" and under the frequency column for Port 3, click the "..." box.

(It is just to the right of the equation.) Change the sign of the value in the right block from "–279 MHz" to "279 MHz" and click OK. This should change the dialog equations for both Port 3 and Port 4. Click "OK" to close the "Port Configuration" dialog.

- *Step 6:* Create a fourth and final channel using the same VNA LO as Trc2 and same millimeter-wave converter LO relationship as in the previous step. Click to highlight Trc2 (which makes it the active trace). Then select Channel > Channel Select > "Add Channel + Trace + Diag Area." This creates a new trace (Trc4). With Trc4 highlighted, Click TRACE > Meas > Wave Quantities, and select "b2 src Port1." It should look the same as Trc2. Click Channel > Mode >"Port Config" and under the frequency column for Port 3, click the "…" box. (It is just to the right of the equation.) Change the sign of the value in the right block from "–279 MHz" to "279 MHz" and click OK. This should change the equations for both Port 3 and Port 4. Click "OK" to close the "Port Configuration" dialog.
- *Step 7:* Create a trace math function to suppress the receiver images from Trc1 and Trc2. With Trc1 selected (highlighted), click Trace > "Trace Select > Add Trace + Diag Area." This creates Trc5. Click "Trace" > "Trace Function" > "User Def Math…" In the right operator column, click on "Min" Then, in the left-hand column, select "Trc1." Move the cursor to just after the comma, and select "Trc2". Click the box "Result is Wave Quantity." Then click "OK" to close the dialog. In the softkey panel, click "Math = User Def" to activate the math function for Trc5.
- *Step 8:* Create trace math function to suppress the receiver images from Trc3 and Trc4. With Trc3 selected (highlighted), click Trace > "Trace Select > Add Trace + Diag Area." This creates Trc6. Click "Trace" > "Trace Function" > "User Def Math…" In the right operator column, click on "Min." Then, in the left column, select "Trc3." Move the cursor to just after the comma, and select "Trc4." Click the box "Result is Wave Quantity." Then click "OK" to close the dialog. In the softkey panel, click "Math = User Def" to activate the math function for Trc6.
- *Step 9:* Create the trace math function to suppress the millimeter-wave receiver images from Trc5 and Trc6. With Trc5 selected (highlighted), click Trace > "Trace Select > Add Trace + Diag Area." This creates new Trc7. Click "Trace" > "Trace Function" > "User Def Math…" In the right operator column, click on "Min." Then, in the left column, select "Trc5." Move the cursor to just after the comma, and select "Trc6." Click the box "Result is Wave Quantity." Then click "OK" to close the

dialog. In the softkey panel, click "Math = User Def" to activate the math function for Trc7.

This completes the image-suppression process for millimeter-wave converters. You should now see only a single spectral line at the center frequency.

To conserve screen area, it is possible to move all the traces to the same diagram area, and use the Trace Manager (Trace > Trace Select > Trace Manager) to turn Traces 1 to 6 off so that only Trc7 is displayed. A more advanced approach would be to eliminate Trc5 and Trc6 altogether by consolidating the software preselection functions into a single step as shown for Trc7:

$$\text{Min}\left(\text{Min}\left(\text{Trc1},\text{Trc2}\right),\text{Min}\left(\text{Trc3},\text{Tr4}\right)\right) \tag{9.10}$$

The images should now be suppressed from Trc7, as shown in Figure 9.14. To further reduce screen clutter, drag all traces to a single diagram area. Then, under TRACE > Trace Config > "Traces" tab, click on "Trace Manager..." and unclick the boxes in the "On" column corresponding to Trc1 through Trc6. You will be left with only the final, image-free trace (Trc7). (The others are labeled "invisible.")

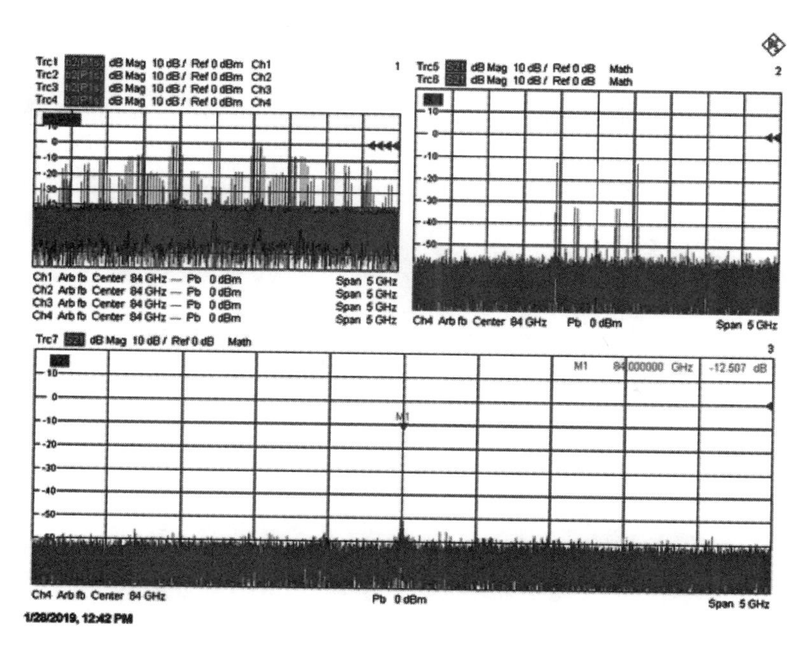

Figure 9.14 Removing images from a mm-wave converter in spectrum analyzer mode. At (a), no image suppression. At (b), after suppressing the VNA IF images. At (c), final result after also suppressing the mm-wave IF images.

While this software preselection procedure does a nice job of suppressing image responses for millimeter-wave converter-based spectrum analysis applications, it does so at the expense of requiring four channels (three additional sweeps).

References

[1] Federal Communications Commission, "Millimeter Wave Propagation: Spectrum Management Implications," *Office of Engineering and Technology*, Bulletin Number 70, July 1997, pp. 5–16.

[2] Li, E. S., et al., "Solving Resonance Problems in Transitions: End-Launch Connector Considerations for Coaxial-to-Microstrip Transitions," *IEEE Microwave Magazine*, Vol. 20, No. 3, 2019, pp. 64–75.

[3] Basu, S., and L. Hayden, "An SOLR Calibration for Accurate Measurement of Orthogonal On-Wafer DUTs," *IEEE MTT-S International Microwave Symposium Digest*, Vol. 3, 1997, pp. 1335–1338.

[4] Rohde & Schwarz GmbH & Co. KG, "R&S ZCxxx Millimeter-Wave Converters Specifications," 2018, p. 5.

[5] Hiebel, M., "Application Note 1EZ55, Millimeter-Wave Measurements Using Converters of the R&S ZVA Family," Rohde & Schwarz GmbH & Co. KG, 2007, pp. 9–10.

[6] Hiebel, M., "Application Note 1EZ56 Multiport Millimeter-Wave Measurements Using Converters of the R&S ZVA Family," Rohde & Schwarz GmbH & Co. KG, 2007, pp. 14–17.

[7] Hiebel, M., "Application Note 1EZ57 Testing Millimeter-Wave Mixers Using Converters of the R&S ZVA Family," Rohde & Schwarz GmbH & Co. KG, 2007, pp. 14–17.

[8] Balanis, C. A., *Advanced Engineering Electromagnetics*, New York: John Wiley & Sons, 1989 p. 357.

[9] Penrose, R., *The Road to Reality: A Complete Guide to the Laws of the Universe*, New York: Vintage Books, 2004, pp. 410–411.

[10] Stone, J. A., and J. H. Zimmerman, "Index of Refraction of Air," https://emtoolbox.nist.gov/Wavelength/Documentation.asp, February, 2001.

10

Signal and Power Integrity Measurements

10.1 Signal Integrity (SI)

Modern electronic systems predominately process digital information. This digital information is typically transmitted as binary information or 1 and 0 logic levels. These two logic levels are usually characterized by two voltage levels, in the case of TTL, 5V for a logic 1 and 0V for a logic 0. In the case of CMOS logic levels, logic 1 may be as low as 1V. While this may seem relatively straightforward, the challenge arises in the transitions of the signals between these two levels. When data rates are low, these transitions times are slow and relatively immune to variations in the channels. These variations include impedance, trace lengths, and electromagnetic effects such as cross coupling and electromagnetic radiation. At higher frequencies, these imperfections induce distortions in the data such as reflections or cross-coupling which may result in erroneous data reception.

In modern digital systems, data rates are very high, in the tens of gigahertz or higher. These systems are particularly sensitive to channel variations. Because of this, channels need to be modeled, characterized, and tested thoroughly to ensure robust, error-free data communications. SI is the study of electrical channels and all of the tools that can be used to optimize these channels to ensure

Table 10.1
Example Devices and Unique Measurement Requirements for SI Measurements

Device/Application	Section	Unique Measurement Requirements (Beyond IL & RL)
2X Thru de-embedding	10.6	Measuring, creating, and applying de-embedding files using 2x thru on SFD
Delta L	10.7	Embedded trace
PCB pro bing	10.8	PCB calibration

error-free data communications. What is unique about these systems in relation to other systems that we have analyzed is that they require wider bandwidths. Most communications systems that we have discussed have bandwidths from 1 octave up to 1 decade. In contrast, digital systems have bandwidths from DC up to 3 to 5 times the maximum clock frequency. This gives bandwidths up to tens of gigahertz, extremely broadband.

While this chapter is not intended to be an exhaustive study of SI, it is intended to give a brief overview of SI and how modern VNAs are used in this discipline.

10.1.1 Steps for Establishing a Four-Port Differential Measurement Baseline

Chapter 8 discussed basic TDR measurements using a VNA, in particular the Rohde and Schwarz ZNB VNA. Chapter 10 will focus on more exhaustive SI measurements. Some of the information from Chapter 8 will be repeated for convenience sake. While some steps are somewhat specific to the Rohde and Schwarz ZNB, the other leading manufacturers have similar capabilities in generating and analyzing an eye diagram.

10.1.1.1 Eye Diagram Setup

- *Step 1:* As always, determine the appropriate interface between the VNA and the DUT. Most commonly in SI measurements, the interfaces will be multipin digital connectors as shown in Figure 10.1. Other interfaces such as vias and embedded traces will also be discussed.

  Obviously digital connectors will not mate with typical RF or microwave connectors such as 3.5 mm or 2.92 mm. To solve this problem a test fixture is required. Often these are purchased from third parties but can also be developed internally. The advantage of purchasing one is that the de-embedding coefficients often are included or a means to easily extract them to move the reference plane from the coax to the digital interface. Figure 10.2 shows an example of a USB 3.1 test fixture. In cases where the coefficients are not included, they can most easily be extracted by using a common extraction tool such as Keysight AFR,

Figure 10.1 High-speed multipin connectors.

Figure 10.2 USB test fixture.

AtiTec ISD, or Packet Microwave SFD. These tools use a common technique such as 2× thru to measure a PCB trace and extract S-parameters from the coupon.

- *Step 2:* Start by connecting a coaxial thru.
- *Step 3:* Preset the instrument.
- *Step 4:* Set up the ZNB for a differential S-parameter measurement as shown in Figure 10.3.
  - a. Trace group;
  - b. Press Meas;
  - c. Balanced ports;
  - d. 2 × Balanced.
- *Step 5:* Display differential insertion loss Sdd21 as shown in Figure 10.4:
  - a. Trace group;
  - b. Press Meas;
  - c. S-parameter;

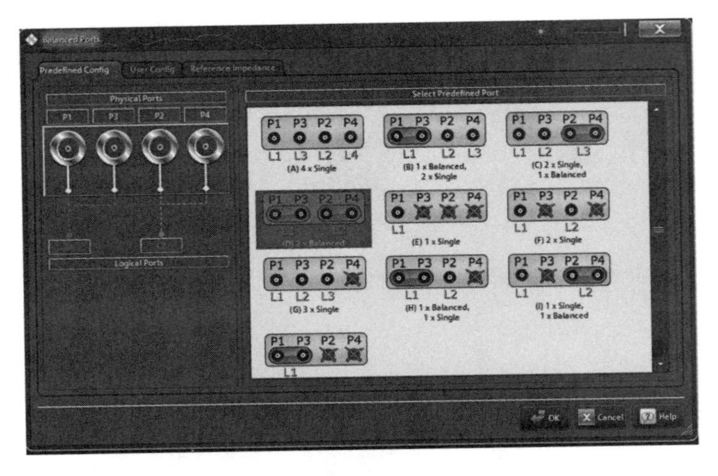

Figure 10.3 ZNB port configuration menu.

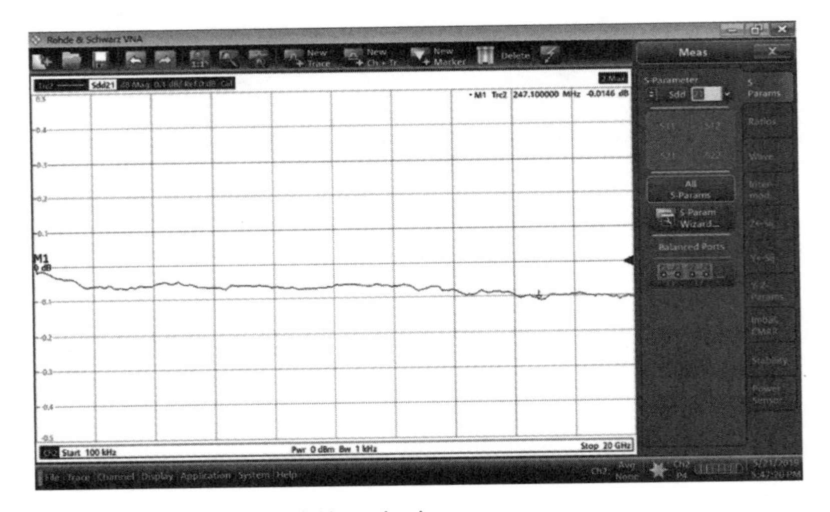

Figure 10.4 Settings for differential insertion loss.

- d. Select Sdd and 21.
- *Step 6:* Select simulation range as shown in Figure 10.5:
  - a. System Group;
  - b. Press Applic;
  - c. Setup;
  - d. Stimulus;
  - e. Set maximum measure delay to 10 ns;
  - f. Rise time to 92.5 ps;

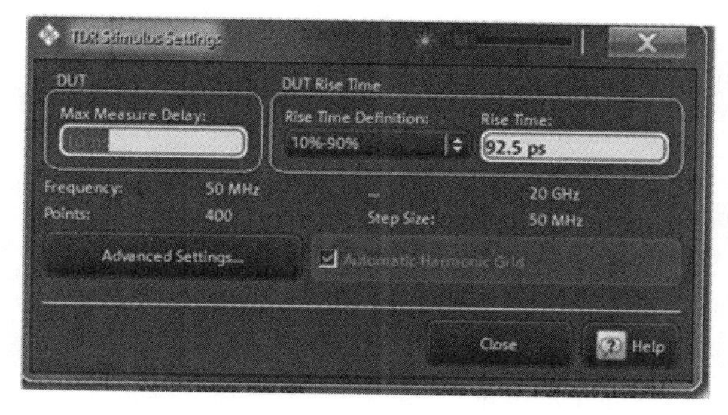

Figure 10.5 ZNB dialog for configuring the TDR stimulus.

- g. This sets the VNA frequency range to 50 MHz to 20 GHz with 400 points.

- *Step 7:* Display eye diagram as shown in Figure 10.6:
  - a. System group;
  - b. Press Applic;
  - c. Menu select;
  - d. TDR;
  - e. Check the Eye Diagram box;
  - f. Press eye diagram;
  - g. Select data rate 5 Gbps;
  - h. Pattern length 2^10-1;

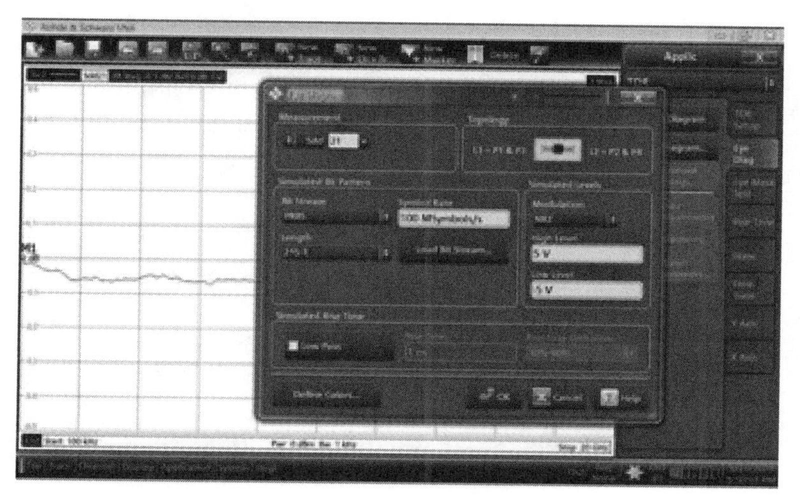

Figure 10.6 ZNB dialog for setting up an eye diagram.

- i. "Zero level" –0.8V. "One" level 0.8V;
- j. Enable lowpass;
- k. Set rise time to 10 ps;
- l. Click OK. Push restart.

- *Step 8:* Display eye diagram sweep as shown in Figure 10.7.
  - a. Trace group;
  - b. Press scale;
  - c. Autoscale;
  - d. Autoscale eye diagram;
  - e. Channel group;
  - f. Sweep control tab;
  - g. Press restart sweep.
- *Step 9:* Set the source level and receiver bandwidth.

  Since most TDR-based SI applications are used to measure passive devices (cables, connecters, PCB traces), the instrument default settings for PWR (–10 ~ 0 dBm) and BW (1 ~ 10 kHz) are sufficient. Wider bandwidths are used for low-loss DUTs and production measurements requiring speed. Narrower bandwidths are used for lossy DUTs requiring higher dynamic range. Typically use an IF bandwidth of 1 kHz and a power level of 0 dBm. Be careful when calibrating to use the recommended input level for the electronic automatic calibration units. Also if measuring any active devices, be aware of the correct input drive level to avoid compression.

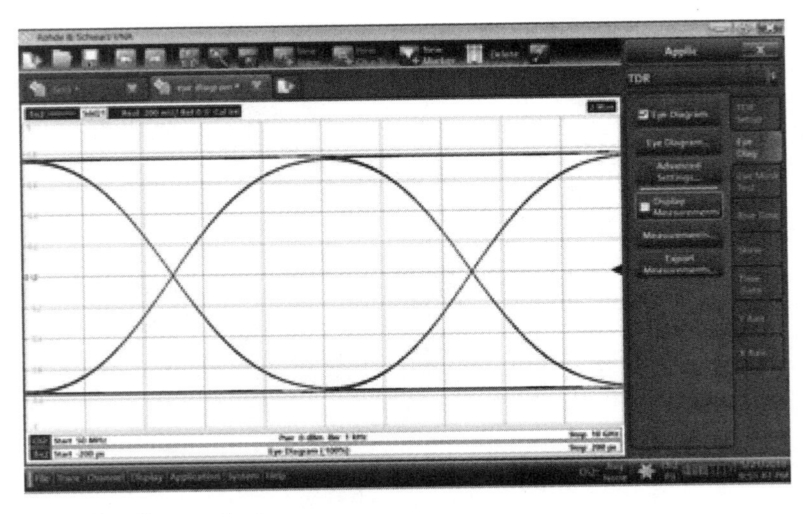

Figure 10.7 Eye diagram display.

- *Step 10:* Perform a two-port or four-port calibration using an appropriate calibration kit or autocal unit (e.g., 3.5-mm calibration kit for SMA or 3.5-mm connectors). Save the calibration and/or the test setup after completion. A TRL calibration or thru-reflect-line may be used but requires designing a zero-length thru, a transmission line between approximately 20° and 160°, and a reflect, usually a short. For wider bandwidths, multiple thrus can be used. The number of standards required is summarized in Table 10.2.
- *Step 11:* After completing the calibration, it is important to independently verify the calibration by installing a thru standard connecting input pairs and output pairs.
- *Step 12:* If the calibration is valid, connect the DUT and configure the traces for the desired measurement.
- *Step 13:* Connect the device with required adapters and interfaces. If not connecting directly to the DUT, be aware that a de-embedding file will be required. As shown in Figure 10.8, S-parameter files may be required to move the reference plane from the coaxial interface to the desired DUT plane. If the test fixture is one that includes an S-parameter file, it can be entered into the instrument. There are likely to be an input and an output file that may be different.
- *Step 14:* Apply de-embedding offsets depending on what tools you are using.
    - a. The simplest method of de-embedding is the offset de-embedding whereby some simple characteristics of the offset are entered such as delay, length, permittivity, and velocity factor. The simplest case would be a trace on a PCB. The offset embed menu is shown in Figure 10.9.

Table 10.2
Number of Lines Required for TRL Calibration

Bandwidth	Number of Transmission Lines
64:1	2
512:1	3
4096:1	4

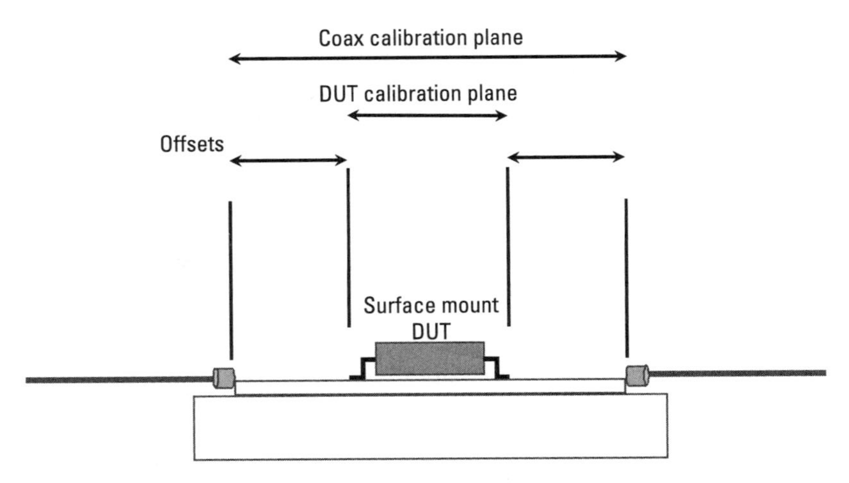

Figure 10.8 Device plus interface.

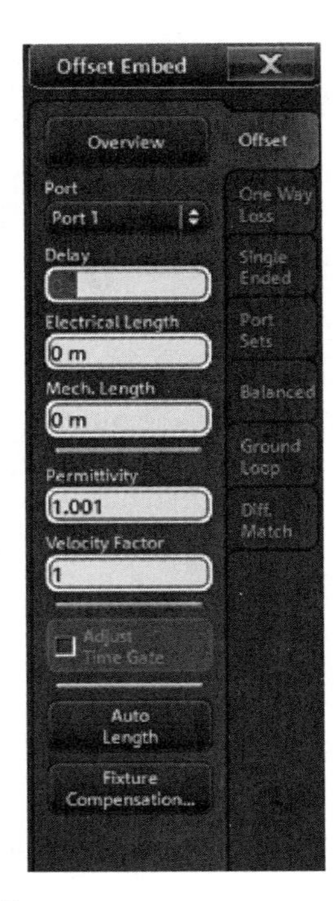

Figure 10.9 ZNB offset embedding menu.

Most VNAs include a feature called fixture compensation that provides some basic extraction capabilities that can be used at lower frequencies as shown in Figure 10.10 to find the offset between the connector interface and the DUT. The fixture extraction menu is used by adding an open and/or short at the end

Figure 10.10 Fixture compensation menu.

of a trace or cable that can be measured to generate the de-embedding file. It is quite simple to use, but is only accurate up to approximately 1 GHz.

The most accurate method of de-embedding is the S-parameter de-embedding whereby an S-parameter file is entered that was either provided with the test fixture or somehow measured such as with probing, modeling or an extraction tool such as ISD, SFD, or AFR. The files are applied port by port giving the ability to have unique structures on each port. Fixture extraction will be discussed later in 2× thru de-embedding.

Now that we have completed the setup of the VNA for time-domain measurements and de-embedded input and output traces using SFD, we are ready to make eye diagram measurements.

10.1.1.2 Eye Diagram Measurements

The extended time-domain option of the Rohde and Schwarz ZNB give the ability to analyze numerous SI characteristics on the eye diagram and give numerical data associated with the eye. Other manufacturers have similar capabilities.

- *Step 1:* Display mixed-mode S-parameters as shown if Figure 10.11:
  - a. Trace group;
  - b. Press Trace Config;
  - c. Add Trace + Diag;
  - d. Add SDD11 and SDD21;
  - e. Traces will be put into a new diagram below the eye diagram (note that you cannot have eye and S-parameters in the same window).

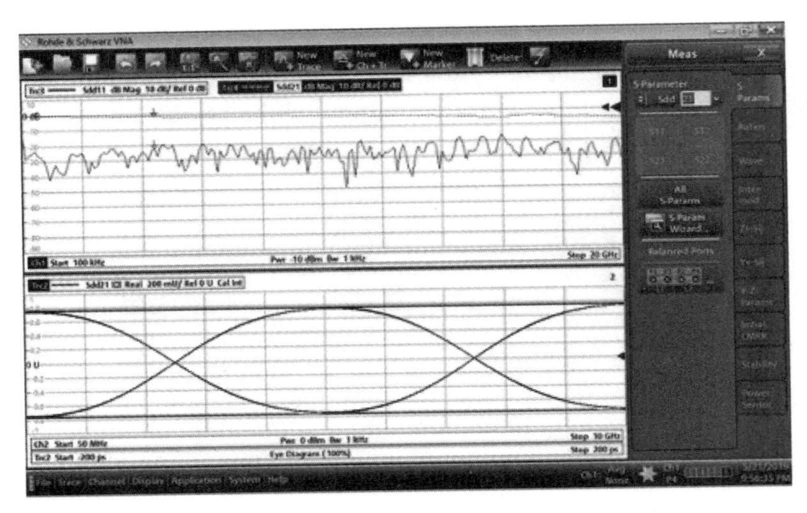

Figure 10.11 Dual display of eye diagram and mixed-mode S-parameters.

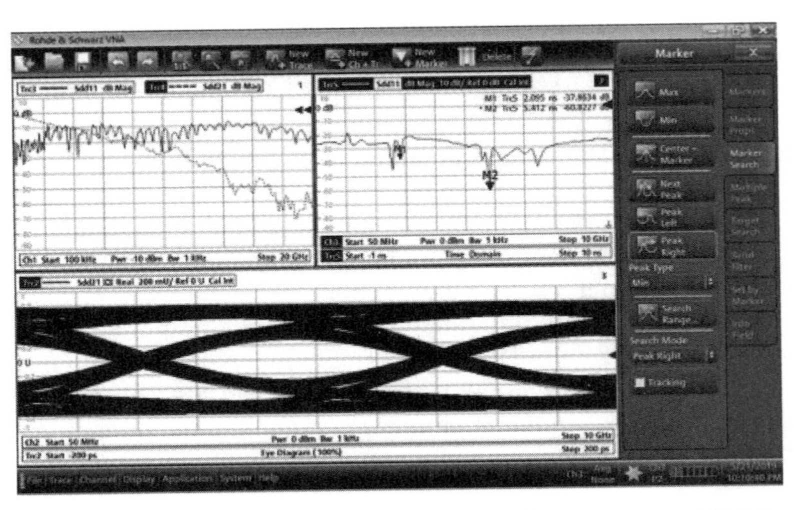

Figure 10.12 Triple display of eye diagram, mixed-mode S-parameters, and TDR Z versus distance.

- *Step 2:* Display differential input impedance as shown in Figure 10.12.
  - a. This impedance versus distance sweep is shown in the upper right window of Figure 10.12.
- *Step 3:* Analyze the TDR Trace as shown in Figure 10.13:
  - a. Double-click on the TDR Window in the top left of Figure 10.12 to get a single display area;

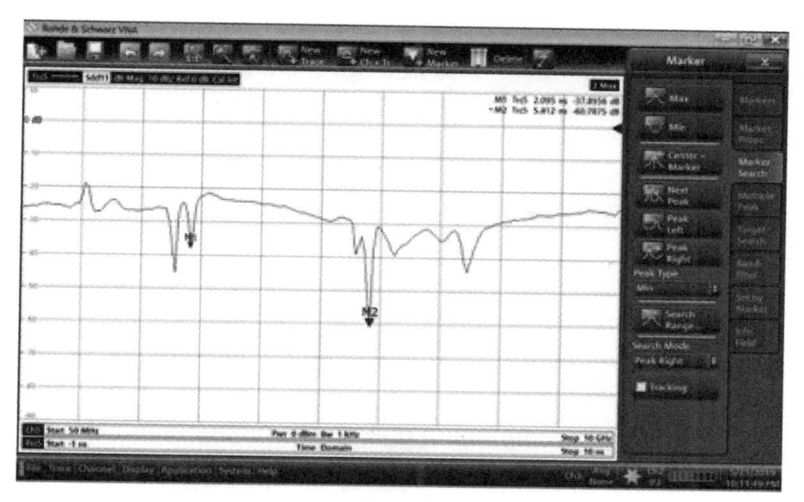

Figure 10.13 TDR display of impedance versus distance of backplane.

- b. Trace group;
- c. Marker;
- d. Marker Search tab;
- e. Select Min;
- f. Trace group;
- g. Marker;
- h. Marker Search;
- i. Peak right until M2 is located in the right dip;
- j. Double-click on the TDR trace to show all display areas.

- Often in SI applications, the traces are the DUTs to be measured in isolation. Figure 10.14 shows a picture of the backplane that we are analyzing. The intent of this measurement is to measure the larger backplane PCB, not the two smaller daughter cards. The marker values shown in Figure 10.13 tell us exactly where to move the reference plane moved to, 2.095 ns at the input and 5.412 ns at the output.
- *Step 4:* Remove influence of daughter card and connectors as shown in Figure 10.15:
  - a.Select the SDD11 trace by selecting the Trace Info Tab at the top right of the screen;
  - b. Trace group;
  - c. Trace Config;
  - d. Time gate enabled;

Figure 10.14 Backplane PCB with coax and multipin connectors.

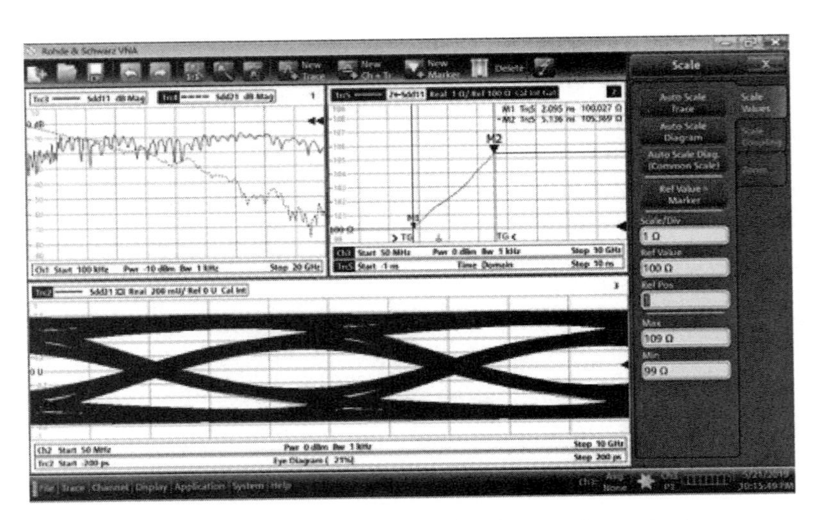

Figure 10.15 TDR response with connectors and daughter cards gated out.

- e. Axis Pair Start Stop;
- f. Start 2.095 ns, and stop 5.412 ns;
- g. Repeat for SDD21.

10.1.1.3 Analyzing Impairments

The key benefit of using an eye diagram in general is the ability to troubleshoot a channel to see how well the system responds to impairments in that channel. These impairments can be simulated by the VNA and the eye will be updated to include the desired impairment. The channel visualizes the effects on the eye and gives numerical results. In addition to impairments, different PRBS sequences can be applied as different sequences have different statistics associated with them. An additional benefit of extended time domain option is the ability to look at the eye diagram at different outputs.

- *Step 1:* Vary the eye diagram by adding impairments in the signal path as shown in Figure 10.16:
  - a. System Group
  - b. Advanced Settings
  - c. Change transparency of the window to improve transparency

There are six controls as well as the measurement point that can be used for analysis, which are shown in the advanced settings display.

10.1.1.4 ZNB Advanced Eye Diagram Settings

Generator

The generator gives the ability to change the bit stream pattern, length, data rate, and logic levels and to simulate rise time effects on the input data stream. The bitstream can also be encoded and scrambled to increase randomness.

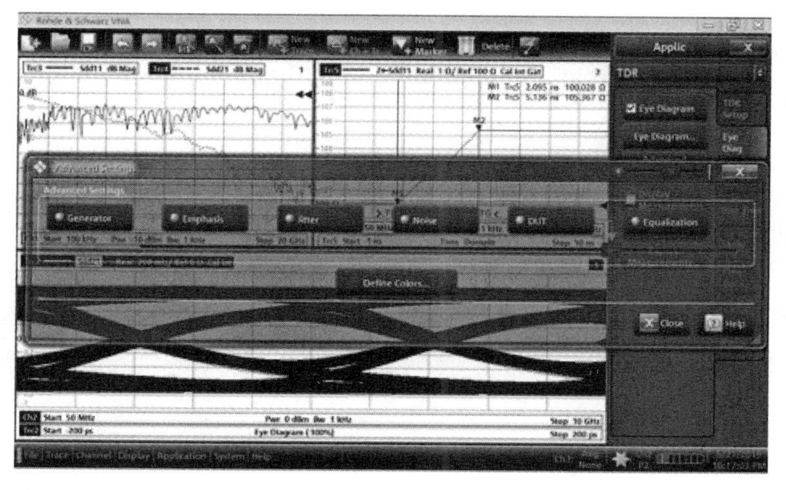

Figure 10.16 Eye diagram advance dialog box.

Emphasis

The emphasis enables the ability to add pre-emphasis filtering to the bitstream that was generated in the generator block. This is commonly used for dispersive channels such as FR4 PCB materials. This predistorts the signal so that the frequency dependence of the channel can be improved.

Jitter

Typically jitter is not measure or analyzed by a VNA. Jitter shows the variations in the zero crossings or horizontal variation of the eye diagram. Jitter is measured by a phase noise analyzer or a high-speed oscilloscope. Jitter typically has statistics associated with it. For example, random jitter is typically Gaussian in nature while periodic jitter usually a spurious-type signal with fixed amplitude, frequency, and phase. Other types of jitter that can be defined include Dirac and user-specified. The user-specified jitter enables the user to measure jitter with a scope and download the measured jitter into the channel model of the ZNB.

Noise

Noise is adding AWGN noise to the channel and shows as vertical fluctuations in the amplitude and zero crossings of the signal.

DUT

DUT give the ability to simply vary the measurement, change the DC value (1 open, 0 load, −1 short), or compare the waveform to an ideal device.

Equalization

Similar to emphasis, equalization will vary the response of the bitstream with a filter, but at the output as opposed to the input eye diagram.

10.1.2 SI Basics

Chapter 8 focused on the basics of time-domain reflectometry (TDR) on high-frequency DUTs. This chapter expands on these TDR measurements that are commonplace in high-speed digital systems. This leads us to the field of SI. The purpose of SI is to determine how to efficiently get high-speed signals from a source to a receiver over some physical channel (physical layer). The channel might include cables, connectors, adapters, backplanes, or other interconnects designed to pass high-speed digital signals. The intent of this chapter is not to thoroughly examine the study of SI, but its relevance to network analysis. Some relevant standards of which to be aware are:

- PCIe;
- USB;
- HDMI;
- Display port;
- Ethernet;
- SATA;
- DVI;
- Thunderbolt.

Most of the modern versions of these standards include S-parameters in their set of compliance tests for which VNAs are optimized.

SI can be summarized as the engineering discipline of analyzing and mitigating the degradation of digital waveforms transmitted over a channel [1]. In analyzing these signals, the problems that arise typically fall into the categories of timing, noise, and electromagnetic interference (EMI). Timing effects are generally not well suited for VNA measurements; they are typically measured with an oscilloscope or phase noise analyzer, but will be briefly addressed in this chapter. EMI is a very challenging topic that deals with electromagnetic radiation and propagation and induction, a topic not covered. Noise is very applicable to being measured with a VNA and will be the primary topic of discussion in this chapter. Noise can consist of reflections, crosstalk, rail collapse, or some other signal quality issues. Rail collapse is a power integrity issue requiring a very low-frequency instrument that will be measuring very low impedances that will be discussed in Section 10.2.

Some of the common SI tests performed with a VNA may include but are not limited to:

- Single-ended S-parameters;
- Differential S-parameters;
- Mixed mode S-parameters;
- Crosstalk;
- Eye diagrams (eye height, eye width);
- Eye mask tests;
- Characteristic impedance variation;
- Dissipation factor;
- Dielectric constant.

When signals are transmitted over any physical medium, the received signal is never identical to the transmitted signal. These changes in the transmitted signal are caused by imperfections in the transmission medium between source and receiver. These imperfections are called impairments or distortions. Trying to quantify these impairments and reduce or eliminate them is a primary goal of SI. The most common distortions seen in SI are:

- Impedance mismatch;
- Noise;
- Jitter;
- Skew;
- Imbalance;
- Limited bandwidth;
- Crosstalk;
- Dispersion;
- Intersymbol interference (ISI).

10.1.2.1 Impedance Mismatch

The fundamental characteristic that is typically measured first by SI engineers is impedance as was discussed in Chapter 8. The 50Ω transmission lines and cables and 100Ω differential lines and cables are the most common. These traces need to maintain reasonable impedance tolerances while staying within the PCB manufacturing constraints. For 100Ω differential transmission lines, maintaining precise length matching is equally critical. Not maintaining these tolerances can cause reflections or reduced common-mode rejection. The VNA is able to measure all of the single-ended and differential transmission line characteristics. Typically transmission lines on PCBs are microstrip or stripline. As was discussed in Chapter 8, impedance is commonly measured using the TDR function of the VNA whereby the impedance versus distance is calculated by the instrument with an inverse FFT operation.

10.1.2.2 Noise

Noise is typically defined as any unwanted signal in an electronic system. It is always present in most real-world systems, does not contain information, and generally degrades system performance. Figure 10.17 shows an ideal sine wave and a sine wave with noise.

There are different types of noise in SI: white noise, spurious noise, ground bounce, switching noise, and jitter. White noise is environmental noise that is

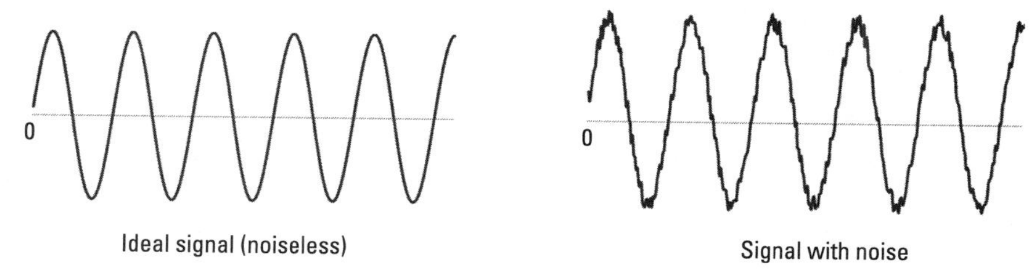

Figure 10.17 Comparison of an ideal sine wave and one containing jitter.

equally distributed across the entire spectrum. Spurious noise typically comes from other clocks, sources, or harmonics of these signals. Ground bounce occurs when there is too much resistance or reactance in the path to ground. Switching noise is a similar phenomenon when in the power rails.

10.1.2.3 Jitter

Jitter is a form of noise. It is the random variation in timing or period of a signal. Looking at the noiseless signal on the left in Figure 10.17, the signal crosses the zero axis at periodic intervals. These are often referred to as the zero crossings. The variation from this exact period is the jitter. Jitter is typically measured with either a high-speed oscilloscope, or a phase noise analyzer, not a VNA and expressed in time. Clock jitter in modern high-speed serial communication systems is measured in picoseconds or femtoseconds.

10.1.2.4 Skew

When using differential signaling, it is critical that the differential trace pairs are of identical length, if not, the two signals arrive at the receiver at different times. The difference in propagation times in a system is defined as skew. The VNA can measure the electrical length of any transmission line extremely accurately. Figure 10.18 shows a skew measurement on a ZNB VNA.

10.1.2.5 Balance

The primary purpose of differential signaling is to reject common mode noise. For this to work effectively, the characteristics of the two lines must be close to identical in impedance as well as length. Balance measures the loss characteristics of two lines; they need to be identical in loss in addition to the impedance.

10.1.2.6 Bandwidth

As digital signals increase in frequency, not only does the frequency content increase, the required bandwidth similarly increases. For an ideal square wave, the even-order harmonics are zero, while the odd harmonics have amplitude values governed by (10.1).

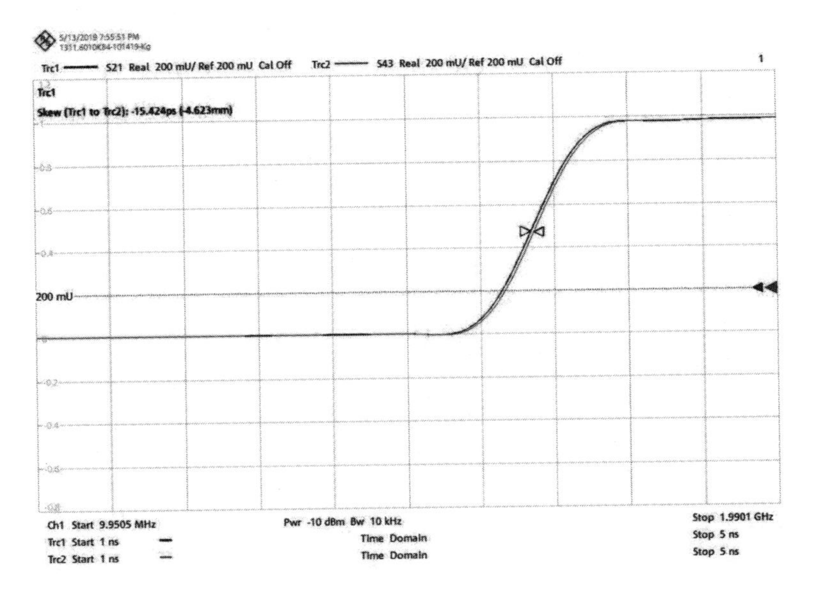

Figure 10.18 VNA skew measurement.

$$A_n = \frac{2}{\pi \cdot n} \tag{10.1}$$

where n is the harmonic number. Therefore, the amplitude of the harmonics decreases inversely with frequency. Given the bandwidth of the signal including harmonic content, we can calculate the rise time and vice versa using (10.2).

$$RiseTime = \frac{0.35}{Bandwidth(GHz)} \tag{10.2}$$

10.1.2.7 Crosstalk

Crosstalk is undesired coupling of signals between transmission lines in proximity. This phenomenon occurs when transmission lines are not effectively shielded or are too close to other transmission lines or conductors. Crosstalk occurs most commonly at the ends of transmission lines, often at the interconnects. SI engineers commonly characterize interconnects measuring near-end crosstalk (NEXT) and far-end crosstalk (FEXT) referring to the inputs and outputs of cable pairs. The crosstalk of one trace to another can be estimated by (10.3) whereby S is the spacing between traces and H is the thickness of the dielectric. Measuring the coupling between signal traces is extremely common in SI.

$$CT_{dB} = 20 \cdot \log\left(\frac{1}{1+\left(\frac{S}{H}\right)^2}\right) \tag{10.3}$$

10.1.2.8 Dispersion

In ideal transmission lines, the velocity of a waveform through that transmission line is frequency-independent. Dispersion is a measure of how frequency-dependent a signal path is. At low frequencies, series resistance is the dominant factor in the transmission line impedance. At high frequencies, the reactance is dominant and can vary with dielectric materials that are frequency dependent and lossy. Because of these frequency-dependent reactances, the different waveforms that encompass the signal do not travel at the same frequency and arrive at the load at different times. This causes pulse distortion as shown in Figure 10.19. As can be seen, the nondispersive signal has a constant or flat group delay versus frequency while the dispersive channel does not.

10.1.2.9 ISI

VNAs are probable the most accurate way of measuring group delay. When measuring group delay versus frequency as shown in Figure 10.19, it is clear that the group delay is frequency-dependent (i.e., not flat). This changing group delay often causes ISI. It is basically caused when arriving symbols of data

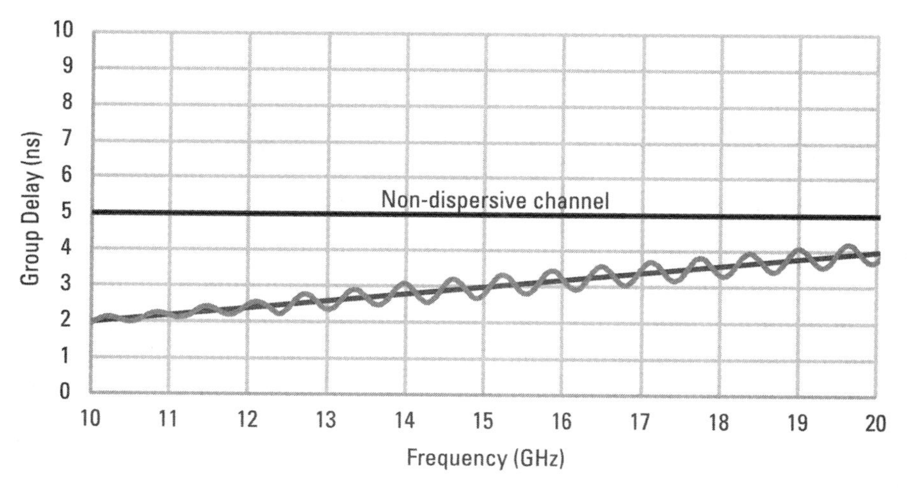

Figure 10.19 Group delay as a measure of dispersion. Nondispersive channel (top) has flat, constant group delay. Dispersive channel (bottom) has both slope and ripple in its group delay response.

interfere with subsequent symbols. This will close the eye diagram as shown in eye diagrams will be explained in the next section.

10.1.3 Eye Diagrams

One of the key tools used for evaluating SI measurements is the eye diagram or data eye as shown in Figure 10.20. The eye diagram is constructed from a digital waveform by folding the parts of the waveform corresponding to each individual bit into a single graph with signal amplitude on the vertical axis and time on horizontal axis diagram. It shows the combined effects of noise and ISI. This is shown in Figure 10.21. It is constructed by taking a data stream as shown in the top of Figure 10.22. As shown, the bit period is approximately 2 divisions wide. Every 2 divisions represent the next bit. Starting from the left, each bit is drawn on top of the previous bit in the lower figure. So the width of the eye diagram is simply the period of 1 bit, in this case, 2 divisions. After a lengthy bitstream is analyzed, the eye show all possible transition possibilities: low to high, high to low, low to low, and high to high, all laid on top of each other. Obviously these signals in real-world systems are not perfect square waves as shown. They look more like the eye diagram shown in Figure 10.20.

Some of the common measurements that can be quantified by the eye diagram include:

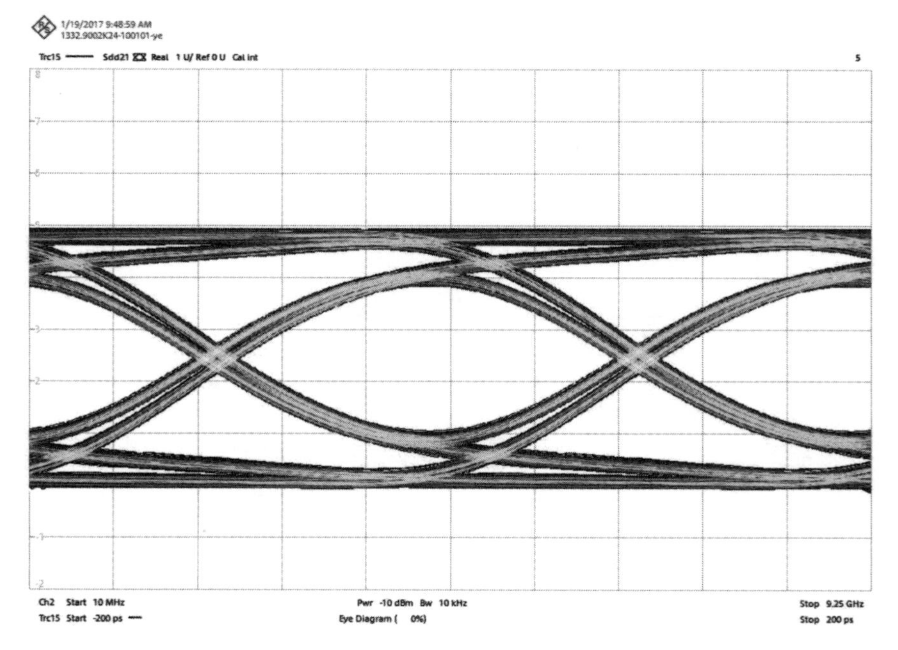

Figure 10.20 VNA eye diagram.

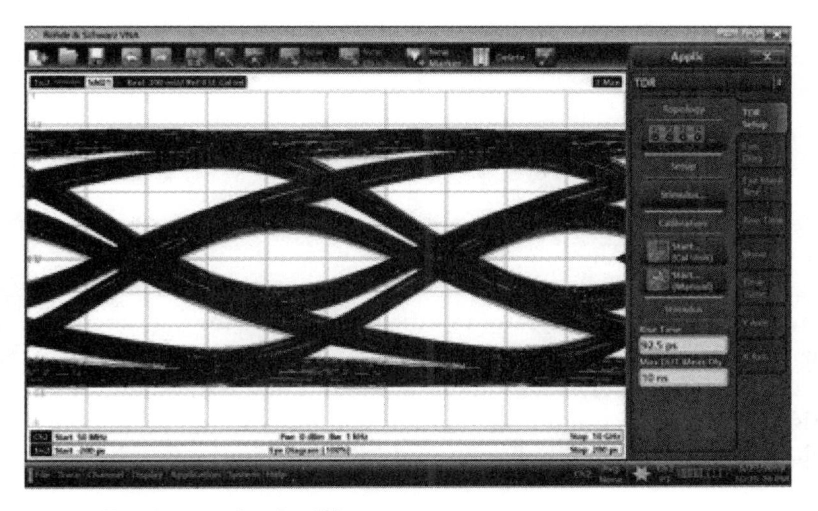

Figure 10.21 Eye diagram showing ISI.

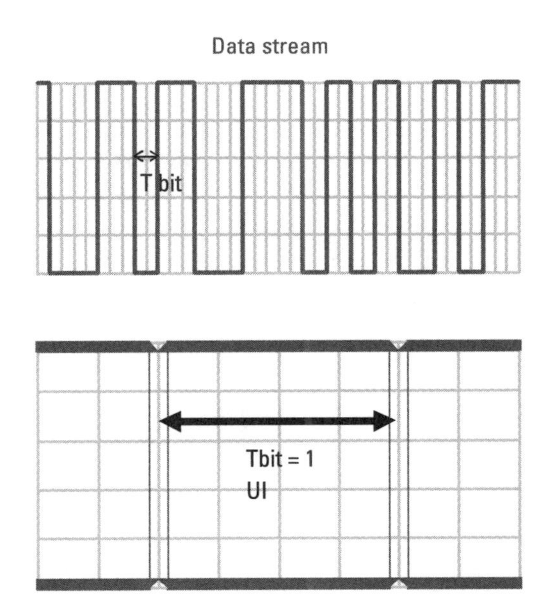

Figure 10.22 How an eye diagram is constructed.

- Eye height;
- Eye width;
- Bit period;
- Amplitude;
- Jitter;

- Eye-crossing percentage;
- Rise time;
- Fall time;
- Zero level;
- One level.

Figure 10.23 shows common measurements performed with an eye diagram. The eye diagram is created in a diagram by showing high and low transitions of a digital waveform in one bit period. It is very useful in quickly showing all of the above parameters in one diagram. The eye diagram will show the statistical average of a very large number of samples with a persistence display of the eye. The following sections describe these measurements.

10.1.3.1 Amplitude

This is the difference between the high and low levels with a 1 a high and a 0 low.

10.1.3.2 Eye Height

The eye height shows the amount of amplitude noise on a digital signal. Less noise is a wide-open eye, and more noise is a closed eye.

10.1.3.3 Eye Width

The eye width is the ideal crossing point of the eye based on the bit period. Variations in this crossing point can be used to calculate the jitter.

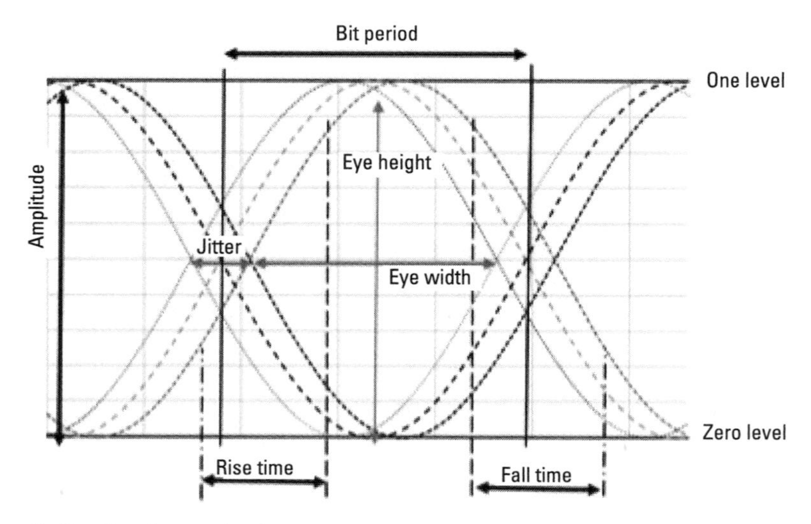

Figure 10.23 Eye diagram basics.

10.1.3.4 Bit Period

The width or horizontal opening of the eye diagram is the inverse of the bit period. When normalized with different data rates, it is called the Unit Interval or (UI), making it easier to compare eye diagrams of different data rates.

10.1.3.5 Jitter

One of the key characteristics in SI measurements is jitter. It is variations in the period of the data signal or deviation from ideal timing. Typically jitter is calculated by looking at the variations in the rising and falling edges of an eye diagram as it crosses from low to high and high to low. This is the midpoint going from 0 to 1 or 1 to 0. Typically, these variations behave statistically in nature. Random jitter usually follows a normal distribution. The width of the histogram is the peak to peak jitter while the RMS jitter is one standard deviation of the histogram.

10.1.3.6 Rise Time

The rise time is the transition time from low to high on the eye diagram. Usually it is from 10% of the one level to 90% of the one level.

10.1.3.7 Fall Time

The fall time is the transition time from high to low on the eye diagram. Usually it is from 90% of the one level to 10% of the one level.

10.1.3.8 Zero Level

The zero level is the mean amplitude of logic zero.

10.1.3.9 One Level

The one level is the mean amplitude of logic one.

10.1.3.10 VNA Eye Diagram

VNAs do not have the ability to source digital signals nor analyze them. So in order to generate an eye diagram, the bit transition need to be calculated rather than measured. This function is provided by most of the major VNA manufacturers. So using a VNA to generate the eye diagram involves three steps:

1. Measure the S-parameters to get the frequency response.
2. Perform an inverse FFT (IFFT) on the measured response to calculate the time-domain response (10.4).
3. Calculate the eye diagram by convolving the time-domain response with a predefined PRBS (pseudorandom bit sequence). This PRBS is typically an ideal bitstream.

The resulting time-domain response is an ideal predefined bit sequence altered by the frequency response of the channel shown in Figure 10.24.

$$h(t) = F^{-1}\left[S(\omega)\right] \tag{10.4}$$

10.2 Power Integrity

Power integrity focuses on delivery DC power to devices predominantly through transmission planes. SI focuses on transmission of signals through transmission

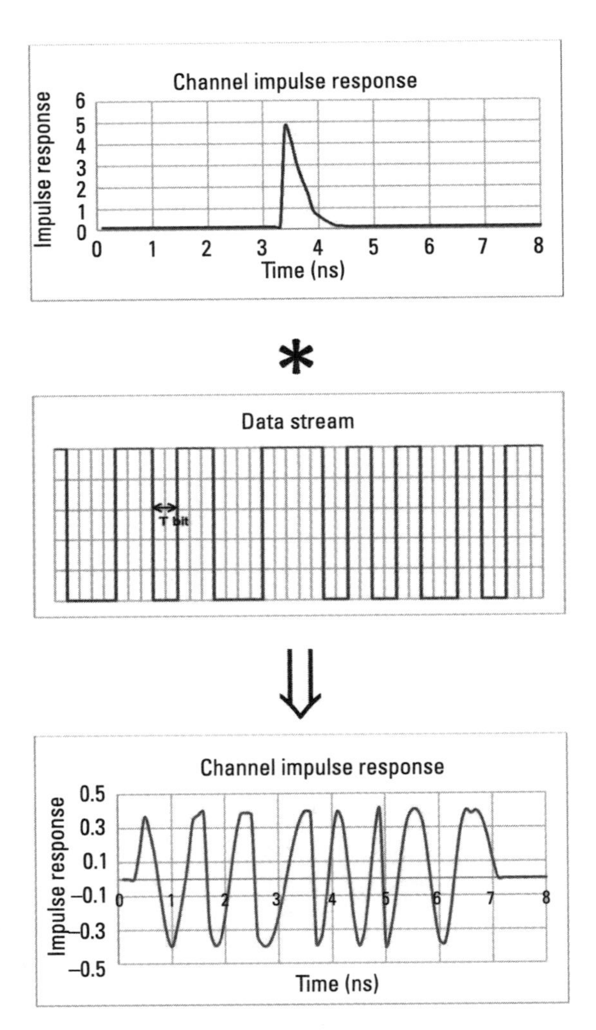

Figure 10.24 Plot of convolution between input data and channel response.

lines, typically 50Ω single-ended or 100Ω differential. Power integrity focuses on very low impedances across frequency and with dramatically varying load. The common issues that are analyzed in power integrity networks are voltage ripple, ground bounce, and maintaining required DC levels.

Most modern digital systems use a myriad of components that are enabled and disabled to save power. The power distribution network (PDN) consists of everything that is connected between the external power source and the power pins of the components. Typically these systems consists of switched or linear voltage regulators, isolated power and ground planes, power traces, ferrites, decoupling capacitors, and series inductors. The primary measurements are to look at impedance versus frequency from the components on the board looking towards the power planes and ground planes. These impedances need to be very low typically in the milliohms. The methods for making these impedances with a VNA are summarized in Table 10.3. The most common method used for low-impedance DUTs for power is the shunt-thru method. In this method, two ports of the VNA are used in parallel, whereby port 1 drives the current through the device and port 2 measures the voltage. This is shown in Figure 10.25.

In this setup, because the two ports are both connected to the DUT, the measured impedance is calculated from (10.5). In Figure 10.26 we see the various effects that are typically being measured with a VNA using this technique.

$$Z_{DUT} = 25 \cdot \frac{S_{21}}{1 - S_{21}} \tag{10.5}$$

The series technique is similarly done but using two ports with a resistor in series. Finally, the reflection method is achieved by making a standard S-parameter measurement but converting the format to Z-parameters in the display.

10.3 Reciprocity

When a device's forward and reverse gain are equal (i.e., $S_{21} = S_{12}$), it is reciprocal. Most passive devices and structures are reciprocal. Nearly all interconnects

Table 10.3
Summary of Power Integrity Measurement

Techniques Method	Impedance Range	Frequency Range	Accuracy
Reflection method	0.5–2K Ω	High-frequency range	10%
Series-thru method	25–1M Ω	Approximately 250 MHz	10%
Shunt-thru method	0.25–25 Ω	—	10%

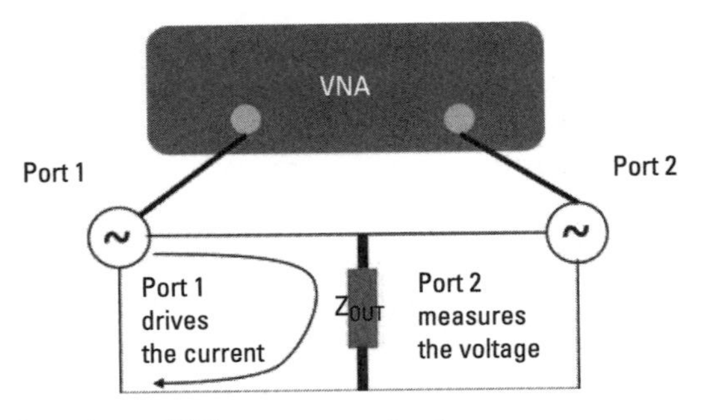

Figure 10.25 Low-impedance VNA measurement setup.

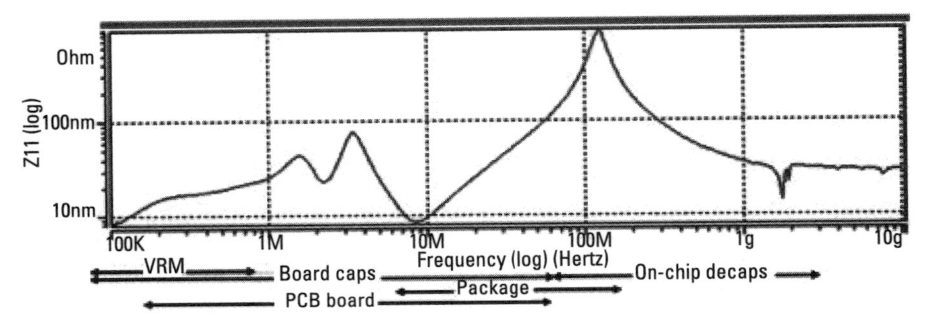

Figure 10.26 Common effects measured with shunt-thru measurement technique.

in digital systems are reciprocal. Typical devices include connectors, adapters, cables, PCB traces, and vias. When reciprocity errors are too high, mismatch errors are measured incorrectly. VNA measurements typically have much lower reciprocity errors than traditional TDRs.

10.4 Passivity

Passive devices cannot have gain and must have loss. When VNAs are calibrated incorrectly or have poor impedance match, passive devices can appear to have gain as there may be a standing wave present. This problem can also occur when incorrect de-embedding coefficients are used, not matching the actual devices or structures. When setting up measurements, always ensure that you have some means of verification as has been discussed in Chapter 9. This is most critical in SI measurements and passivity is a perfect example of this. Finally, ensure that good practices are being used as was discussed.

10.5 Causality

Devices cannot give a response before a stimulus is applied. When causality has been violated, reflections will be seen prior to the input in time looking at S_{11}. Most VNAs do not have a DC response, nor do they integrate up to infinite frequency. Because of this, causality violations are possible and commonly seen. It is very important when making precision measurements to verify causality and quantify it with one of the SI tools discussed.

10.6 2x Thru De-Embedding

Higher bandwidth standards with frequencies in the tens of gigahertz are commonplace today. Simple calibration and fixture de-embedding techniques are not accurate enough to use beyond 1 GHz. Calibrations techniques such as TRL can be used but require multiple structures on the PCB with a fair degree of complexity, and require multiple traces of precise lengths, typically a quarter-wavelength. This makes extraction of low-frequency parameters very challenging.

The 2x thru de-embedding was designed to simplify the PCB calibration process and include more realizable low-frequency structures. Other lengths are possible such as 1x thru and unequal input and output lengths but less common. The test coupon will be a trace on the PCB or alternate PCB. The advantage of being on the same PCB is that the parasitics of the actual PCB will be included in the measurement of the 2x thru. These tools also help to extract some characteristics that may vary with the process such as ε_r or surface roughness.

- *Step 1:* If a structure on the PCB such as a 2x thru is being used, the ZNB will extract the de-embedding files using the SFD tool in Figure 10.27. Add traces for measuring Z<-Sdd11, Z->Sdd22, Sdd11and Sdd22, Sdd12, and Sdd21.
- *Step 2:* Launch de-embedding tool on ZNB as shown in Figure 10.28. Press Channel->Offset & Embedding ->Logical Port.
- *Step 3:* Run SFD de-embedding tool as shown in Figure 10.29:
    - a. Choose the port order.
    - b. Choose the fixture tool (SFD).
    - c. Run tool.
- *Step 4:* Step through the SFD measurement.
- *Step 5:* Measure the coupon with the SFD de-embedding tool as shown in Figure 10.30.

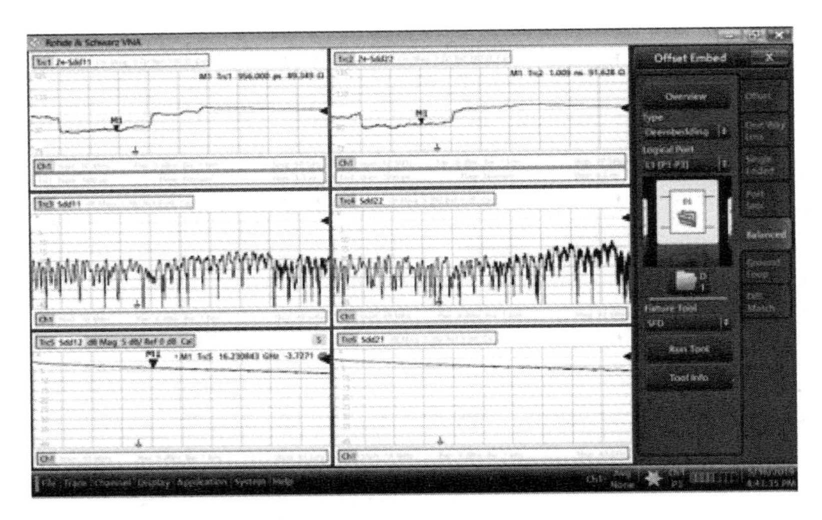

Figure 10.27 De-embedding reference measurements.

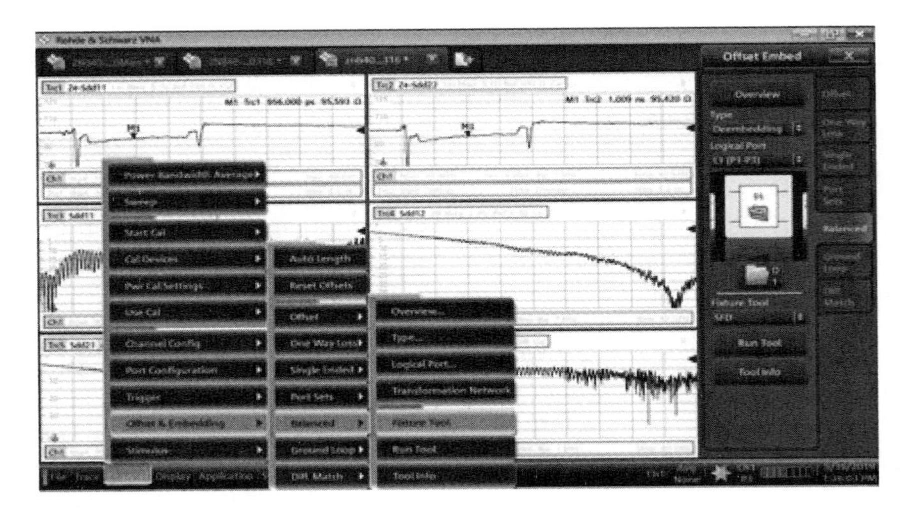

Figure 10.28 SFD launch menu.

- a. Selective active ports.
- b. Choose 2x thru (measure or load the 2x thru).
- c. Measure DUT + fixture.
- d. Selective active ports (reference (4) in Figure 10.30).
- e. Measure DUT + fixture (reference (3) in Figure 10.30).
- f. Save fixture model.

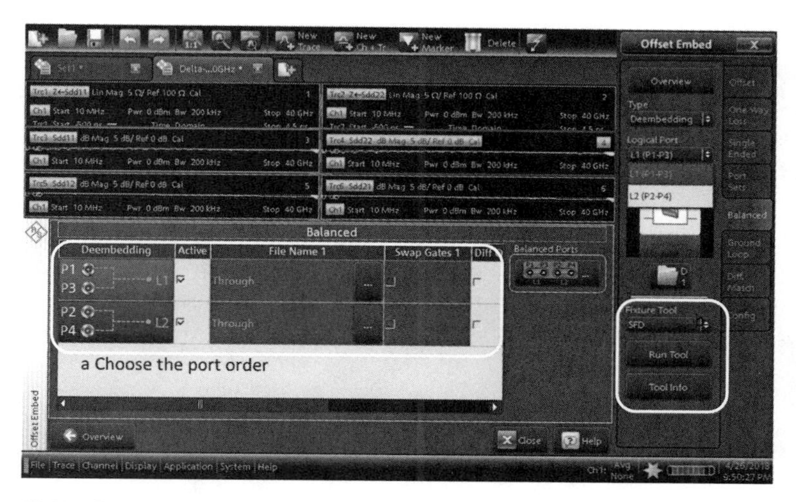

Figure 10.29 Running SFD in the ZNB VNA.

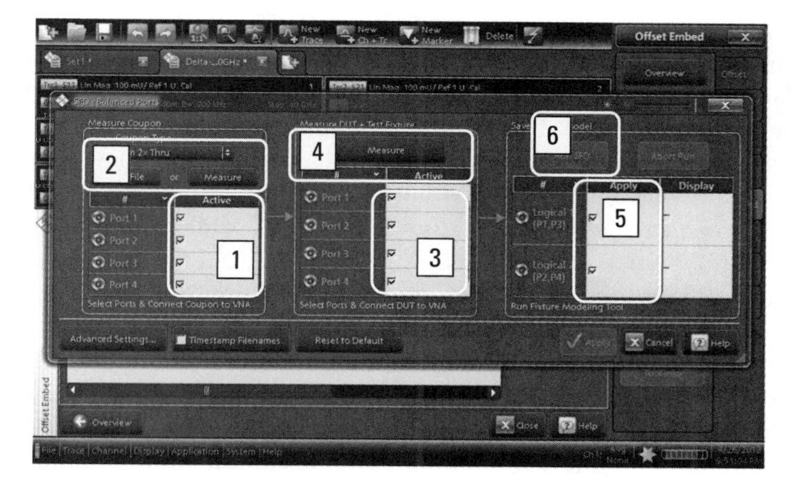

Figure 10.30 Measurement and generation sequence for SFD tool.

- g. Apply the logical port (reference (5) in Figure 10.30).
- h. Run SFD (reference (6) in Figure 10.30).

- *Step 6:* Sequence measure DUT + fixture.
  - a. Select active ports.
  - b. Measure DUT + fixture.

10.7 Delta L

Delta L is a de-embedding technique that was developed by Intel Corporation that is used for production testing of digital backplanes and similar circuits. This technique is used for de-embedding traces that may include vias. The primary advantage is that it is much simpler than most other techniques such as TRL but includes effects of vias. It can be used on microstrip as well as stripline traces. The goal of the Delta L technique is to determine the loss/unit length and apply it to any length of trace with similar via structures. Figure 10.31 shows example traces used for Delta L. Note that it is insufficient for measuring full S-parameter data.

$$\text{dB/unit loss} = [\text{Insertion Loss A} - \text{Insertion Loss B}]$$

10.8 PCB Probing

VNA probing takes a few forms. For many years, high-end wafer characterization systems have existed that are expensive and designed to work with semiconductor wafers. PCBs tend to have much more variations in interfaces, size, and interconnects. Interfaces are less standardized with daughter boards, different orientations, and distant ground vias. Calibration again is one of the challenges; this is where the Delta L and 2x thru calibrations are commonly used. In order to do any calibrations on these PCBs, the boards need to be designed to include calibration traces whether they are Delta L, 2x thru, or TRL.

Delta L was designed for these types of calibrations to be robust, simple, and cost-effective. An example of a larger PCB measurement setup is shown in Figure 10.32. A PCB of this size might be something like a networking device. These are manufacturing probes that have mounting holes on the PCB so technicians can quickly measure the board then move onto the next board.

In contrast, Figure 10.33 shows a very small PCB such as a mobile device. These are more challenging to probe and typically require microscopes,

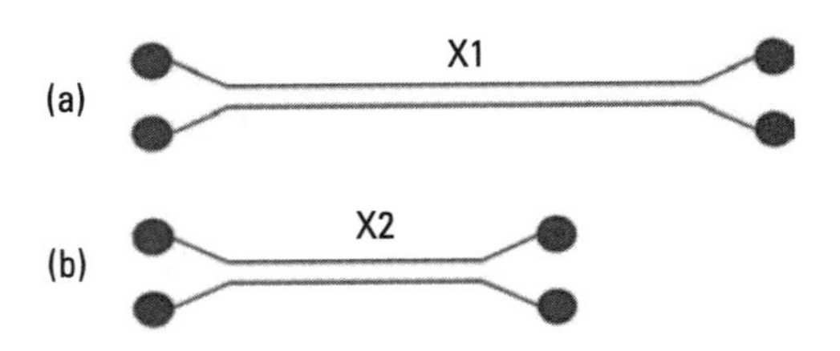

Figure 10.31 Buried stripline.

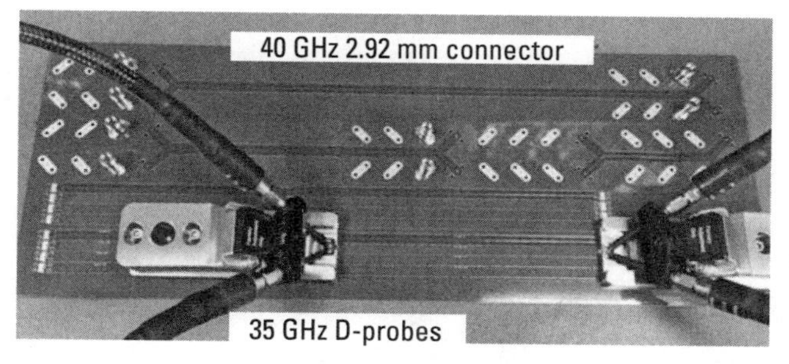

Figure 10.32 Typical larger PCB DUT being probed with manufacturing probes.

Figure 10.33 Typical smaller PCB DUT being measured.

precision positioners, and PCB mounts and are fairly challenging to establish planarity.

Figure 10.34 shows a general-purpose probing system that can be used to accommodate nearly any scenario. This system can be used to measure any size PCB and various pitches of both single-ended and differential signal probes.

- *Step 1:* Mount the PCB and optics as shown in Figure 10.34.
  - a. Screw the holder in to pinch each of the four corners of the PCB and establish the level.
  - b. Position the camera to view the circuits of interest on the PCB such as shown in Figure 10.34.
  - c. View the circuit of interest with appropriate zoom for viewing and manipulation.

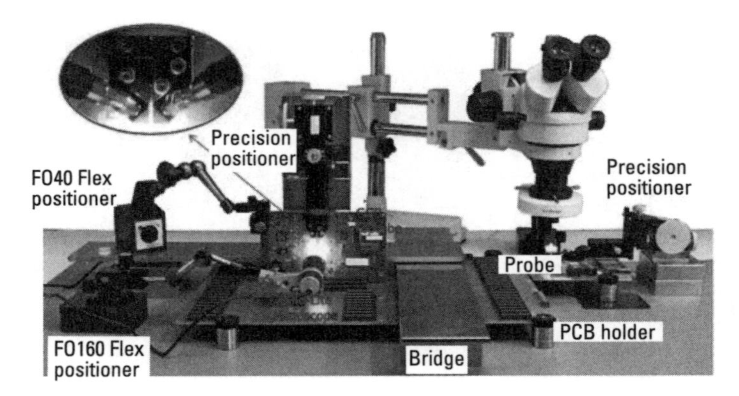

Figure 10.34 Flex PCB probing system for general-purpose PCB probing. (Courtesy Packet Microwave.)

- *Step 2:* Establish planarity with Mylar tape as shown in Figure 10.35. When establishing planarity, you are establishing the flatness of where the probe tips land on the board. As height and angle change, it affects the planarity. This is most easily accomplished by using Mylar tape on your PCB.
  - a. Place a small piece of Mylar tape on the PCB near the structure of interest.
  - b. Land the probes and observe the probe tip footprint; you should see equal indentations from both probe tips.
  - c. If the indentations are not equal, raise the probe tips and adjust planarization knob.

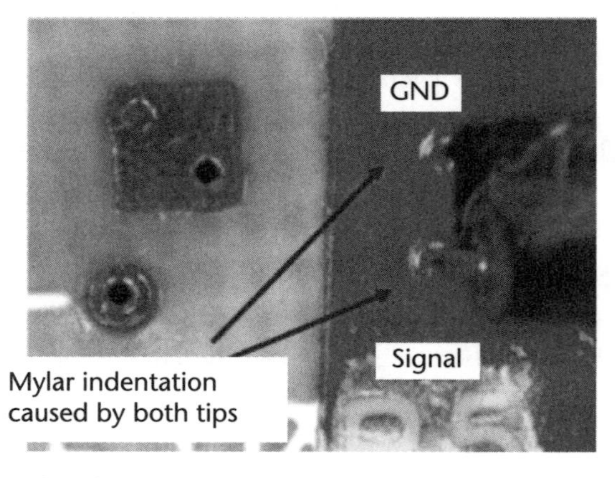

Figure 10.35 Probing planarity test.

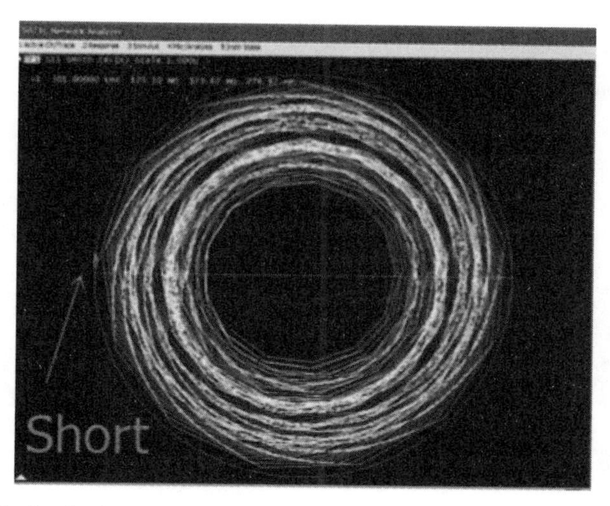

Figure 10.36 Probe tip short measuring S11.

- d. Iterate steps c and d until the indentations are equal and planarity is established.
- *Step 3:* Verify short as shown in Figure 10.36:
  - a. If planarity was accomplished, you will see clockwise rotation on the Smith chart.
  - b. If measurement does not resemble Figure 10.36, verify probes are landed and, if so, check planarity again.
- *Step 4:* De-embed or calibrate probes:
  - a. Many of the PCB probes will include de-embedding files that can just be entered in the offsets menu previously discussed.
  - b. If de-embedding, establish coaxial calibration at the connector, then apply files and verify short, open, and load with probe tips.

Reference

[1] https://en.wikipedia.org/wiki/Signal_integrity.

About the Authors

Neil Jarvis is a graduate of the Naval Postgraduate School in systems analysis. He has an M.S. in technology management from Pepperdine University, a B.A. in math/economics from the University of California Santa Barbara, and a B.S.E.E. in engineering from San Jose State University. After earning his degrees, he enlisted in the Naval Reserves as an avionics technician and later as an aeronautical engineering duty officer. He holds multiple patents in RF and microwave.

Mr. Jarvis has been a founder in two wireless communications start-up companies in the Silicon Valley with over 25 years of experience in testing, managing, developing, and manufacturing wireless and RF products. His experience ranges from high-end military sensors to RF semiconductors. Most recently, Mr. Jarvis has been an applications engineer at Rohde and Schwarz USA Inc., a leader in test and measurement equipment.

Mr. Jarvis is married with two children and lives in Silicon Valley.

Greg Bonaguide is a senior applications engineer for Rohde & Schwarz, specializing in spectrum analyzers and vector network analyzers. He received an M.S.E.E. in 1996 and an M.S.E.M. in 1989 from the University of South Florida, and a B.S.E.E. in 1982 from the University of Hartford. He has worked in the RF and microwave field since 1982 in roles involving design, systems engineering, and test and measurement. His experience includes both commercial and aerospace/defense work for companies including Cubic Corporation, E-Systems, GTE Spacenet, Raytheon, and Philips Semiconductors.

Mr. Bonaguide has published in journals such as *QST*, the *ARRL Handbook*, *Microwaves & RF*, *Test & Measurement World*, and *Evaluation Engineering*. He has also authored and delivered MicroApps presentations at past IEEE-MTT conferences. He served on the 2016 EDI CON Technical Advisory Committee, was the cochairman for the 2014 International Microwave Symposium (IMS) Microwave Application Seminars (MicroApps), and served as the chairman of the IEEE Florida West Coast Section MTT/AP/ED Societies (1999 to 2000).

Mr. Bonaguide is married with two children and lives in the Boston area.

Index

Artech House Microwave Library

Behavioral Modeling and Linearization of RF Power Amplifiers, John Wood

Chipless RFID Reader Architecture, Nemai Chandra Karmakar, Prasanna Kalansuriya, Randika Koswatta, and Rubayet E-Azim

Control Components Using Si, GaAs, and GaN Technologies, Inder J. Bahl

Design of Linear RF Outphasing Power Amplifiers, Xuejun Zhang, Lawrence E. Larson, and Peter M. Asbeck

Design Methodology for RF CMOS Phase Locked Loops, Carlos Quemada, Guillermo Bistué, and Iñigo Adin

Design of CMOS Operational Amplifiers, Rasoul Dehghani

Design of RF and Microwave Amplifiers and Oscillators, Second Edition, Pieter L. D. Abrie

Digital Filter Design Solutions, Jolyon M. De Freitas

Discrete Oscillator Design Linear, Nonlinear, Transient, and Noise Domains, Randall W. Rhea

Distortion in RF Power Amplifiers, Joel Vuolevi and Timo Rahkonen

Distributed Power Amplifiers for RF and Microwave Communications, Narendra Kumar and Andrei Grebennikov

Electric Circuits: A Primer, J. C. Olivier

Electronics for Microwave Backhaul, Vittorio Camarchia, Roberto Quaglia, and Marco Pirola, editors

EMPLAN: Electromagnetic Analysis of Printed Structures in Planarly Layered Media, Software and User's Manual, Noyan Kinayman and M. I. Aksun

An Engineer's Guide to Automated Testing of High-Speed Interfaces, Second Edition, José Moreira and Hubert Werkmann

Envelope Tracking Power Amplifiers for Wireless Communications, Zhancang Wang

Essentials of RF and Microwave Grounding, Eric Holzman

Frequency Measurement Technology, Ignacio Llamas-Garro, Marcos Tavares de Melo, and Jung-Mu Kim

FAST: Fast Amplifier Synthesis Tool—Software and User's Guide, Dale D. Henkes

Feedforward Linear Power Amplifiers, Nick Pothecary

Filter Synthesis Using Genesys S/Filter, Randall W. Rhea

Foundations of Oscillator Circuit Design, Guillermo Gonzalez

Frequency Synthesizers: Concept to Product, Alexander Chenakin

Fundamentals of Nonlinear Behavioral Modeling for RF and Microwave Design, John Wood and David E. Root, editors

Generalized Filter Design by Computer Optimization, Djuradj Budimir

Handbook of Dielectric and Thermal Properties of Materials at Microwave Frequencies, Vyacheslav V. Komarov

Handbook of RF, Microwave, and Millimeter-Wave Components, Leonid A. Belov, Sergey M. Smolskiy, and Victor N. Kochemasov

High-Efficiency Load Modulation Power Amplifiers for Wireless Communications, Zhancang Wang

High-Linearity RF Amplifier Design, Peter B. Kenington

High-Speed Circuit Board Signal Integrity, Second Edition, Stephen C. Thierauf

Integrated Microwave Front-Ends with Avionics Applications, Leo G. Maloratsky

Intermodulation Distortion in Microwave and Wireless Circuits, José Carlos Pedro and Nuno Borges Carvalho

Introduction to Modeling HBTs, Matthias Rudolph

An Introduction to Packet Microwave Systems and Technologies, Paolo Volpato

Introduction to RF Design Using EM Simulators, Hiroaki Kogure, Yoshie Kogure, and James C. Rautio

Introduction to RF and Microwave Passive Components, Richard Wallace and Krister Andreasson

Klystrons, Traveling Wave Tubes, Magnetrons, Crossed-Field Amplifiers, and Gyrotrons, A. S. Gilmour, Jr.

Lumped Elements for RF and Microwave Circuits, Inder Bahl

Lumped Element Quadrature Hybrids, David Andrews

Microstrip Lines and Slotlines, Third Edition, Ramesh Garg, Inder Bahl, and Maurizio Bozzi

Microwave Circuit Modeling Using Electromagnetic Field Simulation, Daniel G. Swanson, Jr. and Wolfgang J. R. Hoefer

Microwave Component Mechanics, Harri Eskelinen and Pekka Eskelinen

Microwave Differential Circuit Design Using Mixed-Mode S-Parameters, William R. Eisenstadt, Robert Stengel, and Bruce M. Thompson

Microwave Engineers' Handbook, Two Volumes,
Theodore Saad, editor

Microwave Filters, Impedance-Matching Networks, and Coupling Structures, George L. Matthaei, Leo Young, and E. M. T. Jones

Microwave Imaging Methods and Applications, Matteo Pastorino and Andrea Randazzo

Microwave Material Applications: Device Miniaturization and Integration, David B. Cruickshank

Microwave Materials and Fabrication Techniques, Second Edition,
Thomas S. Laverghetta

Microwave Materials for Wireless Applications, David B. Cruickshank

Microwave Mixer Technology and Applications, Bert Henderson and Edmar Camargo

Microwave Mixers, Second Edition, Stephen A. Maas

Microwave Network Design Using the Scattering Matrix,
Janusz A. Dobrowolski

Microwave Power Amplifier Design with MMIC Modules,
Howard Hausman

Microwave Radio Transmission Design Guide, Second Edition,
Trevor Manning

Microwave and RF Semiconductor Control Device Modeling,
Robert H. Caverly

Microwave Transmission Line Circuits, William T. Joines,
W. Devereux Palmer, and Jennifer T. Bernhard

Microwaves and Wireless Simplified, Third Edition,
Thomas S. Laverghetta

Modern Microwave Circuits, Noyan Kinayman and M. I. Aksun

Modern Microwave Measurements and Techniques, Second Edition, Thomas S. Laverghetta

Modern RF and Microwave Filter Design, Protap Pramanick and Prakash Bhartia

Neural Networks for RF and Microwave Design, Q. J. Zhang and K. C. Gupta

Noise in Linear and Nonlinear Circuits, Stephen A. Maas

Nonlinear Microwave and RF Circuits, Second Edition, Stephen A. Maas

On-Wafer Microwave Measurements and De-Embedding, Errikos Lourandakis

Passive RF Component Technology: Materials, Techniques, and Applications, Guoan Wang and Bo Pan, editors

Practical Analog and Digital Filter Design, Les Thede

Practical Microstrip Design and Applications, Günter Kompa

Practical Microwave Circuits, Stephen Maas

Practical RF Circuit Design for Modern Wireless Systems, Volume I: Passive Circuits and Systems, Les Besser and Rowan Gilmore

Practical RF Circuit Design for Modern Wireless Systems, Volume II: Active Circuits and Systems, Rowan Gilmore and Les Besser

Production Testing of RF and System-on-a-Chip Devices for Wireless Communications, Keith B. Schaub and Joe Kelly

Q Factor Measurements Using MATLAB®, Darko Kajfez

QMATCH: Lumped-Element Impedance Matching, Software and User's Guide, Pieter L. D. Abrie

Radio Frequency Integrated Circuit Design, Second Edition, John W. M. Rogers and Calvin Plett

Reflectionless Filters, Matthew A. Morgan

RF Bulk Acoustic Wave Filters for Communications, Ken-ya Hashimoto

RF Design Guide: Systems, Circuits, and Equations, Peter Vizmuller

RF Linear Accelerators for Medical and Industrial Applications, Samy Hanna

RF Measurements of Die and Packages, Scott A. Wartenberg

The RF and Microwave Circuit Design Handbook, Stephen A. Maas

RF and Microwave Coupled-Line Circuits, Rajesh Mongia, Inder Bahl, and Prakash Bhartia

RF and Microwave Oscillator Design, Michal Odyniec, editor

RF Power Amplifiers for Wireless Communications, Second Edition, Steve C. Cripps

RF Systems, Components, and Circuits Handbook, Ferril A. Losee

Scattering Parameters in RF and Microwave Circuit Analysis and Design, Janusz A. Dobrowolski

The Six-Port Technique with Microwave and Wireless Applications, Fadhel M. Ghannouchi and Abbas Mohammadi

Solid-State Microwave High-Power Amplifiers, Franco Sechi and Marina Bujatti

Spin Transfer Torque Based Devices, Circuits, and Memory, Brajesh Kumar Kaushik and Shivam Verma

Stability Analysis of Nonlinear Microwave Circuits, Almudena Suárez and Raymond Quéré

Substrate Noise Coupling in Analog/RF Circuits, Stephane Bronckers, Geert Van der Plas, Gerd Vandersteen, and Yves Rolain

System-in-Package RF Design and Applications, Michael P. Gaynor

Technologies for RF Systems, Terry Edwards

Terahertz Metrology, Mira Naftaly, editor

TRAVIS 2.0: Transmission Line Visualization Software and User's Guide, Version 2.0, Robert G. Kaires and Barton T. Hickman

Understanding Microwave Heating Cavities, Tse V. Chow Ting Chan and Howard C. Reader

Understanding Quartz Crystals and Oscillators, Ramón M. Cerda

Vertical GaN and SiC Power Devices, Kazuhiro Mochizuki

The VNA Applications Handbook, Gregory Bonaguide and Neil Jarvis

For further information on these and other Artech House titles, including previously considered out-of-print books now available through our In-Print-Forever® (IPF®) program, contact:

Artech House Publishers
685 Canton Street
Norwood, MA 02062
Phone: 781-769-9750
Fax: 781-769-6334
e-mail: artech@artechhouse.com

Artech House Books
16 Sussex Street
London SW1V 4RW UK
Phone: +44 (0)20 7596 8750
Fax: +44 (0)20 7630 0166
e-mail: artech-uk@artechhouse.com

Find us on the World Wide Web at: www.artechhouse.com